Palgrave Studies in Economic History is designed to illuminate and enrich our understanding of economies and economic phenomena of the past. The series covers a vast range of topics including financial history, labour history, development economics, commercialisation, urbanisation, industrialisation, modernisation, globalisation, and changes in world economic orders.

More information about this series at
http://www.palgrave.com/gp/series/14632

Palgrave Studies in Economic History

Series Editor
Kent Deng
London School of Economics
London, UK

Renate Pieper ·
Claudia de Lozanne Jefferies ·
Markus Denzel
Editors

Mining, Money and Markets in the Early Modern Atlantic

Digital Approaches and New Perspectives

Editors
Renate Pieper
Institute of History
University of Graz
Graz, Austria

Claudia de Lozanne Jefferies
Department of Economics
City, University of London
London, UK

Markus Denzel
Historisches Seminar Lehrstuhl für
Universität Leipzig
Leipzig, Germany

Palgrave Studies in Economic History
ISBN 978-3-030-23893-3   ISBN 978-3-030-23894-0   (eBook)
https://doi.org/10.1007/978-3-030-23894-0

Cover illustration: Robert Kawka/Alamy Stock Photo

This Palgrave Macmillan imprint is published by the registered company Springer Nature Switzerland AG
The registered company address is: Gewerbestrasse 11, 6330 Cham, Switzerland

Acknowledgements

We would first like to thank Professor Ken Deng, editor of *Palgrave Studies in Economic History*, for accepting our book as part of the series.

This volume owes a great deal to Werner Stangl, Harry Vossen and Harald Kleinberger-Pierer for their meticulous editorial input. We are also grateful to Laura Pacey, Ruth Noble, Clara Heathcock, Yuvaraj Krishnan and all the other Palgrave staff engaged in the transformation of our manuscript into a book.

Last but not least, our gratitude goes to the Karl-Franzens University of Graz, the University of Leipzig and the Economics Department of City, University of London for their generous financial support.

Renate Pieper
Claudia de Lozanne Jefferies
Markus Denzel

Contents

Notes on Contributors

Peter Bernholz is a retired Full Professor at the University of Basel in Switzerland. He has published the monographies *Monetary Regimes and Inflation* and *Totalitarianism, Terrorism and Supreme Values*.

Jérôme Blanc is Full Professor in Economics at Sciences Po Lyon, France. His works deal with money and the plurality of its forms and practices, mainly analysed through socioeconomic viewpoint and history of ideas.

José Enrique Covarrubias is a researcher at the Instituto de Investigaciones Históricas of the National Autonomous University of Mexico (UNAM). His most important research themes include economic history of Mexico, history of political and sociological ideas in Mexico, and travellers in Mexico in the nineteenth century.

Claudia de Lozanne Jefferies is Senior Lecturer in Economics at City, University of London, UK. Her main fields of research are monetary and financial history.

Markus Denzel is Professor and Chair of Social and Economic History at the University of Leipzig, Germany. His main fields of research are international payments, the role of money and bills, and currency history.

Ludovic Desmedt is Professor in Economics at the University of Burgundy and member of the Laboratoire d'Economie de Dijon (LEDi). He is co-editor of the *Revue d'histoire de la pensée économique*. His research interests are focused on the history of economic thought and the evolution of banking practices.

Andrés Calderón Fernández is a researcher at the National Autonomous University of Mexico (UNAM). His main lines of investigation are the history of economic institutions and prices in early modern Mexico.

Johan García Zaldúa is currently an associated researcher at the Centro de Investigação Transdisciplinar «Cultura, Espaço e Memória» (CITCEM). He has a Ph.D. in Text and Event in Early Modern Europe by the University of Porto and the University of Kent.

Alfredo Garcia-Hiernaux is a Contracted Professor at the Universidad Complutense de Madrid. His research focuses on early modern economic history and econometric computational methods.

Rafael Dobado González is Professor of Economic History at the Universidad Complutense de Madrid. His research deals mainly with early modern era international economic history (silver production and circulation, prices, wages, consumption patterns, numeracy and heights).

Domenic Hofmann is a researcher at the University Graz, Austria. He has a Ph.D. with special focus on early modern and economic history. His research focus is the Atlantic world in the sixteenth and seventeenth centuries.

Harald Kleinberger-Pierer is a researcher at the University of Graz. His main lines of investigation are the history of science and visual culture.

Peter Paul Marckhgott-Sanabria is a researcher at the Austrian Centre for Digital Humanities of the Austrian Academy of Sciences. His research focuses on the history of administration from the seventeenth century to the nineteenth century.

Manuel González-Mariscal is a Contracted Professor at the Universidad de Sevilla. His main lines of investigation are history of prices, standards of living, history of consumption, historical demography and agrarian history.

Renate Pieper is Professor at the University of Graz, Austria. Her main fields of research are early modern economic and cultural history, the Spanish Empire and its European connections, especially cultural exchange, communication media and networks, mining, prices and state finances.

Amélia Polónia is Associate Professor at the University of Porto, where she is the Scientific Coordinator of CITCEM (Transdisciplinary Research Centre Culture, Space and Memory). Her current research interests include transoceanic transference patterns and the environmental impacts of colonial dynamics.

Werner Stangl is a researcher at the University of Graz, Austria. His research focuses on social and economic history and historical geography of Spanish America, employing methods of digital and spatial humanities.

List of Figures

List of Tables

Part I

Introduction

1

Mining, Money and Markets in the Early Modern Atlantic: Digital Approaches and New Perspectives

Renate Pieper, Claudia de Lozanne Jefferies
and Markus Denzel

In 1635, in Antwerp, the Flemish diplomat and court painter Peter Paul Rubens designed and constructed an "Arch of the Mint" (Fig. 1.1) for the ceremonial entry of Cardinal Infante Ferdinand,[1] the new governor

[1]Ferdinand of Spain and Portugal (Spanish: Fernando de Austria; 1609/10–1641), the third son of Philipp III of Spain, since 1619 administrator of the archbishopric of Toledo and cardinal, since 1620 archbishop of Toledo, therefore known as Cardinal Infante; 1632–1633 Viceroy of Catalonia, 1633–1634 governor of the duchy of Milan; 1633–1641 governor of the Spanish Low Countries.

R. Pieper (✉)
Institute of History, University of Graz, Graz, Austria
e-mail: renate.pieper@uni-graz.at

C. de Lozanne Jefferies
School of Arts and Social Sciences,
City, University of London, London, UK
e-mail: claudia.jefferies.1@city.ac.uk

M. Denzel
Historisches Seminar, Universität Leipzig, Leipzig, Germany
e-mail: denzel@rz.uni-leipzig.de

© The Author(s) 2019
R. Pieper et al. (eds.), *Mining, Money and Markets
in the Early Modern Atlantic,* Palgrave Studies in Economic History,
https://doi.org/10.1007/978-3-030-23894-0_1

Fig. 1.1 From mining to money (*Source* Peter Paul Rubens, The Arch of the Mint, 1635. (c) KMSKA. Photo Rik Klein Gotink)

of the Spanish Low Countries, who the year before had won a resounding victory at the battle of Nördlingen. For this *Joyeuse Entrée*, Rubens composed a collection of allegorical images linking mining and money on both sides of the Atlantic: miners, resembling those in the woodcuts of Georg Agricola, supposed to represent indigenous workers at the silver-rich mountain of Potosí in modern Bolivia; Vulcan forging coins and medals, fuelling commerce in the city of Antwerp; American parrots and monkeys; and Jason and Medea seeking the Golden Fleece.

Inspired by Rubens, this volume brings together a collection of articles on subjects previously dispersed across the breadth of economic history literature. The resulting mosaic traces the entwinement of mining and money across the transatlantic world. Some pieces hitherto missing from the mosaic will inspire further research on the entanglement between these deeply interrelated fields of craft and economy.

The "Arch of the Mint" bound the European commercial centre of Antwerp into the transatlantic trade network, a key step in the process of globalisation (Flynn 2003; Martínez Ruiz 2018). In the same vein, this book focuses on globalisation and the entanglement of the Americas with Western Europe in a "transatlantic world". In particular, it focuses on the territories of the Spanish Habsburgs—carriers of the Golden Fleece—as well as some of their competitors in continental Europe: the northern Netherlands, the German territories and France.

Rubens designed his arch in 1635, at the end of a long period of intensive silver mining in Spanish America (TePaske 2010, p. 20; Garner 2007). In order to capture the wide-ranging effects of mining on the transatlantic world, the studies assembled here examine not only that first silver boom around 1600 but the whole period from the beginning of the American mining industry in the early sixteenth century, through its decline and recovery in the second half of the seventeenth century and its zenith in the second half of the eighteenth century to the transformations, driven by the emergence of nation states in the Americas, which took place during the early nineteenth century.

As Rubens's allegory suggests, precious metals—gold and silver—from the Americas (e.g. Lopezosa Aparicio 1999; Hausberger and Ibarra 2014) played a vital role in pre-modern economies, whether they were traded as commodities in the form of bullion or fuelling merchant networks as specie. Due to its relative abundance, silver was the most frequently exchanged metal across the early modern Atlantic. This book tracks the trajectory and transformations of silver from its origin in ores on both sides of the Atlantic up to its arrival in the financial centres of Central and Western Europe, where it was traded mainly as currency. As a point of comparison, it also examines the mining and monetary use of copper. Mining and its products entangled both sides of the Atlantic and, as well as economics, heavily influenced the technical, social, political, administrative, theoretical and cultural aspects of early modern society. This volume scrutinises the fluctuations and interrelationships associated with silver and copper mining in the early modern Atlantic world, but excludes gold and the gold-producing regions of Africa and the Americas.

Given the current state of the literature, this book aims to bring the interrelated fields of mining history, minting history and monetary history into a common discourse (cf. Kellenbenz 1981; Fischer et al. 1986, vol I; Van Cauwenberghe 1989, 1991). Recent literature has largely centred on the Pacific, highlighting the connections between Spanish America and the Asian world (e.g. Depeyrot and Flynn 2016; Denzel 2014; Hirzel and Kim 2008; Kim 2013; Flynn 1996, 1997, 2003; Von Glahn 2003). In contrast, this volume focuses on the Atlantic—roughly two-thirds of American bullion exports in the early modern period were shipped east, across the Atlantic (cf. Pieper 1990, 1992; TePaske 2010).

Further, our analysis emphasises the technical, geographical and economic aspects of mining and minting. Recent literature tends to engage with social rather than technical issues (Povea Moreno and Zagalsky 2017; Barragán Romano 2017; Bakewell 1997, 2011; Robins 2011) and those studies which do deal with technical questions are dated or—with few exceptions (Sánchez Gómez and Pieper 2000)—stress a unidirectional European impact on Spanish American mining industries (Brown 2012; Platt 2012; Hausberger 1997; Sánchez Gómez 1995, 2017; Sánchez Gómez and Mira Delli-Zotti 2000; Lacueva Muñoz 2010). In contrast, we investigate how the combination of indigenous and European methodologies led to innovation on both sides of the Atlantic so that, for instance, American mining experts were sought after for European ventures and the extent to which mining in the

Americas drove technical innovation in eighteenth-century Europe. We engage with social and political questions as well, but do so with respect to human geography using digital approaches. In the realm of economic questions, we draw upon recent archival research on prices, interest and exchange rates. Other publications leveraging similar economic indicators rely mostly on time series published some time ago (Munro 2002, 2003).

The impact of American precious metals on European concepts and monetary theories is still the subject of debate which began in the sixteenth century and re-emerged at the end of the twentieth century (Bernholz 1992; de Lozanne Jefferies 1997, 2014; Denzel 2004; Lucassen 2007). This book focuses on the role played by precious metals in the form of specie in the development of early modern state administration and the emergence of theoretical debates in early modern economic treatises, which still persist in modern monetary theory. Finally, this volume also engages with the cultural imaginaries—evident in Rubens's painting—associated with silver, both as bullion and as specie. Up until now, specie, in contrast to diamonds, has not generally been considered a cultural issue worthy of study in itself (Siebenhüner 2018).

The following contributions rely mainly on the methodologies of economics, economic history and history. In the last two decades, economic history in particular has experienced an integration of computerised statistical tools as state of the art, but the implementation of digital geographical tools (GIS) as an instrument for research is still in its infancy. Therefore, several chapters discuss the use of this relatively novel software as a tool for researching and explaining economic histories.

To analyse the trajectory of silver and copper from the mines to their ultimate cultural impact, we cover a broad transatlantic area and time span, from the commercial revolution of the late Middle Ages to the Industrial Revolution. Further, we employ a number of theoretical approaches. Some chapters apply economic theories, and others centre on legal, political or cultural concepts. Given the variety of methodological approaches, geographical entities and political changes, especially around 1800, a common terminology would obfuscate the differences in approach without aiding in comprehension. Some chapters use terms such as "empire" for the Holy Roman Empire as well as for the Spanish Empire, which encompassed the American viceroyalties, Spain and the southern Netherlands and for some time the kingdoms of Portugal and Naples. Other contributions avoid the concept of "empire", pointing to post-colonial theories, and prefer to use the term "transnational" even before the existence of nation states. Most texts dealing with Spanish America use "colony" and "colonial" for those territories before their political independence, although such terminology was not employed by contemporaries. They rather spoke of the Viceroyalty of New Spain, encompassing the modern south-western United States, Mexico and Central America north of Panama, and the Viceroyalty of Peru, encompassing Spanish America south of Costa Rica up to the early eighteenth century and then splitting up into three viceroyalties over the course of that century. Spanish American elites stressed this terminology for a long time as it underlined that they were ruled by *viceroys* like the kingdoms of Naples, Valencia, Aragón and Portugal, not by *governors* like the southern Netherlands. The vernacular only shifted towards the use of "colony" with the independence movement and the emergence of new elites in the early nineteenth century.

Irrespective of their chosen terminology, the chapters of this volume are united in stressing the agency of all the inhabitants of the Atlantic world, whether they resided in a colony or a nation state. This book shows how individuals from both sides of the early modern Atlantic actively exchanged ideas and participated in the mining, minting and marketing of silver and copper.

To ground "from mining to money" with a theoretical approach, Peter Bernholz assesses the importance of precious metals as commodities and as money in the development of complex economies. Arguing from an economic theory perspective, he focuses on money and currency from Antiquity to the Middle Ages, establishing that information, transaction and transport costs would have been much higher without the use of valuable commodity money. It seems obvious that with higher costs, many goods would not have been produced and brought to market, but this does not mean that commodity money was invented for that purpose: rather, with the development of a decentralised economic system, valuable imperishable objects became an effective means of payment and therefore supported further economic growth. Analysing later European monetary history, Ludovic Desmedt and Jerôme Blanc interpret monetary reform attempts and monetary debates through the lenses of three monetary systems: a dualist monetary system (articulating an "imaginary money" with "real currencies"), a metallic system (fully fledged metallic money with invariable legal tender) and a credit-money system (wherein the issued currency is backed by a real asset in which it can be repaid or is defined with reference to such an asset). These three systems refer to distinct ways of conceiving the quality of money and, thus, what can be "good money". Finally, Renate Pieper finds that silver was presented in two distinctive ways in still lives and vanitas paintings in the northern and southern Low Countries and in Spain during the Golden Age of the seventeenth century. In the form of bullion, of a commodity for trade or of manufactured artefacts, silver signified wealth and conspicuous consumption, but in the form of coins, it connoted human vice, gambling and warfare. This dichotomous representation underlay contemporary descriptions of the arrival of Spanish silver fleets and the further distribution of silver within Europe in handwritten newsletters and printed newspapers. Money needed to be backed to a certain extent by (American) silver and the real value of bullion, but *as money*, silver was used as a representation of evil in its most tangible form.

Tracing the trajectories of silver and copper, Part II of this book examines specific aspects of mining activity, particularly in early modern Mexico (New Spain) and Spain between the sixteenth and nineteenth centuries. Werner Stangl begins with a geographical approach, using GIS with three different examples to establish how situating our sources in the proper geographic context improves our understanding of early modern mining in the Americas. As a further example of geographical contextualisation, Amélia Polonia and Johan García present a study of copper mining in Michoacán, demonstrating that European cultural patterns were not the only ones responsible for altering landscapes; indigenous techniques, combining both local and European methods, were crucial for the success of the private copper mining industry in the sixteenth century. The situation in private American silver mining enterprises was similar, as demonstrated in Domenic Hofmann's contribution, and so the Spanish Crown often employed mining experts born in the Americas to promote a precious metal mining sector in mainland Spain. Even though these efforts met with little success, the Crown continued to invest in the Spanish mining sector, because relying on the influx of American silver was deemed too risky. Peter Marckhgott-Sanabria shows that Spanish silver mining ventures continued into the late seventeenth century with the purpose of training officials for the quicksilver mines providing Spanish America with mercury. By then, the Council of Finances had become the central administrative institution of the "public" Spanish mining sector and pursued extensive information on mining-related issues. This section establishes that the transformation of the landscape and local geography was considerable and entangled mining activities on both sides of the Atlantic, regardless of whether they were private or crown-administrated enterprises.

Part III focuses on the transformation and interaction between bullion and specie. Claudia Jefferies investigates the controversy between mine owners and merchants in early modern New Spain. The results suggest that the allegedly high premium charged for exchanging bullion and specie on-site fell within the usual range of charges for similar financial services that linked Spanish colonial mining towns with European financial centres. The story of copper currency in the first decades of nineteenth century-Mexico, told by Enrique Covarrubias, illustrates the problems arising from decentralisation of the minting process, the financial and technical decay of the old Mexico City mint and the failure of contemporary Mexican politicians and intellectuals to establish a monetary order, at least in comparison with their efforts to reorganise and strengthen the treasury. Finally, Harald Kleinberger-Pierer takes a specialised technical approach: technical documentation, expertise and literature were intimately connected to minting. As the methodologies and the scope of technical drawings of minting machines developed, so did the requirements and demands for technical expertise to be met by the court and its administration. This section on minting technologies in early modern times focuses on the costs of minting: the authorities aimed to maximise the seigniorage and minimise the overheads. When this was finally achieved, it became apparent that minting could be a more profitable affair than selling bullion.

The final section deals with the impact of minted and unminted silver and copper money on commodity and financial markets in Europe and Mexico from the sixteenth to the early nineteenth centuries. Manuel González-Mariscal reopens the debate on the price revolution of the sixteenth century. He presents a new price series, including the prices of accommodation in Seville, the most important and fastest growing transatlantic trading port of the period. His findings show that, when using commodity baskets, the price level for Seville was much higher than previously thought. For some periods, they can establish the impact of American silver on general price levels. Andrés Calderón Fernández, Rafael Dobado González and Alfredo García-Hiernaux explore the main financial market of New Spain/Mexico. With improved series of silver production and outflow, they demonstrate that there was no relationship between silver production and interest rates. Comparing the situation of the credit market before and after Mexican independence in different contexts, such as the profile of creditors and debtors or the average and median size of loans, they establish that interest rates remained almost stable with only small increases after political independence. Nonetheless, the authors identify a structural break in the credit markets, a consequence of the church losing its status as a provider of credit. Focusing on credit and money in Europe, Markus A. Denzel states that a clear distinction must be made between commodity markets, where current money was used and the markets of high finance for bills of exchange, where crown and state financing participated. The commodity markets existed in continuous dependence on the availability of silver and thus on the price of silver at fairs; the financial markets needed silver or gold as security for their transactions and the price of silver interested bankers only in relation to the prevailing money of account. This section analyses the impact of precious metal flows on markets where they could have caused inflationary or—in the case of shortage—deflationary tendencies, accompanied by the emergence of money of account.

To our minds, the results of the studies presented in this book could inspire further research:

- first, the digitalisation and visualisation of mining and minting geographies;
- second, rethinking price revolutions with reference to the combination of mining, minting and prices;
- third, testing the hypothesis of the stability of nominal interest rates in early modern economies based on precious metals and without inflation expectations.

Like Rubens's arch of the *Joyeuse Entrée,* we have composed a mosaic which brings the Atlantic world together, linking mining to minting to money on both sides of the ocean. We conclude with the motto of the Order of the Golden Fleece in Rubens's banderole:

Pretium laborum non vile!

Bibliography

Primary Source

Rubens, Peter Paul, 1635. *Erepoort van de Munt.* Sheat from the rear, Antwerp. Koninklijk Museum voor Schone Kunsten Antwerpen, Inventory 317.

Secondary Sources

Bakewell, Peter, ed., 1997. *Mines of Silver and Gold in the Americas.* Aldershot: Ashgate.
Bakewell, Peter, 2011. *Miners of the Red Mountain: Indian Labor in Potosi, 1545–1650,* 2nd ed. Albuquerque: University of New Mexico Press.
Barragán Romano, Rossana, 2017. "Potosí's silver and the Global World of Trade (Sixteenth to Eighteenth Centuries)." In: *On the Road to Global Labour History: A Festschrift for Marcel Van der Linden,* edited by Karl Heinz Roth. Leiden and Boston: Brill: 61–92.
Bernholz, Peter, 1992. "The Discovery of the New World and the Development of the Purchasing Power Parity Theory." In: *Economic Effects*

of the European Expansion, edited by José Casas Pardo. Stuttgart: Steiner: 234–261.

Brown, Kendall W., 2012. *A History of Mining in Latin America from the Colonial Era to the Present*. Albuquerque: University of New Mexico Press.

de Lozanne Jefferies, Claudia, 1997. *Geldtheorie und Geldpolitik im frühneuzeitlichen Spanien*. Saarbrücken: Verlag für Entwicklungspoliitk.

de Lozanne Jefferies, Claudia, 2014. "La pensée monétaire en Castille au 17eme siècle." In: *Les Pensées Monétaires dans l'Histoire. LÉurope, 1517–1776*, edited by Jérôme Blanc and Ludovic Desmedt. Paris: Garnier: 743–784.

Denzel, Markus A., 2004. "The System of Cashless Payment as a Basis for the Commercial Integration of Europe and the World." In: *From Commercial Communication to Commercial Integration, Middle Ages to 19th Century*, edited by Markus A. Denzel. Stuttgart: Steiner: 199–248.

Denzel, Markus A., 2014. "Die Wirtschaftsbeziehungen zwischen China und Europa: Seehandel und Zahlungsverkehr von 1702 bis 1914." In: *Jahrbuch für europäische Überseegeschichte* 14: 101–152.

Depeyrot, Georges; Flynn, Dennis O., 2016. *From Underground to End-Users: Global Monetary History in Scientific Context*. Wetteren: Moneta.

Fischer, Wolfram; McInnis, R. Marvin; Schneider, Jürgen, eds., 1986. *The Emergence of a World Economy, 1500–1914*. Wiesbaden: Franz Steiner Verlag.

Flynn, Dennis O., 1996. *World Silver and Monetary History in the 16th and 17th Centuries*. Aldershot: Ashgate.

Flynn, Dennis O., 1997. *Metals and Monies in an Emerging Global Economy*. Aldershot: Ashgate.

Flynn, Dennis O., 2003. *Global Connections and Monetary History*. Aldershot: Ashgate.

Garner, Richard, 2007. "Dataset: Spanish-American Silver Registrations." In: *Inside My Desk: History Data Desk*. http://www.insidemydesk.com. Accessed October 1, 2019.

Hausberger, Bernd, 1997. *La Nueva España y sus metales preciosos: La industria minera colonial a través de los 'libros de cargo y data' de la Real Hacienda, 1761–1767*. Frankfurt am Main: Vervuert.

Hausberger, Bernd; Ibarra, Antonio, eds., 2014. *Oro y plata en los inicios de la economía global: de las minas a la moneda*. México: El Colegio de México, Centro de Estudios Históricos.

Hirzel, Thomas; Kim, Nanny, eds., 2008. *Metals, Monies, and Markets in Early Modern Societies: East Asian and Global Perspectives*. Berlin and Münster: Lit.

Kellenbenz, Hermann, ed., 1981. *Precious Metals in the Age of Expansion*. Stuttgart: Klett-Cotta.

Kim, Nanny, 2013. *Mining, Monies, and Culture in Early Modern Societies: East Asian and Global Perspectives*. Boston: Brill.

Lacueva Muñoz, Jaime J., 2010. *La plata del rey y sus vasallos. Minería y metalúrgia en México (siglos XVI y XVII)*. Sevilla: CSIC, Universidad de Sevilla, Diputación de Sevilla.

Lopezosa Aparicio, Concepción, dir., 1999. *El oro y la plata de las Indias en la epoca de los Austrias: exposición*. Madrid: Fundación ICO.

Lucassen, Jan, 2007. *Wages and Currency: Global Comparisons from Antiquity to the Twentieth Century*. Bern: Lang.

Martínez Ruiz, José Ignacio, ed., 2018. *A Global Trading Network: The Spanish Empire in the World Economy (1580–1820)*. Sevilla: Editorial Universidad de Sevilla.

Munro, John H., 2002. *Prices, Wages, and Prospects for 'Profit Inflation in Brabant, England, and Spain, 1501–1670: A Comparative Analysis*. Toronto: University of Toronto, Department of Economics and Institute for Policy Analysis.

Munro, John H., 2003. "The Monetary Origins of the 'Price Revolution': South German Silver Mining, Merchant Banking, and Venetian Commerce, 1470–1540." In: *Global Connections and Monetary History, 1470–1800*, edited by Dennis O. Flynn, Arturo Giráldez, and Richard von Glahn. Aldershot: Ashgate: 1–34.

Pieper, Renate, 1990. "The Volume of African and American Exports of Precious Metals and his Effects in Europe, 1500–1800". In: *The European Discovery of the World and Its Economic Effects on Pre-industrial Society, 1500–1800*, edited by Hans Pohl. Stuttgart: Steiner: 97–117.

Pieper, Renate, 1992. "American Silver Production and West European Monetary Supply in the Sixteenth and Seventeenth Century." In: *Economic Effects of the European Expansion*, edited by José Casas Pardo. Stuttgart: Steiner: 77–98.

Platt, Tristan, 2012. "Container Transport: From Skin Bags to Iron Flasks: Changing Technologies of Quicksilver Packaging Between Almadén and America, 1788–1848." In: *Past and Present* 214, 1: 205–253.

Povea Moreno, Isabel M.; Zagalsky, Paula C., eds., 2017. "Dossier 'Conflictos y violencia en los distritos mineros de América española (siglos

XVI–XVIII)'." In: *Revista Historia y Justicia* 9. https://doi.org/10.4000/rhj.1096.

Robins, Nicholas A., 2011. *Mercury, Mining and Empire: The Human and Ecological Cost of Colonial Silver Mining in the Andes*. Bloomington: Indiana University Press.

Sánchez Gómez, Julio, 1995. "La minería en el siglo de Oro." In: *La Ciencia Española en Ultramar*, edited by Alejandro Díaz Torre, Daniel Pacheco Fernández, Tomás Mallo Gutiérrez, and Angeles Alonso Flecha. Madrid: Ediciones Doce Calles: 107–111.

Sánchez Gómez, Julio, 2017. "De la nada a la cúspide. La minería en el siglo XVI Hispano." In: *Modernidad de España. Apertura Europea e integración atlántica*, edited by Antonio Miguel Bernal. Madrid: Marcial Pons: 553–567.

Sánchez Gómez, Julio; Mira Delli-Zotti, Guillermo, 2000. *Hombres, Técnica, Plata. Minería y sociedad en Europa y en América, siglos XVI–XIX*. Sevilla: Ediciones Aconcagua.

Sánchez Gómez, Julio; Pieper, Renate, 2000. "Tras las huellas de un espejismo: La minería en Nueva España y Europa Central en la segunda mitad del siglo XVIII." In: *Jahrbuch für Geschichte Lateinamerikas* 37: 49–72.

Siebenhüner, Kim, 2018. *Die Spur der Juwelen. Materielle Kultur und transkontinentale Verbindungen zwischen Indien und Europa in der Frühen Neuzeit*. Köln and Weimar: Böhlau.

TePaske, John J., 2010. *A New World of Gold and Silver*, edited by Kendall W. Brown. Leiden: Brill.

Van Cauwenberghe, Eddy H. G., ed., 1989. *Precious Metals, Coinage and the Changes of Monetary Structures in Latin-America, Europe and Asia (Late Middle Ages—Early Modern Times)*. Leuven: University Press.

Van Cauwenberghe, Eddy H. G., ed., 1991. *Money, Coins, and Commerce: Essays in the Monetary History of Asia and Europe (From Antiquity to Modern Times)*. Leuven: University Press.

Von Glahn, Richard, 2003. "Money Use in China and Changing Patterns of Global Trade in Monetary Metals, 1500–1800." In: *Global Connections and Monetary History, 1470–1800*, edited by Dennis O. Flynn, Arturo Giráldez, and Richard von Glahn. Aldershot: Ashgate: 187–205.

Part II

Theory

2

Money in History Based on Precious Metals

Peter Bernholz

The Importance of Money in a Decentralised Market Economy

In a pure barter economy without money, the exchange of goods and services is difficult or impossible for a number of reasons. First, in any bilateral exchange between two persons, it is necessary that each of them can offer the precise goods desired by the other (a "double coincidence" of demand and supply) and that the two persons seeking that precise exchange are able to find one another. This imposes significant search costs over time and the wastage of resources, for example because of the necessity of travelling to different places in search of exchange partners. This is true even if we assume that both partners agree on the relative value of the goods to be exchanged; even then, the quantity of goods offered and demanded may be different. This limits the amount

P. Bernholz (✉)
University of Basel (Emeritus), Basel, Switzerland
e-mail: peter.bernholz@unibas.ch

© The Author(s) 2019
R. Pieper et al. (eds.), *Mining, Money and Markets
in the Early Modern Atlantic*, Palgrave Studies in Economic History,
https://doi.org/10.1007/978-3-030-23894-0_2

of goods to be exchanged to the smallest quantity in the lowest offer or demand. More importantly, even if all possible bilateral exchanges of this kind take place, no exchange requiring the participation of more than two people could be realised.

How can the informational and search-related costs in a barter economy be reduced? One measure could be the introduction of local markets at pre-agreed times and locations. After a corresponding announcement, everybody would know where to find potential partners for their desired exchange of goods. However, this innovation does not immediately do away with the requirement for a double coincidence of demand and offer. In order to ensure the existence of market shops for all pairs of goods which could be exchanged, in a market with n goods, it would be necessary to create $n*(n-1)/2$ shops. For $n = 1000$ goods, for instance, the market would require 49,950 shops, which would be prohibitively expensive. Given these logistical constraints, it is not surprising that no such institutions developed in the past to enable bartering among all possible pairs of goods. Indirect exchange using a generally accepted means of payment, for instance commodity money in the form of gold or silver coins, is a way out of this dilemma. In an economy with commodity money market, participants only need to be convinced that the means of payment will retain their value so that they can be used by the new owner for further purchases. The use of precious metals helps to provide that certainty.

If indirect exchange using money is possible then the double coincidence, that is the search for someone who offers exactly the good one needs and demands the good one offers, is no longer necessary. As a result, barter is separated into two phases: first, the offered good is sold for money and then the desired good can be purchased from someone who sells it. The use of money eliminates the high search and information expenses associated with a pure barter economy. If we once again consider a local market on certain days, the saving of resources is obvious: creating one thousand market shops for one thousand goods is much less expensive than creating 49,950 and keeping informed about one thousand shops is substantially less demanding than the impossible task of staying informed about 49,950. Fewer resources are needed to inform oneself about one thousand shops. People who want a given product only need to know which shop sells it. Similarly, the supplier of a product only needs to be informed about where he should go to sell it. Certainly, costs in such an economy are increased by the fact that two transactions—sale and purchase—are necessary instead of one, but in a market with many goods that increase is more than compensated by the savings associated with the use of money.

The example below illustrates this point (Fig. 2.1). It can easily be extended to situations of multilateral exchange with more than three people. In Fig. 2.1, we assume that the people involved have already agreed on the relative value of the three goods to be exchanged. We further assume that the amounts offered and demanded correspond to the wishes of the respective persons. The three potential participants A, B and C are all offering goods demanded by one of the others but, unfortunately, not by the person offering the goods they seek in exchange. In this situation, the use of commodity money like gold or silver coins with intrinsic value is a means to solve their problems. For instance, A is prepared to sell shoes to C for money, since he knows that B is ready to accept it for the 3 kg of wheat he wants. Similarly, B takes the

O: Two pairs of shoes; D: 3 kg wheat

A

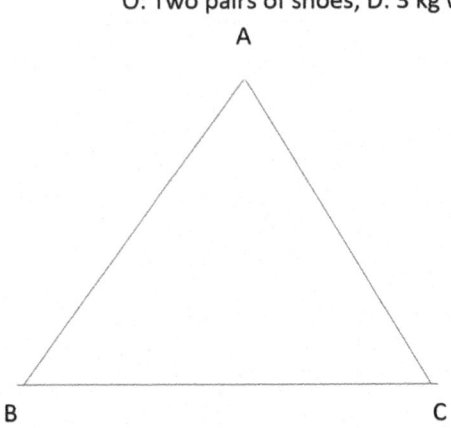

B

C

O: 3 kg wheat; D: One suit   O: One suit; D: Two pairs of shoes

Fig. 2.1 Trilateral barter. O: Offer of good. D: Demand of good

money to buy the suit he desires from C. The existence of commodity money enables the separation of one complicated multilateral barter exchange (where all participants would have to know their offers and wishes and to meet in one place for their exchanges) into three separate bilateral transactions.

There is one condition: the person executing the first payment must have a sufficient amount of commodity money. It has to be available somewhere in their society. Now, as Adam Smith observed, this is a waste of resources because, for example, the silver for coins has to be mined, refined and minted.

On the other hand, the search and information-gathering costs associated with bartering are saved by the use of commodity money. As the number of goods that can be offered and demanded grows, the search and information costs in a barter economy increase exponentially; the expenses for using commodity money are swiftly dwarfed in comparison.

The use of money, further, eliminates the need for all participants to come together for their transactions since money enables them to buy and sell goods bilaterally. Participants in the economy only need to know which shops are selling which goods, as well as the time and location of market days. All multilateral exchanges can now be separated into individual bilateral acts of buying and selling. The associated costs are reduced dramatically, so that transactions become possible which would have been prohibitively expensive before the introduction of commodity money.

The costs associated with producing commodity money by mining, refining and minting gold and silver coins directly suggest future developments: first, a partial substitution of commodity money with banknotes convertible into gold or silver at a fixed rate; second, a further substitution of banknotes or coins with convertible bank accounts; and finally, the introduction of inconvertible banknotes, bank accounts or electronically transferable claims, that is the total abolition of commodity money. However, the savings associated with moving away from expensive-to-manufacture commodity money have often turned out to be very expensive, since governmental monetary policies have frequently led to high or hyperinflation. By contrast, the debasement of commodity money caused by reduction of precious metal content or coin weight is limited to about 8% annual inflation, as evident for instance in the history of the Roman Empire in the fourth century A.D.

The Creation and Use of Coins in Lydia and Ancient Greece

The history of using indirect means of payment begins with the Sumerians and Akkadians in Mesopotamia during the third millennium B.C. These people used silver, in the form of rings or coils, and barley for transactions, demonstrating the historical importance of commodity money as a means of enabling cheaper economic transactions. As there were no silver deposits in Mesopotamia, it had to be imported from Asia Minor, which required a surplus in the balance of trade and also imposed significant transport costs.

Further, the minting of silver had not yet been invented, and so it had to be weighed to determine its value. Loans in silver were available at an interest rate of 20%, whereas loans of barley were given at a rate of 30%, demonstrating the superiority of silver as a means of payment (Van de Mierop 2014). It seems that the first coins were minted at the end of the seventh century B.C., as reported by the historian Herodotus, who grew up in Halicarnassus, now Bodrum, in Asia Minor:

> The Lydians were the first humans of whom we know that they minted gold and silver coins. (Herodotus, Histories 94,1)

Herodotus's report has been confirmed by the discovery of coins below the ruined temple of Artemis in Ephesus, dating to the final decades of the seventh century B.C.

The Lydians were in close contact with the Ionian Greeks, who had settled on the coasts of Asia Minor over several centuries. Over time the Lydians integrated into all the Greek cities except Mycenae. Given the innovative talents of the Greeks, it is not surprising that the development of coinage spread rapidly throughout most Greek territories including mainland Greece, Sicily and other Greek settlements.

Fig. 2.2 Didrachm from Lucania (Sybaris), 550–510 B.C (*Source* http://gams.uni-graz.at/o:numis.35)

Table 2.1 Weights of Athenian and Corinthian silver coins in grams

Athens						
Year (B.C.)	575–525	525–500	460–450	393–300	167–166	
Didrachm	8,429					
Tetradrachm		16,949	17.14	17.19	16.89	
Corinth						
Year (B.C.)	570–550	525–500	515–500	460	420	320
Stater	8.55	8.66	8.61	8.67	8.59	8.61

Source Jenkins and Küthmann (1972)

Over time the Attic silver coin—the drachma—developed into an internationally recognised form of money, especially in the Aegean Sea but also in Naucratis, a Greek trading port in Egypt. Another important coin was issued by Corinth and both monies remained stable for centuries, except for a limited period during the Peloponnesian War when Athens lost control over its silver mines, but returned to the old parity of its coins after the war (Fig. 2.2 and Table 2.1).

The long stability of Greek silver coins is remarkable given the fact that politicians are usually characterised by an inflationary bias. This bias is mainly caused by the demand for expenditure on extraordinary

Table 2.2 Stable money and debasement in the late Middle Ages, first row of each location: weight in grams, second row: value in relation to ca. 1300 (=100%)

Value of gold florin in different currencies	c. 1300	c. 1400	c. 1500	Currency
Castille	5.8	66	375	Maravedi
	100	1137.93	6465.52	
Cologne	6.67	42	112	Schilling
	100	630.00	1680.00	
Flanders	13.13	33.50	80	Groot
	100	255.24	609.52	
Austria	2.22	5	11	Schilling
	100	225.00	495.00	
France	10	22	38.75	Sou
	100	220.00	387.50	
Hanse	8	10.5	31	Schilling (of Lübeck)
	100	131.25	387.50	
Rome	34	73	130	Soldo
	100	214.71	382.35	
Florence	46.50	77.92	140	Soldo
	100	167.56	301.08	
Bohemia	12	20	30	Groschen (of Prague)
	100	166.67	250.00	
Venice	74	93	124	Soldo
	100	125.68	167.57	
Aragon	11.50	12.71	16	Sueldo
	100	110.51	139.13	
England	2.67	3	4.58	Shilling
	100	112.50	171.88	

Source Spufford (1986, Table I)

and ambitious projects or for financing wars. In such situations, politicians are either unable or unwilling to cover the additional costs through tax increases or cuts in other spending, so they turn to inflation to raise funds.

This happened in the Roman Empire during the third and fourth centuries, which saw continuous reductions of the silver content of circulating coins. Later, a gold currency was introduced that remained stable for a long period in the eastern empire, while the west was conquered by Germanic tribes.

Despite a currency reform introduced by Charlemagne around 800 A.D., Western Europe until the twelfth century saw the use of hundreds of different nominal silver penny coins, each of which had evolved separately from the Carolingian denier. Their circulation was generally limited to their immediate neighbourhood. There was also an intermediate denomination, the soldo, derived from the Byzantine gold solidus, which was established as a unit of twelve denarii during the transition from a gold-based currency to a silver-based in the seventh and eighth centuries (Spufford 2014, p. 231).

In the late Middle Ages, the Italian city states flourished, and Venice and Florence introduced gold coins which remained stable for about three centuries (Table 2.2). As evident in the table, the Fiorino d'Oro of Florence held its value for about three centuries. The same was true for the Venetian gold ducat, whereas the silver coins of both states slowly lost their value, albeit to a lesser extent than the silver coins issued by other states. For Florence, the average annual inflation during these 200 years amounted to less than 1%. For Venice, it was even less.

The Introduction of Commodity-Based Paper Money

The crucial financial and monetary innovations of the early modern period originated in north-western Europe, and it was this region that benefitted most from the shift in economic balance and wealth from the Mediterranean to the North Sea and Atlantic which took place in the wake of the European overseas expansion between the 1490s and 1620s. The central financial and goods markets within this system, which eventually developed into the global financial markets of their age, were, of course, Antwerp, followed by Amsterdam, which was in turn gradually replaced during the eighteenth century by London (Denzel 2014, p. 253).

For our purpose of studying the further development of commodity money—that is of silver and gold coins—we are mainly interested in the creation of banknotes convertible into gold and silver as a substitute for still-circulating coins, connected by a fixed exchange rate.

In the pound sterling, England had what was arguably one of Europe's most stable currencies, increasingly based on gold from the eighteenth century on. Pound sterling was used in London and Amsterdam alike and became, after the Dutch currency, Europe's most important means of foreign exchange. There were two elements crucial to the process that we call a "Financial Revolution," which characterised England's commerce and trade from the late seventeenth century. First, a stable currency (the pound sterling) that was now increasingly backed by gold and, second, the Bank of England—a private venture with royal monopoly and permission to issue notes in lieu of cash (Denzel 2014, p. 263).

With the introduction of banknotes convertible at a fixed parity into gold coins, the first step away from commodity-based money had been taken. In fact, during the Napoleonic wars the convertibility of banknotes of the Bank of England was abolished for several years, entirely abandoning commodity-based currency, though afterwards it was restored at the old parity. All the leading European powers abolished banknote convertibility into gold at the beginning of the First World War, and again, once and for all, during the Great Depression of the 1930s. In 1935, President Roosevelt forbade the possession of gold for private American citizens. Finally, when in 1971 President Nixon abolished the last remaining convertibility of the dollar into gold at a fixed parity for central banks, the world had moved to intrinsically worthless money. The costs implied by the use of commodity money, which Adam Smith had complained about, were now saved. That saving, however, came at the price of significant fluctuations between exchange rates, higher inflation and at least 30 instances of hyperinflation with more than 50% per month since the First World War in several countries.

Historical Consequences of Introducing Valuable Commodity Money

We have already seen that without the introduction of valuable commodity money, the information, transaction and transport costs of economic activity would have been much higher than commodity money allowed them to be. It seems also obvious that with such prohibitively high costs many goods would never have been produced and entered the market. This does not, however, mean that commodity money was invented for the purpose of reducing overheads and allowing new goods into the market. Rather, with the development of a decentralised economic system, intrinsically valuable imperishable objects were used as means of payment and supported further economic growth. This hypothesis is supported by the observation that decentralised polities like Sumer and Akkad as well as the classical Greek city states and late medieval countries not only showed the most impressive economic developments but also excelled in many other fields. The development of writing and many other artefacts and techniques were developed in Babylonia; mathematics, poetry and arts in Greek and Hellenistic states; and the accomplishments of the Renaissance and of early modern times built upon and further enhanced those successes.

Bibliography

Bagnall, Roger S., 1985. *Currency and Inflation in Fourth Century Egypt.* [s. l.]: Scholars Press.

Denzel, Markus A., 2014. "Monetary and Financial Innovations in Flanders, Antwerp, London and Hamburg, Fifteenth to Eighteenth Century." In: *Explaining Monetary and Financial Innovation*, edited by Peter Bernholz and Roland Vaubel. [s. l.]: Springer: 253–282.

Deutsche Bundesbank, 1980. *Antike Goldmünzen in der Münzsammlung der Deutschen Bundesbank.* Frankfurt am Main: Deutsche Bundesbank.

Jenkins, G. K.; Küthmann, Harald, 1972. *Münzen der Griechen.* München: Ernst Battenberg.

Spufford, Peter, 1986. *Handbook of Medieval Exchange.* London: Royal Historical Society.

Spufford, Peter, 2014. "The Provision of Stable Monies by Florence and Venice, and North Italian Financial Innovations in the Renaissance Period." In: *Explaining Monetary and Financial Innovation*, edited by Peter Bernholz and Roland Vaubel. [s. l.]: Springer: 227–251.

Van de Mierop, Marc, 2014. "Silver as a Financial Tool in Ancient Egypt and Mesopotamia." In: *Explaining Monetary and Financial Innovation*, edited by Peter Bernholz and Roland Vaubel. [s. l.]: Springer: 17–29.

3

Debating Sound Money in Early Modern Europe: From Dualist to Metallic Monetary Systems

Jérôme Blanc and Ludovic Desmedt

Introduction: The Quality of Money as a Thread Running Through the History of Monetary Ideas in the Early Modern Period

If there is a single question that runs through monetary discourse from the European Renaissance through to the late eighteenth century, it is that of "sound money".[1] Numerous writers bemoan the parlous state of the system they write about or from which they develop their ideas of what a sound monetary system might be. Accordingly, the notion of the "quality of money" is central to that idea. Of course, the overarching question is how to define that "quality" and what precisely is meant

---

[1] This text draws on contributions to Blanc and Desmedt (2014b), namely Louis Baeck, Niall Bond, José Luís Cardoso, Jean Cartelier, Lucien Gillard, Gilles Jacoud, Claudia de Lozanne Jefferies, Antoin Murphy, Anders Ögren, Sevket Pamuk, Danila Raskov, Joël-Thomas Ravix, Leif

---

J. Blanc (✉)
Triangle, Sciences Po Lyon, Lyon, France
e-mail: jerome.blanc@sciencespo-lyon.fr

© The Author(s) 2019
R. Pieper et al. (eds.), *Mining, Money and Markets in the Early Modern Atlantic*, Palgrave Studies in Economic History,
https://doi.org/10.1007/978-3-030-23894-0_3

by "sound". Depending on when, where and who is writing, this may relate to a monetary circulation that avoids surfeit and shortage alike; it may focus on the diversity of specie or the ways and means in which it is issued and articulated, from small change of no intrinsic value for the workaday dealings of the common folk to valuable metal coinage suitable for storing wealth and settling international transactions; it may relate to the question of monetary stability, endangered by the machinations of the monarch and the glut of specie; or it may centre on the problem of counterfeiting and the material quality of coinage.

In order to understand the debates of the time, we must focus on the "quality of money", because this was the primary theme common to monetary discourse across all of Europe. This does not mean that questions of quantity are to be overlooked, though, for at least two reasons. First, the quantitative questions of monetary glut or shortage, alongside the problems of legal tender and foreign exchange, relate to matters of quality by way of the inadequate supply of money to the economy, the search for alternative or supplementary means of payment, the switch to other units of account and so on. Second, dealing with questions of quantity (for monetary aggregates) first requires defining homogenous classes of means of payment—an initial approach to defining their quality.

The question of quantity is, therefore, subordinate to that of quality and may appear as a "false problem" (Aglietta and Orléan 1982, p. 145) masking the broader issue of the quality of money. It has to be acknowledged that the question of "quality" and what is "sound" cannot but give rise to controversy. It is tied in with value systems and leads to major political debates regarding the societal project associated with

L. Desmedt
Laboratoire d'Economie de Dijon (LEDi),
Université de Bourgogne, Dijon, Franche-Comté, France
e-mail: Ludovic.Desmedt@u-bourgogne.fr

Runefelt, André Tiran and Carl Wennerlind. Obviously, we thank them for their contributions to this line of thinking, which is not binding on them. We also thank readers of earlier versions of this paper and in particular Kuroda Akinobu, Jean Cartelier, Bruno Théret, Claudia de Lozanne Jefferies, Patrice Baubeau, Anders Ogren and Pierre Alary.

the monetary system. As Christine Desan puts it, "[a]s a practice, money allows great capacity to the stakeholder and those assisting that authority" (Desan 2014, p. 7). This, in particular, is where the ethical dimension of public faith lies (Aglietta and Orléan 1998). Theoretical divergences among authors are related to their political positions and, because of the performativity of ideas, to monetary institutions.

The topic of quality is that much more difficult to address in economics because the monetary system in which these issues arose was very remote from the monetary system of the industrial era. Luigi Einaudi, attempting to shed light on the idea of imaginary money, rightly underscores the extent to which living in a given system leads to the internalisation of ideas which are difficult to understand for later commentators who, through their own ignorance, initially find nothing but confusion in the writings of the time:

> Prior to the French Revolution, the monetary system of most European countries was based on altogether different principles. Contemporary authors could take these principles for granted and did not have to explain them to others. Their strange terminology causes us, who live in another world, to wander for a while in a dark forest. (Einaudi 1953, p. 235)

The difficulty associated with understanding contemporary authors is intensified in a historical context because "mercantilist" and more broadly "pre-classical" writings have largely been dismissed and considered anathema (Blanc and Desmedt 2014b). Under no circumstances should "pre-classical" be taken to mean "pre-analytical"; on the contrary, while classicism contributed to the development of thinking based on the dichotomy between "real" and "monetary" spheres, many earlier authors developed formulations which combined economic and monetary analyses. This passage from Mark Blaug is a fine summary of the anti-mercantilist steamroller that from the outset discredited most of the authors we shall examine:

> Once upon a time everyone believed that national prosperity depended upon the accumulation of gold bullion resulting from a favourable balance of payments. A man called Adam Smith denied this and later another man called David Ricardo actually proved that free trade and leaving the balance

of payments to take care of itself produced the best of all possible worlds. Thereafter, every schoolboy could expose the fallacies of 'mercantilism.' Members of the German Historical School protested that the mercantilists had been misunderstood, but in the 1930s the big guns of international scholarship went one better than the schoolboy: the mercantilists were not only poor theorists but also poor historians, writing in almost total ignorance of the economies in which they lived. (Blaug 1964, p. 111)

This chapter presents a fresh reading of the monetary debates of the early modern period in the light of the changes in the monetary institution, rejecting an ex-post reading that would cast those authors back into the darkness of "pre-analytical" understanding. Instead, we suggest that the supposed ignorance of authors of the period with respect to the economy of their own age (and therefore with respect to their monetary system) is instead a reflection of the ignorance of contemporary commentators with respect to the economy (and therefore the money) of Europe in the early modern period.

We begin by presenting the analytical framework of the paper, which is constructed around a typology of monetary systems that highlights the circumstances in which money was created. Next, we examine the long-lived dualist system, its principles, its difficulties and the doctrines associated with it. Its crises led to the rise of metallist proposals and the eventual triumph of their vision of what makes a "sound" monetary system. The dualist system was superseded by a metallic system which, despite its triumph, appears untenable because the stabilisation it brought about made the conditions for creating money unmanageably rigid. A fifth section concludes the paper.

An Analytical Framework: Diversity Among Monetary Systems

To get a better grasp of the debates surrounding money in early modern Europe, it is necessary to characterise the prevailing monetary systems on which those debates turned. We begin here with the conceptual framework used by Thomas (1977, pp. 9–94), reworked by Aglietta and Orléan (2002, pp. 123–166) and then by Théret (2007, pp. 64–66).

On the basis of their studies, we can identify four monetary systems in the Western world in the early modern period: the dualist, metallic, convertible and self-referential systems.[2] This paper focuses on the transformations of the monetary system in early modern Europe and the way in which scholarly writings and debates criticised or promoted those transformations, covering ideas, policies and institutions.

In the dualist system, the value in units of account of specie in circulation is defined by monetary proclamations, but is not marked on the coins themselves. Some commentators in the early modern period sought to safeguard the dualist system, despite its difficulties, while others advocated for a metallic system. This paper focuses on the intellectual debates which surrounded the shift from the dualist to the metallic system. When the metallic system was implemented, some authors proposed introducing convertible moneys or even making them the nexus of a new system. Others went so far as to suggest the abandonment of metal and the introduction of a self-referential monetary system (without any intrinsic value). This part of the story will not be addressed by this paper, as it engages with developments that would require a longer study. The fact is that the monetary system Adam Smith (1776) lived in was very different from the one that Copernicus (1526) wrote about. A far-reaching shift came about with the advent of issuing banks, whose credit freed money from an age-old dependence on metal. The "rationale of the seal", characteristic of the merchants' way of creating money (that is minting), was gradually superseded by the "rationale of the signature" characteristic of capitalist minting (Orléan 1998).

It should be specified, however, that the purpose here is not to propose an evolutionary view of monetary systems which would reduce history into stages progressively superseded by the inevitable emergence of some higher level. It is, rather, to shed light on the difficulties of monetary systems and their interrelationships and the evolving forms of pluralistic money. One theoretical challenge for the reasoning that follows is to position these categories as systems. A system is a coherent set of

---

[2]The question of the universal character of these four systems will not be addressed here. Consequently and for the sake of caution, it is not a question of asserting that every creation of money necessarily refers to one of these four systems, but of using this interpretation to cover Europe in the early modern period.

interdependent parts. With historical depth, systems can be envisaged as dynamic phenomena, evolving through the interplay of the parts that compose them (what is coherent at one point in time may subsequently become incoherent) but also through outside shocks (such as the black plague pandemic of the mid-fourteenth century or the *diluvio* of silver from the New World in the second half of the sixteenth century, see Hamilton 1934). An important point is to recognise, in each of the monetary systems to be investigated, the birth of elements that initially seem to be compatible with or even functionally vital to the system in place, or insignificant because they are marginal to it which then, by changing and becoming generalised, transform the system from the inside until they completely overthrow it.

In order to account for these potential tipping points, it is useful to identify for the dualist and metallic monetary systems those elements that are dismissed as external or incompatible (in particular counterfeit money) and those that are tolerated as harmful but compatible (e.g., coins of a bad quality). Both contribute in their own way to weakening and transforming systems; both rely on plurality, whether endured or accepted, inside and outside the system.

We may begin with a definition of money as a system of payment (Cartelier 2013; Cartelier 2018; Aglietta and Cartelier 1998). That system rests upon a set of rules which includes three necessary components: a "nominal unit of account", a minting process (understood as the ability to create means of payment other than by selling goods) and a "procedure for settling monetary balances". On this basis, we shall define a minting regime as the predominant and most stable form of access to means of payment. The evolution of a monetary system may, therefore, stem in particular from a transformation in the minting regime.

The Dualist System and Monetary Policy

It should be specified from the outset that, from the Middle Ages onwards, monetary systems never relied on a single homogenous means of payment. Monetary circulation in sovereign areas was neither uniform nor exclusive of all others. Because the lack of specie was recurrent,

[...] it was delusional to prohibit unlawful specie and foreign specie for so long as local money remained in short supply. The governing powers realised it was more convenient to tolerate such specie and even to make it legal tender (from the second Half of the fifteenth century in France) in order to attract it and potentially channel it towards the royal mints. (Favier 1981, p. 177)

The uniform and exclusive characteristics of money that we are familiar with today are historically specific constructions that only emerged after the end of the dualist system.

Definition of Principles and Historical Deployment

The dualist system provided a means for combatting the "scarcity of money" through the possibility of manipulating coinage. It brought together an "imaginary money" (a unit of account defined by a given weight of precious metal) and "real moneys" (coins struck from precious metals of variable degrees of purity, especially silver, which was more abundant in Europe than gold and copper with the notable exceptions of Sweden and later Russia). Einaudi (1953) illustrates the confusion among authors of the time regarding the notion of "imaginary money": the contemporary definitions he produces appear to be highly diverse and ultimately of little relevance. In addition to those he lists, we might add for example the eighteenth-century definition by Ferdinando Galiani: "what we call *imaginary* money [*immaginiaria*] is money that is not made from a whole piece of metal" (Galiani 1751, p. 169). Galiani specifies that money can be both "ideal" ("which measures prices", p. 147) and "real" (which can be used "to buy things", p. 125).

The fact is, then, that "imaginary money" can only be understood within the dualist system.[3] This does not just come down to there being a unit of account that is separate from a means of payment: monetary systems in which the unit of account is a particular type of means

---

[3]This means that "imaginary money" of the dualist system cannot be likened to a mere unit or money of account, whether official but defined without reference to physical quantities such as a weight of metal or informal because arising from practices of the ordinary people.

of payment in actual circulation are rare. Nor is it just a unit of account for which there is no material form corresponding to unity (unlike, e.g., the Euro, since there is a 1 Euro coin). Under the dualist system, the value, in units of account, of specie in circulation was defined by monetary proclamations but was not marked on the coins themselves. The population was informed of these proclamations by crying the money up or down. In terms of monetary policy, this system provided the possibility of manipulating the value of money. The value of specie might vary easily with enhancements or cry-ups, a kind of manipulation by which the sovereign's monetary authorities could alter their legal rate. Debasement, another kind of manipulation which was less common as it involved recasting existing coins or creating new types of coin, added to the general complexity of monetary circulation.[4] Such manipulations were a way of combatting the scarcity of money, adjusting gold/silver ratios (this is the position of Einaudi 1953, p. 245), and of deriving monetary income for the monarch.

In a dualist system, the metallic definition of the unit of account (e.g., the *livre tournois* in France) was obtained indirectly, by calculating the implicit metallic content of each coin. The difficulty of obtaining a metallic definition of the unit of account meant that the metal was secondary and that it was the *unit of account*, constituted politically, that sat at the centre of the system. It goes without saying that this type of calculation, the basis of much arbitrage and speculation, was feasible only for a very limited section of the population—essentially money changers and merchants (both capable of developing banking activities) as well as sufficiently astute thinkers and, of course, officers of the mint. This is what Jehan Cherruyt de Malestroit (1566/1934), officer of the chamber of accounts in France, set out to show the king with his two paradoxes that, contrary to common belief, no more was paid (in metal) then than three hundred years earlier, but that the beneficiary

---

[4]Though their differences are often blurred in the English-speaking literature on money of the early modern period, we chose here to clearly distinguish between debasements, as real manipulations of the coin (related to changes in their material content), and enhancements or cry-ups, as nominal manipulations (related to changes in the legal value of the coin). See the discussion in Blanc and Desmedt (2010), drawing on Gould (1970).

of a fixed income like a rent received less (in metal) than before. This is also what Jean Bodin (1568) did in his *Réponse* to Malestroit: both wrote on the dualist system and sought to explicate its effects. However, they diverged not only in the choices made for their reasoning (giving a glimpse of the complexity of the phenomenon) but also in their conclusions as to what should be done.

Under the dualist system, money was created by striking metals, under the sovereign's control and on conditions laid down by the sovereign, in a mint. The minting process enabled the sovereign to earn an income in seigniorage. Economically, the fees for seigniorage were the compensation for a service consisting in making available metal discs of uniform sizes and certified weights and metal contents that were recognised and accepted as instruments of payment in transactions. However, seigniorage was more than that: it was a manifestation of political power. It should be added that, under the conditions of the dualist system, real or nominal manipulations were actions in the domain of minting, since they altered the nominal value of money in circulation.

Historically, Einaudi (1953) situates the dualist era between the time of Charlemagne and the French Revolution: a thousand-year reign. Boyer-Xambeu et al. (1994) emphasise that the Carolingian system, built around the *livres, sous* and *denier* (pounds, shillings and pence) and the striking of the silver *denier* (*denarius*, penny) as the keystone of the system,[5] became gradually detached from measures of weight, that the units were soon no longer represented by any coin and that the £/s/d arrangement spread under various names to all of Christendom between the twelfth and thirteenth centuries. The result was that, in any given country, each coin in circulation invariably represented a fraction or multiple in the single system of account by 12 and 20 (and quarter and half for the smallest divisions) (Boyer-Xambeu et al. 1994). Above all, the introduction of gold coins into an area in which the precious metal struck in the mints and in circulation had been silver since the time of Charlemagne gave the system its classical characteristic.

---

[5]It should be recalled that the pound (*livre*) was divided into twenty shillings (*sous*) and the shilling into twelve pennies (*deniers*) in many European countries—with the then Muslim Iberian peninsula being an exception.

The emergence of gold and silver coinage introduced a new constraint on adjustment, which drove the development of new enhancements:

> And so alongside real manipulations, which continued to be possible and practised, a second purely nominal kind of manipulation found its place. In other words [...] a true system of money of account was created, related to the actual coins only by a perpetually variable equivalence: an 'imaginary' money as it was readily called from the sixteenth century on (with, it would seem, initially at least, a nuance of hostility). The pound, according to Turquam's picturesque expression,[6] became comparable to stirrup straps that could be shortened – or lengthened – at will. At the same time it was a tremendous means of action, the political scope of which has perhaps not always been sufficiently appreciated, that was placed in the hands of the governing powers. (Bloch 1954, p. 44)

The twelfth and thirteenth centuries may be thought of as the time when Europe definitively took up the dualist system.

Difficulties Specific to the Dualist System

The way that the dualist system articulated the unit of account and the means of payments provided many possibilities for adjustment in the form of governmental manipulation, but it was also a source of major difficulty whenever the tools for making adjustments were inadequately or unfairly employed. Manipulations had direct effects on distribution: if there was an enhancement between the time a loan was contracted and the time it was repaid, the creditor would receive less metal than he had advanced. It is not difficult to imagine why an indebted monarch might resort to practices of this sort, and history is full of such practices. This was how the pound sterling of 1793 came to be defined by just 35.7% of its weight in gold of 1278 and the *livre tournois* of 1789 by 3.6% of its weight in gold of 1266 (see Aglietta 2002) (Table 3.1).

---

[6]Thomas Turquam was an important officer of the French Cour des monnaies (Currency court) during the troubled period of the years 1570s.

Table 3.1 Depreciation of units of account in weight of fine gold, England and France

France			England		
Reigns and dates of measures	Milligrammes fine gold in the livre tournois	Residual percentage value of the livre tournois	Reigns and dates of measures	Milligrammes fine gold in the pound sterling	Residual percentage value of the pound sterling
Louis IX (1266)	8270	100	Edward I (1278)	20.5	100
Philippe le Bel (1311)	4200	50.7	Edward III (1350)	17.4	84.8
Louis XI (1480)	2040	24.6	Henry VII (1489)	15.47	75.5
Henri IV (1600)	1080	13.1	Henry VIII (1535)	9200	44.9
Louis XIII (1640)	621	7.5	Elizabeth I (1560)	7750	37.8
Louis XIV (1700)	400	4.8	George III (1793)	7320	35.7
Louis XVI (1789)	300	3.6			

Source Excerpt from Cailleux (1980, pp. 253–254)

As Maurice Lagueux put it:

> the obsessive fear of arbitrary devaluations that had darkened the monetary history of the Ancien Régime probably contributed to bringing about this instinctive distrust of an Authority whose interests in monetary matters did not always coincide with those of the citizens. (Lagueux 1990, p. 91)

Three major difficulties coincided with the problem of long-term depreciation of the unit of account by manipulation.

First, the simultaneous use of different metals for the striking of coinage led to a problem with the ratio of values between those metals, especially for the ratio of gold to silver, even if copper may have played a role in some instances. Here lay the problem of bimetallism or monometallism. Enhancements (or nominal manipulations) were a tool for regulating the gold/silver ratio, but if misunderstood or misused, they were rejected, leading many observers to believe that the gold/silver ratio should be fixed definitively and that all manipulations should cease. This was Jean Bodin's (1593) position in particular.

Second, debasements (or real manipulations) complicated monetary circulation and could give rise to the famous mechanism Thomas Gresham is generally credited with identifying. In a letter to Queen Elizabeth I in 1558, he pointed out the dangers that debasement might hold for the country's metal reserves (Gresham 1558). He observed that debasement could lead to the best coin in circulation being exported. Excessive debasement would mean that coins with a lower content of fine metal had the same payment power as older coins with a "richer" composition. The older coins would therefore be exported or melted down to be exchanged abroad, while only the baser coins would remain within the country. Therefore, in the event of debasement, and if the quantity of bad money exceeded current needs for transactions, "bad money drives out good".[7] In other words, when a divergence arose

---

[7]On this condition, see de Roover (1949, p. 93): "Gresham [...] does not state that bad money necessarily drives out good. On the contrary, he shows that bad money may be greatly overvalued and will not drive out the better coins, provided that the baser coins are issued only in limited quantity and not in excess of the needs of trade".

between the legal value and the market value of different means of payment, the one with the highest market value would be exported while coins of lower precious metal content would be used in the settlement of debts. It is worth noting that Gresham was not the originator of this idea; rather, he was merely repeating a widely held opinion.[8] Bimetallism compounded the difficulty by causing a to and fro between gold and silver coin depending on the fluctuations in the gold/silver ratio. The focus of manipulations on copper coinage, like the Spanish *vellón* in the seventeenth century, meant that high-quality gold and silver coins all but vanished (Hamilton 1943, p. 493).

Third, real moneys were themselves dual. While some coins were made of precious metals that formed the basis of their worth and so commanded a high value in units of account and were used in expensive transactions, for international trade and for saving, others were largely fiduciary, of little worth, and were used in day-to-day transactions (see, e.g., Fantacci 2005). While in the first case coins had a legal value comparable with the market value of the metal from which they were made (which does not mean the two values were identical, especially as they were not determined in the same way), in the second case the legal value was much higher than the market value of the metal in the coins. Under these circumstances, weighing and precisely determining the metallic contents by experts of monetary circulation were meaningless as a means of defining the value of a coin. Nevertheless, manipulating the metal content could spark crises because the quality of the metal content could build trust. In Castile, in a context of chronic crisis of pure copper small coinage, *vellóns* made up almost the entirety of the coinage used in making payments; more than 92% of spending in the 1650s was facilitated with copper *vellóns*, according to Hamilton (1943, p. 478). Generally, small change was an intractable problem both theoretically (to what rule should its issue be subjected) and technically (in what form should it be issued), and many monetary crises in Europe

---

[8]"It is the phenomenon that was already vaguely perceived in the ancient world [...] until MacLeod had the odd idea of attributing it to Gresham" (Laurent 1933, p. 9). Hauser speaks of "the famous law discovered by Copernicus and for which the honour went to Gresham" (Hauser 1932, p. xli).

developed over the problem of small change, of which there was either too much or too little.[9]

The Excesses of the Dualist System

Under the dualist system, the royal unit of account and therefore the monarch was at the centre of the monetary world. Counterfeiting therefore consisted in the striking of coins without the sovereign's say-so. However, potential for waywardness in the dualist system was not limited to just this type of forgery. The idea of false money as deployed by writers of the time covered several types of situation[10] (see Table 3.2): enhanced and debased money (through allowing a legal value that was higher in units of account than their metal content, by nominal or real manipulation, respectively), degraded money (by a metal content that did not comply with royal proclamations) or counterfeit money (coins made by individuals who did not hold the sovereign right to strike money). Each of these instances involved falsification, but for different reasons: poor control and abuse (enhanced and debased money), criminal offence (degraded money) and even the crime of lèse-majesté (counterfeit money). Criminals faced death everywhere and could be roasted and boiled in France. As for those rulers who engaged in abuses, they risked being called counterfeiters, the fate that befell Philippe the Fair of France in Dante's *Divine Comedy*, whose extraordinary manipulations have gone down in history. Some decades after that episode, Nicole Oresme considered that "every change [i.e. manipulation] of money, except in the very rare cases which I have mentioned, involves forgery and deceit, and cannot be the right of the prince" (Oresme 1355, p. 24); as to how to characterise such practices, "I doubt whether it should not rather be termed robbery with violence or fraudulent extortion" (Oresme 1355, p. 28).

---

[9]Sargent and Velde (2002) form part of the debate suggesting a new view of the fairly chaotic historical pathway towards what they term the "standard formula" that supposedly settled the problem.

[10]For a discussion of these forms of "falsification", see Blanc (2007) and Blanc and Desmedt (2010).

Table 3.2 Three forms of monetary falsification in the dualist system

Type of falsisifcation	Origin	Definition	Appropriate legal category
Enhancement and debasement	Sovereign	Increase of the legal value above the value of metal content. This discrepancy depended on the minting and seigniorage at the time of issue	Breaking contract and danger for public trust
Degradation	Officers of the mint or subjects	Reduction of the fineness or the weight of the metal content below the royal proclamations, because of diversion or poor enforcement of proclamations by officers of the mint, clipping or natural wear and tear of coins	Transgression of law (crime)
Counterfeit money	Subjects	Forgery by individuals with no sovereign right to strike coins	Transgression of law (crime of lèse-majesté)

In this way (nominal and real), monarchical enhancement and debasement, coin clipping and counterfeiting by subjects were bound together, as was the circulation of foreign coin whenever it was of poor quality. An author who denounced monetary manipulations could therefore create a diversion: it was possible, in order to spare the sovereign's sensibility, to revile false money in general and to call for a monetary reform that would prevent all forms of monetary falsification (Blanc and Desmedt 2010).

Dualist Doctrines

The dualist system was developed in the context of Thomist philosophical realism. Medieval monetary doctrine was not opposed to rulers manipulating the rates of specie in circulation as necessity dictated. Insofar as striking coinage was a privilege of the ruler, seigniorage, which etymologically relates to that privilege and which was technically increased by enhancement and debasement, was considered a legitimate source of income. The successors of Thomas Aquinas defended monetary nominalism "by [which] coins were signs endowed with a certain value by the prince and relatively independently of their metal content" (Bichot 1984, p. 49). The ruler's legitimacy in manipulating the currency did not persist in all circumstances, however, and debate occasionally arose over whether or not a prince had overreached his legitimate right to manipulate money. More specifically in the late Middle Ages, jurists supported the prince and canon lawyers were more hesitant; Nicole Oresme, the translator of Aristotle and a philosophical nominalist, was far more critical and leaned towards monetary metallism. For him, manipulations were only acceptable if they were accepted by the community from which the prince came. They had to be warranted by special circumstances and had to remain temporary.[11]

---

[11]Eisenstadt identifies three perceptions and conceptions by which manipulations might be allowed: the position of the legist, generally in support of rulers; the more hesitant position of the canon lawyer; and the Aristotelean position that was critical and based occasionally on arguments of natural law (Weber 1968, p. 101, note).

The nominalist medieval monetary approach, which was potentially critical of rulers' manipulations, had a major influence on the monetary ideas of early modern Europe. Let us look at three instances of monetary controversy, each of which sees a confrontation between views regarding the viability of the dualist system. In the Duchy of Saxony around 1530, a debate (with weighty implications for local politics) cast the supporters of the Albertine branch in opposition to those of the Ernestine branch of the ruling dynasty over the reform of the monetary system. The Ernestines thought of manipulation as a reasonable way to increase the prince's resources; the Albertines opposed them and won the day (see Bond 2014; Perrotta 2000). In France, the monetary crisis of the 1560s gave rise to a heated exchange that is often neglected on this side of the debate.[12] Malestroit (1567) proposed a solution that was not foreign to the dualist system but, on the contrary, operated within it. In order to establish an unchangeable ratio of 1–12 between gold and silver, he proposed reducing the metal content of the unit of account by manipulating specie. To do so, he increased seigniorage and announced that it could provide revenue of 100,000 ecus per year (Tourette 1567, p. 145), a departure from his *Paradoxes*, which places the origin of the sensation of dearness in manipulations (Malestroit 1566). Not only was Malestroit an author within the dualist system, he worked to shore it up, not to overturn it. This was not the case for his opponent Jean Bodin. In England, the monetary crisis of 1694–1695 prompted William Lowndes (1695), Secretary of the Treasury, to propose a solution that was similarly consistent with the dualist system: a rise in the legal value of money (that is an enhancement or nominal manipulation) in order to stem the leakage and melting of specie. His opponent, John Locke, would come up with an external solution that would contribute to transforming the dualist system.

---

[12]It is generally known only for Jean Bodin's foundation of the quantity theory of money, although a highly debateable point.

The Metallic Pathway: Towards "Immutable" Moneys

First Types of Criticism

The dualist system came with a number of problems, such as excessive manipulation, the difficulty of managing the gold/silver ratio and similarly of managing the ratio between small change and valuable coin. Meanwhile, monetary innovations under the dualist system included improved bookkeeping practices and above all the appearance of paper money. All this led to the emergence of new theoretical frameworks which required the dualist system to evolve or be left behind.

A few decades after the episode of Philippe the Fair, and in the context of a series of extraordinary manipulations, Nicole Oresme explicitly positioned the sovereign as one whose actions in monetary matters ought to serve their community:

> Furthermore, it was ordained of old, with good reason, and to prevent fraud, that nobody may coin money or impress an image or design on his own gold and silver, but that the money, or rather the impression of its characteristic design, should be made by one or more public persons deputed by the community to that duty, since, as we have said, money is essentially established and devised for the good of the community. And since the prince is the most public person and of the highest authority, it follows that he should make the money for the community […]. Although it is the duty of the prince to put his stamp on the money for the common good, he is not the lord or owner of the money current in his principality. (Oresme 1355–1370, pp. 10–11).

Oresme was walking a tightrope: he considered the temporary manipulation of money legitimate, provided it was done in exceptional circumstances and in accordance with the opinion of the Estates General. In doing so, he sought to justify the manipulations made necessary by the ransom to be paid to the English for the release of Jean le Bon but also to close the door on the sovereign's arbitrary power to alter currencies.

In the south-west of the Holy Roman Empire, the theologian Gabriel Biel (1516) denounced monetary manipulations as false money, while recognising the advantages of this form of taxation. Nevertheless, he, too, did not take the step that would have led him to imagine a new monetary system. In East Prussia, Nicolaus Copernicus (1526), one of the first great monetary writers of early modern Europe, was highly critical of the manipulations and their disastrous effects:

> Although there are countless scourges which in general debilitate kingdoms, principalities, and republics, the four most important (in my judgment) are dissension, [abnormal] mortality, barren soil, and debasement of the currency. The first three are so obvious that nobody is unaware of their existence. But the fourth, which concerns money, is taken into account by few persons and only the most perspicacious. For it undermines states, not by a single attack all at once, but gradually and in a certain covert manner. (Copernicus 1526)

Copernicus did consider it justifiable, though, to debase money on two conditions: that the notables agreed and that the former specie was withdrawn from circulation to avoid confusion.

It was other proposals, therefore, that led to the way out of the dualist system. The metallic proposal was to subject money fully to a rule as to its metal content, which could go so far as to abolish seigniorage. As well as eliminating the problem of counterfeiting by nefarious individuals, the metallic proposal would also remove all possibility of falsification by the sovereign (Blanc and Desmedt 2010). The metallic pathway came with the major advantage of preventing princes from further manipulations.

Proposals to Escape Dualism

Within the context of the dualist system, the denunciation of manipulations was not exceptional, since they were subject to a principle of justice by which the sovereign himself was bound. In fact, the inclusion of money among the symbols of sovereignty was a relatively recent construction. In the Kingdom of France, efforts to establish a royal monopoly over minting date to the mid-fourteenth century (the time when Oresme was writing) and were only achieved in the course of the seventeenth century (Rigaudière 1993).

Moreover, many authors who were critical of the dualist system did not propose (either explicitly or implicitly) to abandon it. Their solutions operated within the system; they sought to limit the sovereign's ability to alter moneys by imposing a close tie between the legal value of coins and their metallic content.

The Spanish canonist Azpilcueta provided a clear illustration of the tipping point from the medieval dualist system to a metallic system: "the intrinsic quality of money is not the price the state attributes to it, it is the quality of the material from which it is made" (Azpilcueta 1556, p. XXVIII; see Tortajada 1988). Two centuries later, Ferdinando Galiani summarised the metallist position:

> For a thing to be accepted by all, four qualities are required, to my mind. First, that it has some intrinsic and real value, and that at the same time everyone agrees on its estimation. Second, that it is easy to ascertain the true value. Third, that it is difficult to commit fraud with respect to it. Fourth, that is can be kept for a long time. (Galiani 1751, p. 137)

The metallic pathway that Azpilcueta glimpsed (because he was a man of the dualist system) and Galiani observed (because he was a man of the metallic system) was travelled in three stages.

The first stage was primarily concerned with the minting regime. This stage consisted of abolishing any discrepancy between the legal value of coins and their metallic content (at the price of metal) and only striking coins from pure metal, as proposed by Jean Bodin in France. All seigniorage, therefore, ought to be abolished, and there was discussion about *brassage*, which covered the costs of minting and any profit for the minter. The question of small change for the populace remained on hold.

The second and more radical stage concerned the definition of the unit of account. The French Thomas Turquam (1578) proposed abolishing the duality between real and imaginary moneys by taking the unit of account as the central component of the monetary system. That was expected to lead to the unit of account being defined strictly by a weight of fine metal and to the solidification of an invariable relationship between unit of account and unit of payment. This model may be characterised as a "purely" metallic or "immutable" money (Bloch 1954, p. 49). As such, this was no longer a dualist system, which by definition is "mutable" through enhancements (i.e. nominal manipulations).

The final step was inscribing a legal value on the coins in circulation so that it was irreversibly fixed. Although certain coins in the United Provinces bore an indication of their value in the unit of account as early as 1694, this practice was not widely implemented for coins with intrinsic value before the nineteenth century.

There were a number of authors who, within the context of the dualist system, argued that it should be transformed or even abandoned in favour of a metallic system in which real and nominal manipulations were no longer tolerated: Jean Bodin and Thomas Turquam in France, Juan de Mariana in Castile, Youri Krizhanich in Russia, Gaspar Scaruffi in Reggio Emilia, Ferdinando Galiani in Naples and John Locke in England. We concentrate on three "Johns" who wrote between the 1560s and 1690s: in France, Jean Bodin; in Castile, Juan de Mariana; and in England, John Locke.

In France, the legal scholar Jean Bodin, writing a formal response to Malestroit on the subject of the "dearness of all things" in the difficult monetary context of the 1560s, objected to falsification in any shape or form, including falsification by the prince. He explicitly cast the sovereign right to revalue moneys against public confidence that had to be safeguarded at all costs. In his *Réponse* of 1568, he articulated an imagined ideal system which he went on to supplement and refine. By 1578, his system contained the following recommendations: first, there should remain only fine gold coins and fine silver coins, small change being made of silver as well. Their legal value should be immutable and correspond exactly to their intrinsic value, which meant abolishing seigniorage and *brassage* (the costs of minting money). The ratio between the values of gold and silver should be set definitively at one to twelve. Money creating activities should be concentrated in a single mint in one city in the kingdom, and finally, hammering out coins should be replaced by casting them (Bodin 1593, book VI, chapter 3). One of the advantages of this system was, for him, that everyone down to the "coarsest and most ignorant" should be immediately able to recognise the quality of the coins through simple use of the senses: "to the eye, by their sound, by their weight, without fire, without a burin, without a touchstone" (Bodin 1593, p. 124).[13] As it was easy to tell the true from the false and the good from the bad, the various abuses would be fewer; everyone would be watched over by everyone else. Because Bodin's plan was so radical and because of his involvement in the prior debates, some people thought he was responsible for the monetary reform of 1577. That reform introduced France to a metallic system centred on a single coin, the gold *écu*, and thus broke with the logic of the dualist system and stabilised moneys for some time to come. In actual fact, Bodin was not involved and it was the *Cour des monnaies*, in particular Thomas Turquam, who thought up and carried out this metallist reform—which, ultimately, failed (Blanc 2011).

At the beginning of the seventeenth century, the Jesuit Juan de Mariana's freedom of speech cost him his freedom of movement.[14] Advocating ideas like those developed by Oresme, he denounced absolute sovereignty and the monarch's ability to revalue money without the subjects' consent in a context in which the issue of debased copper coins had pushed Castile into the long crisis of *vellóns*. Monetary manipulations were viewed as monarchical falsification, and only temporary, reversible manipulations in exceptional circumstances could be justified: money, like weights and measures, was the foundation of commerce and "cannot be changed without danger and harm to commerce" (Mariana 1609, p. 19). Manipulations were "against right reason and the natural law herself – it is a sin to change [coins]" (Mariana 1609, p. 38). He clearly perceived that the shortage of money was a terrible problem and that manipulations could be used to counter it, but every manipulation, he claimed, made for dearer prices. Without the subjects' consent, manipulations were unfair; with their consent,

---

[13]A touchstone was an instrument only professional money changers possessed. It enabled them to determine the metallic quality of coins by the colour of the trace they left.

[14]He was imprisoned and then held in confinement for several years.

they were generally fatal (Mariana 1609, p. 61). In any event, coins had to have some intrinsic value, even small change. Accordingly, Mariana was in favour of a pure copper coinage whose legal value matched their intrinsic value. The ideal system of a "well-constituted republic" was a metallic system in which the (extrinsic) legal value equalled the (intrinsic) value of the metallic content, plus the minting cost (Mariana 1609, p. 15).

At the other end of the same century, John Locke, in England, published several chapters and books directly concerning monetary matters. While his contemporary Barbon asserted that "nothing can have an Intrisick Value" (Barbon 1696, p. 6), Locke thought that intrinsic value pre-existed trade. In the primitive state of the world, he suggested, exchange took place without any monetary mediation.[15] Locke asserted in modern theoretical terms the primacy of trade and so discredited any hierarchical or political intervention in the emergence of forms of money. The sovereign's intervention with legitimate money was not required: "Silver, i.e. the quantity of pure silver separable from the alloy, makes the real value of money" (Locke 1691, pp. 146–147). Whereas many of his contemporaries held different positions, Locke was to work on the far-reaching English monetary reform of 1696. His goal was to do away with manipulations and counter the proliferation of counterfeit coins: counterfeiting was private appropriation of the right to strike money. As the chosen solution was to restore the integrity of the coins in circulation, it would involve mobilising immense resources and became just as important as winning the 1689 war begun against France. Both efforts were fights for national identity and sovereign authority. An authority that could not ensure respect for the seal it placed on coins was a weakened, symbolically mutilated authority and that was a serious problem for monetary sovereignty. Under Locke's guidance, the state won back its monopoly over the issue of monetary signs—though the harsh currency shortage that followed led to the rise of numerous private emergency issues.

---

[15]"Thus in the beginning all the world was America [...]; for no such thing as money was anywhere known". Locke (1690), "The Second Treatise", paragraph 49.

Did Stabilisation Mean Monetary Revolution?

It was by moving towards the metallic pathway, formulated in different times and under different circumstances by Bodin, Mariana, Locke and others that money was stabilised in Europe over the course of some fifty years, beginning in the late seventeenth century.[16] This was the case in the United Provinces between 1681 and 1694, in England in 1696 (for silver) and 1717 (for gold), in France in 1726 (gold and silver) under the aegis of Cardinal Fleury and in Spain between 1725 and 1735.[17] By the 1730s, a third of the way through the eighteenth century, most European moneys had been stabilised (Vilar 1976).

In this way, from being a "political object subject to a will, money became an economic object governed by the law of value, the determination of which escapes from all willpower" (Benetti 1994, p. 1186). Marc Bloch spoke of the "monetary regime of the capitalist era", which began with "the foundation of the new regime" and "the advent of stable money" in England in 1717 and in France in 1726 (Bloch 1954, p. 79). In his view, "however remote the premises of [modern capitalism] [may] have been, it only flourished in the eighteenth century – and with its advent the frequent manipulations temporarily disappeared" (Bloch 1954, p. 77). Monetary stability was a necessary precondition for an effective and sound system of credit to develop, which proved to be a precondition for the development of industrial capitalism. What has been called the "monetary revolution" associates this stabilisation with the emergence of convertible/bank moneys:

> With the growth of private credit, in which the rise of capitalism had its origins, mistrust of the money of account hindered the productive utilisation of savings. [...] The monetary revolution preceded the industrial revolution by a good half-century. The former created the economic and social structures within which the latter was able to take place. [...] However, the most important factor was the institution of a system in

---

[16]That is, the implicit weight in precious metals of units of account was stabilised.

[17]E. J. Hamilton speaks of "stability" in Spain; see also Tucci (1984) and Tabatoni (1999).

which a (private) bank issued a currency, trust in which was maintained by convertibility into a high-quality metal currency constituting a monetary base which was itself linked to the unit of account via a ratio decreed by the sovereign. (Aglietta 2002, pp. 41–42)

The monetary revolution was, therefore, the confluence of the transformation of metallic minting and of the use of paper, which involved trial and error and was not without its crisis points.

It may seem strange that the advent of this metallic system was not identified as such by Thomas (1977) and then by Aglietta and Orléan (2002): for them, the dualist system yielded to a system of convertible money. It seems intuitive to us to identify and separate this metallic system from both the dualist system and the later convertible system. The minting process is indeed very different. Moreover, the shortcomings of the metallic system do not lead inevitably to the deployment of convertible paper: the British case, on which most historical reconstructions of the evolution of the monetary system are based, may be considered something of an exception. Finally, the authors who have sought to go beyond the dualist system with the advent of a metallic system have not caught sight of the disastrous effects of the rigidifying of the supply of metallic money. The connection between the reform of metallic money and the emergence of paper money was not foreseen, it was an emergent impact unnoticed until after the event.

A Centralised and Rigidified Money Supply as the Price of Stabilisation

In this emerging system centred on metal, monetary sovereignty was reshaped in a way that largely ruled out the possibility of manipulations. It was also centralist: simpler and sounder monetary circulation presupposed greater control, through the mints, of what was issued in the country—issuing baronies had to be integrated into the system of royal mints, which was itself simplified—and what came into the country. For Bodin and Locke, metallism required reducing and even centralising the striking of coins. The far-reaching overhaul of the minting regime meant redefining the unit of account to be characterised by a weight of metal that, if not immutable, was at least stable over the long term. The system therefore implied the unification of monetary circulation and the centralisation of minting. However, this unifying force failed to withstand the necessities of everyday circulation and the needs of economic growth.

Under the metallic system, false money could no longer be blamed on the ruler. In England Isaac Newton, appointed master of the Royal Mint in 1696, hunted down counterfeiters even into the country's taverns at a time when royal minting was growing increasingly sound (see Wennerlind 2004). In 1764, François-André Abot de Bazinghen, one of the few authors to have written on the *Cour des monnaies* in the Kingdom of France, identified nine types of false money that corresponded to forms of the crime of lèse-majesté (Abot de Bazinghen 1764, vol. 1, pp. 499–505). Falsification by manipulation was no longer an issue. Better still, no entry in his dictionary concerns the forms of sovereign monetary manipulations: there was nothing related to enhancement, adulteration, debasement or the like. The entry headed "*Altérer la monnaie*" (altering the coins) covered fraudulent practices in the mints or by individuals, in particular clipping (Abot de Bazinghen 1764, vol. I, p. 51). Abot de Bazinghen, in this respect, was an author rooted in the metallic system.

The advent of stable monetary systems consistent with the metallist outlook was not without its problems. The strengthening of money and the implementation of policy restricting the striking of small coins had disastrous social effects. Regarding the particular case of France, Marc Bloch wrote: "Significantly [...] strengthening was constantly unpopular with the lower and middle categories of urban dwellers. [...] Urban riots were consistently caused by the return to strong currency" (Bloch 1954, p. 73).

The implementation of a stable metallic monetary system could only occur at the cost of a more rigid money supply. Although the dualist system had fallen well short of providing an elastic supply for the requirements of circulation, this problem was ameliorated by the interplay of enhancements (nominal manipulations) and, less frequently, debasements (real manipulations). Under the metallic system, minting could not keep pace with the needs of circulation. Only an alternate source of monetary issue could offset the depressive effects of the new regime. The English case is interesting in this respect, but exceptional: Lockean monetary stabilisation went hand in hand with the founding of the Bank of England and the emergence of credit money. This success, however, was far from common: it was not the case in Ireland, and on the continent, banks did not develop as quickly or as early as in Great Britain. Further, even in Britain the elasticity of the supply of bank money was not enough to make up for the inelasticity of the supply of metallic money. As a result, Britain faced numerous issues of emergency coins and counterfeiting during the eighteenth century until private tokens and tradesmen's tickets exploded in the last fifteen years of the century, in a context of industrialisation that required ever more coin in small denominations to pay workers (Mathias 1979; Dykes 2011). In America, both French and English colonies had to cope with serious shortages of metallic specie. Having been denied the right to strike money by their respective crowns, they were forced to innovate with card money in Quebec and provincial notes in the British colonies (see Dimand 2008). It could reasonably be argued that the development of credit money was critical to the rise of the British Empire.

Conclusion

In early modern Europe, the dualist monetary system provided a degree of flexibility in the money supply at the cost of allowing discretionary action by princes, which were called into question in times of major crises. The crises of the dualist system were marked by inflation, chaotic monetary circulation involving a variety of currencies of diverse quality and ultimately a loss of trust in the sovereign's capacity to control money (Table 3.3 gives a summary comparison of the characteristics of the dualist and metallic systems).

The elimination of sovereign's manipulations brought about the emergence of a metallic monetary system, but it also imposed a more rigid supply of royal specie causing a deflationary dynamic that was only partly offset by the emergence of new means of payment. Some new means of payment were completely outside the sovereign monetary

Table 3.3 Dualist and metallic systems compared

	Dualist system	Metallic system
Minting	Striking metal (logic of the seal)	Striking metal for coins of full intrinsic value (relativized logic of the seal)
Advantages	Possibility of adjusting money supply and its legal value. Possibility of devaluing debts (including sovereign debts)	Monetary stability. Conservation of value of monetary assets
Disadvantages	Real and nominal monetary manipulations	Inelasticity of money supply
Basis for monetary quality	Sovereign authority	Precious metals ('nature')
Bad money	'False money' (sovereigns, clippers, counterfeiters)	Non metal and base metal coins, counterfeit specie
Crises and solutions	Inflation, simultaneous circulation of coins of all qualities, mistrust of sovereign's capacity to manage money → stabilizations (Europe, eighteenth century)	Deflation and emergency private tokens → credit money (England, late seventeenth century)
Theorists	Medieval jurists and canon lawyers, including Thomas Aquinas; Ernestines, Malestroit, Lowndes…	Sixteenth and seventeenth centuries: Bodin, Mariana, Locke, Scaruffi, Genovesi, Leibniz, Krizhanich…

system (multiple tokens and emergency coins); others fit into the metallic system rooted in sovereignty. In the case of convertible paper moneys, the means of payment was linked more or less tightly with a recognised form of wealth, primarily metal coins. Others were outside the realm of metal but within the sphere of sovereignty: in the case of self-referential money (or fiat money), issuing operations were based on debt and no definition or conversion into a real asset was guaranteed.

Monetary stabilisation was a precondition for the development of convertible moneys and self-referential moneys, and both were a condition for the survival of the metallic system until the tipping point when it was transformed by the proliferation of credit moneys. It was indeed the credit-based system that transformed the system of payment by moving away from the conditions of minting and also focusing on the conditions for settling monetary balances. The gold standard was a system of convertible money that became established, because of the failings of the metallic system, as early as 1717 in England and more widely in the early nineteenth century (Amato and Fantacci 2012).

Bibliography

Primary Sources

Abot de Bazinghen, François-André, 1764. *Traité des monnoies, et de la juridiction de la Cour des monnoies, en forme de dictionnaire*, vol. 2. Paris: Guillyn.

Azpilcueta, Martin de (= Doctor Navarro), 1556/1978. *Comentario resolutorio de cambios* (1556). French translation in *Or et monnaie chez Martin de Azpilcueta,* edited by Bernard Gazier. Paris: Economica.

Barbon, Nicolas, 1696. *A Discourse Concerning Coining the New Money Lighter*. London: Chiswell.

Biel, Gabriel, 1516/1930. *Tractatus de potestate et utilitate monetarum*. English translation in *Treatise on the Power and the Utility of Moneys*, edited by Robert Belle Burke. Philadelphia: University of Pennsylvania Press.

Bodin, Jean, 1568/1578. *Response to the Paradoxes of Malestroit*. Bristol: Thoemmes, 1997.

Bodin, Jean, 1593. *Les six livres de la République*, vol. 6. Paris: Fayard, 1986.

Copernicus, Nicolaus, 1526. Monete Cudende Ratio, translated by Edward Rosen. http://copernicus.torun.pl/en/archives/money/4/?view=transkrypcja&lang=en. Accessed December 9, 2018.

Galiani, Ferdinando, 1751/2005. *De la Monnaie. Della Moneta*, translated by André Tiran and Anne Machet. Paris: Economica.

Gresham, Thomas, 1558. "Letter to Queen Elizabeth." In: *The Life and Times of Sir Thomas Gresham*, edited by John William Burgon, vol. 1. London: E. Wilson: 483–486.

Locke, John, 1690. *Two Treatises of Government*. London: Awnsham Churchill.

Locke, John, 1691/2011. "Some Considerations of the Consequences of the Lowering of Interest, and Raising the Value of Money." In: *Écrits monétaires*, edited by A. Tiran, French translation by F. Briozzo. Paris: Classiques Garnier.

Locke, John, 1695a/2011. "Short Observations on a Printed Paper Intituled, For Encouraging the Coining Silver Money in England, and After for Keeping it Here." In: *Écrits monétaires*, edited by A. Tiran, French translation by F. Briozzo. Paris: Classiques Garnier: 209–234.

Locke, John, 1695b/2011. "Further Considerations Concerning Raising the Value of Money." In: *Écrits monétaires*, edited by A. Tiran, French translation by F. Briozzo. Paris: Classiques Garnier: 235–363.

Lowndes, William, 1695. *A Report Containing an Essay for the Amendment of the Silver Coins*. London: Charles Bill.

Malestroit, Jehan Cherruyt de, 1566/1934. "Les paradoxes du seigneur de Malestroict, Conseiller du Roi et maistre ordinaire de ses comptes, sur le faict des monnoyes présentez a sa majesté, au mois de mars, MDLXVI." In: *Ecrits notables sur la monnaie*, edited by Le Branchu, vol. 1. Paris: Félix Alcan: 49–68.

Malestroit, Jehan Cherruyt de, 1567/1937. "Mesmoires sur le faict des monnoyes, proposez et leues par le maistre des comptes de Malestroit au Conseil privé du Roi." In: *Paradoxes inédits du Seigneur de Malestroit touchant les monnoyes avec la response du Président de la Tourette*, edited by Luigi Einaudi. Turin: Giulio Einaudi: 99–130.

Mariana, Juan de, 1609/2002. "A Treatise on the Alteration of Money." In: *Journal of Markets and Morality* V, 2: 523–593.

Oresme, Nicole, 1355/1989. "Traité sur l'origine, la nature, le droit et les mutations des monnaies." In: *Traité des monnaies (Nicolas Oresme) et autres écrits monétaires du XIVe siècle (Bartole de Sassoferrato, Jean Buridan)*, edited by Claude Dupuy. Lyon: La Manufacture: 47–92.

Tourette, Alexandre de la, 1567/1937. "Response du Sr de la Tourette aux paradoxes du Sr de Malestroit et touchant les monnoyes." In: *Paradoxes inédits du Seigneur de Malestroit touchant les monnoyes avec la response du Président de la Tourette*, edited by Luigi Einaudi. Turin: Giulio Einaudi: 131–145.

Turquam, Thomas, 1578. *Advis de M. Thomas Turquam… afin d'abolir le compte à sols et à livres*. Paris: Dalier et Roffet.

Secondary Sources

Aglietta, Michel, 2002. "Whence and Whither Money?" In: *The Future of Money*. Paris: OECD: 31–72.

Aglietta, Michel; Cartelier, Jean, 1998. "Ordre monétaire des économies de marché." In: *La monnaie souveraine*, edited by Michel Aglietta and André Orléan. Paris: Editions Odile Jacob: 129–157.

Aglietta, Michel; Orléan, André, 1982. *La violence de la monnaie*. Paris: PUF.

Aglietta, Michel; Orléan, André, eds., 1998. *La monnaie souveraine*. Paris, France: Odile Jacob.

Aglietta, Michel; Orléan, André, 2002. *La monnaie entre violence et confiance*. Paris: Odile Jacob.

Amato, Massimo; Fantacci, Luca, 2012. *The End of Finance*. Cambridge, UK: Polity Press.

Benetti, Carlo, 1994. "Troc, bons d'achat et monnaie: La conception de Ferdinando Galiani." In: *Revue économique* 45, 5: 1177–1187.

Bichot, Jacques, 1984. *Huit siècles de monétarisation. De la circulation des dettes au nombre organisateur*. Paris: Economica.

Blanc, Jérôme, 2007. "Beyond the Quantity Theory. A Reappraisal of Jean Bodin's Monetary Ideas." In: *Money and Markets. A Doctrinal Approach*, edited by Alberto Giacomin and Maria Cristina Marcuzzo. London: Routledge: 135–149.

Blanc, Jérôme, 2011. "La réforme monétaire française de 1577: une expérience radicale et son échec." In: Workshop *La souveraineté monétaire et la souveraineté politique en idées et en pratiques: identité, concurrence, corrélation?* Paris: Sciences Po, décembre 8–9, 2011.

Blanc, Jérôme; Desmedt, Ludovic, 2010. "Counteracting Counterfeiting? Bodin, Mariana and Locke on False Money as a Multidimensional Issue." In: *History of Political Economy* 42, 2: 323–360.

Blanc, Jérôme; Desmedt, Ludovic, eds., 2014a. *Les pensées monétaires dans l'histoire: l'Europe, 1516–1776*. Paris: Classiques Garnier.

Blanc, Jérôme; Desmedt, Ludovic, 2014b. "In Search of a 'Crude Fancy of Childhood': Deconstructing Mercantilism." In: *Cambridge Journal of Economics* 38, 3: 585–604.

Blaug, Mark, 1964. "Economic Thought and Economic History in Britain, 1650–1776." In: *Past and Present* 28: 111–116.

Bloch, Marc, 1954. *Esquisse d'une histoire monétaire de l'Europe*. Paris: Librairie Armand Colin.

Bond, Niall, 2014. "La monnaie et le monstre." In: *Les pensées monétaires dans l'histoire: l'Europe, 1516–1776*, edited by Jérôme Blanc and Ludovic Desmedt. Paris: Classiques Garnier: 811–869.

Boyer-Xambeu, Marie-Thérèse; Deleplace, Ghislain; Gillard, Lucien, 1994. *Private Money & Public Currencies: The 16th Century Challenge*. Armonk, NY: M.E. Sharpe.

Cailleux, Alain, 1980. "L'allure hyperbolique des dévaluations monétaires." In: *Revue de Synthèse*, 99–100: 251–266.

Cartelier, Jean, 2013. "Beyond Modern Academic Theory of Money: From 'Flat Money' to 'Payment System'." In: *New Contributions to Monetary Analysis: The Foundations of an Alternative Economic Paradigm*, edited by Faruk Ülgen. London: Routledge: 155–171.

Cartelier, Jean, 2018. *Money, Markets and Capital: The Case for Monetary Analysis*. London and New York: Routledge/Taylor & Francis.

Desan, Christine, 2014. *Making Money: Coin, Currency, and the Coming of Capitalism*. Oxford, UK: Oxford University Press.

Dimand, Robert, 2008 "David Hume on Canadian Paper Money." In: *David Hume's Political Economy*, edited by Margaret Schabas and Carl Wennerlind. London: Routledge: 168–180.

Dykes, David, 2011. *Coinage and Currency in Eighteenth-Century Britain. The Provincial Coinage*. London: Spink.

Einaudi, Luigi, 1953. "The Theory of Imaginary Money from Charlemagne to the French Revolution." In: *Enterprise and Secular Change: Readings in Economic History*, edited by Frederic C. Lane and Jelle C. Riemersma. London: George Allen and Unwin: 229–261.

Fantacci, Luca, 2005. "Complementary Currencies: A Prospect on Money from a Retrospect on Premodern Practices." In: *Financial History Review* 12, 1: 43–61.

Favier, Jean, 1981. "Etat et monnaie." In: *La Moneta nell'economia europea secoli XIII–XVIII*. Firenze: Le Monnier: 171–184.

Gould, John Denis, 1970. *The Great Debasement: Currency and the Economy in Mid–Tudor England*. Oxford: The Clarendon Press.

Hamilton, Earl J., 1934. *American Treasure and the Price Revolution in Spain, 1501–1650*. Cambridge, MA: Harvard University Press.

Hamilton, Earl J., 1943. "Monetary Disorder and Economic Decadence in Spain, 1651–1700." In: *Journal of Political Economy* 51, 6: 477–493.

Hauser, Henri, 1932. "Introduction." In: *La response de Jean Bodin à M. de Malestoit*. Paris: A. Colin.

Lagueux, Maurice, 1990. "À propos de Montesquieu et de Turgot: peut-on encore parler de la monnaie comme d'un signe?" In: *Cahiers d'économie politique* 18: 81–96.

Laurent, Henri, 1933. *La loi de Gresham au Moyen Âge. Essai sur la circulation monétaire entre la Flandre et le Brabant à la fin du xive siècle*. Brussels: Éditions de la Revue de l'université de Bruxelles.

Mathias, Peter, 1979. "The People's Money in the Eighteenth Century: The Royal Mint, Trade Tokens and the Economy." In: *The Transformation of England: Essays in the Economic and Social History of England in the Eighteenth Century*. London: Methuen: 190–208.

Orléan, André, 1998. "La monnaie autoréférentielle: réflexions sur les évolutions monétaires contemporaines." In: *La monnaie souveraine*, edited by Michel Aglietta and André Orléan. Paris: Odile Jacob: 359–386.

Perrotta, Cosimo, 2000. "Einleitung zum Münzstreit der sächsischen Albertiner und Ernestiner um 1530." In: *Die drei Flügschriften, über den Münzenstreit*, edited by Bertram Schefold. Düsseldorf: Wirtschaft und Finanzen: 101–156.

Rigaudière, Albert, 1993. "L'invention de la souveraineté." In: *Pouvoirs* 67: 5–20.

Roover, Raymond de, 1946. "The Medici Bank Organization and Management." In: *The Journal of Economic History* 6, 1: 24–52.

Roover, Raymond de, 1949. *Gresham on Foreign Exchange*. Cambridge, MA: Harvard University Press.

Sargent, Thomas; Velde, François, 2002. *The Big Problem of Small Change*. Princeton: Princeton University Press.

Smith, Adam, 1776. *An Inquiry into the Nature and Causes of the Wealth of Nations*, edited by Edwin Cannan. London: Methuen, 1904.

Stigler, George, 1983. "Nobel Lecture: The Process and Progress of Economics." In: *Journal of Political Economy* 91, 4: 529–545.

Tabatoni, Pierre, 1999. *Mémoire des monnaies européennes, du denier à l'euro*. Paris: PUF.

Théret, Bruno, 2007. "La monnaie au prisme de ses crises d'hier et d'aujourd'hui." In: *La monnaie dévoilée par ses crises*, edited by Bruno Théret, vol. I. Paris: Editions de l'EHESS: 17–74.

Thomas, Jean-Gabriel, 1977. *Inflation et nouvel ordre monétaire*. Paris: Presses universitaires de France.

Tortajada, Ramon, 1988. "Gains de change et seigneuriage: Remarques sur Monnaie privée et pouvoir des princes. L'économie des relations monétaires à la Renaissance." In: *Revue économique* 39, 2: 461–476.

Tucci, Ugo, 1984. "De la modernité du XVIe siècle au sévère mais riche XVIIe siècle." In: *Etudes d'histoire monétaire*, edited by John Day. Lille: Presses universitaires du Septentrion: 335–351.

Vilar, Pierre, 1976. *A History of Gold and Money 1450–1920*. London: NLB.

Weber, Max, 1968. *On Charisma and Institution Building*, edited by Shmuel Noah Eisenstadt. Chicago: University of Chicago Press.

Wennerlind, Carl, 2004. "The Death Penalty as Monetary Policy: The Practice and Punishment of Monetary Crime. 1690–1830." In: *History of Political Economy* 36, 1: 131–161.

4

Re-presenting Silver in Early Modern Europe

Renate Pieper

Introduction

Almost every autumn or winter for more than a hundred years, from the second half of the sixteenth century until well into the eighteenth century, galleons loaded with silver from Spanish America arrived at Seville on the southern coast of the Iberian Peninsula. Thanksgiving Masses were sung, and letters were sent to princes, merchants, and other well-to-do individuals to inform them of the arrival of the "treasure fleet". Impressed by the fleet and the celebrations, Alonso Sánchez Coello, a painter at the Spanish court, captured the arrival of the fleet in a monumental painting at the end of the 1570s (Sánchez Coello 1576).

The silver arriving in Seville served a dual purpose in early modern Europe: it was used both as money and as a luxury commodity. Precious metals, especially silver, could perform both functions without changing their external form: silver coins were used as medals for decoration

R. Pieper (✉)
Institute of History, University of Graz, Graz, Austria
e-mail: renate.pieper@uni-graz.at

© The Author(s) 2019
R. Pieper et al. (eds.), *Mining, Money and Markets in the Early Modern Atlantic*, Palgrave Studies in Economic History,
https://doi.org/10.1007/978-3-030-23894-0_4

(Pieper 1995, pp. 180–181) and, in inventories, silver buttons were assessed, like specie, according to their metal content ("Inventory" 1741). As a result, the economic and cultural valuations of silver, which have historically been studied as separate matters (cf. Lacueva Muñoz 2010; Philipps et al. 2004), were closely related. Both silver money and silver artefacts received at least part of their value through the worth of their metal essence (cf. Bernholz; Desmedt and Blanc in this volume). Despite their similarity, the cultural and economic connotations of money and art differed. This chapter discusses the connections and disjunctions in the cultural appreciation of silver in its different forms and analyses the relationship between them.

The interdependence of economic and cultural assessments of silver will be studied in paintings and news media. Images of silver items illustrate the connotations which surrounded the precious metal, and so the first purpose of this chapter is to analyse such paintings. More diffuse, but nevertheless consistently observable, the imaginary associated with silver is evident in written accounts of the arrival of the silver fleets from Spanish America. The narratives and representations of the fleets' arrival in newsletters and newspapers will be the second major topic of interest in this study.

Spanish America was the most important silver-producing region in the early modern world. In order to examine the connections between the imagining and understanding of silver objects and specie, this chapter will focus on the first period of silver boom: silver production doubled within two decades after the introduction of a new metallurgical process in today's Mexico and especially in modern Bolivia, peaking between 1580 and 1635. From 1635 onwards, Spanish American silver production slowly decreased. By the 1650s, it fell beneath the levels of the 1580s and only recovered in the eighteenth century (Te Paske 2010).

American silver had a great influence on the general European imaginary in the decades around 1600, particularly because it financed Western European trade with Asia and Africa, which provided Europeans with a plethora of marvellous things (Brook 2007; Siebenhüner 2018). Territories intensely linked to overseas trade contributed significantly to the emerging cultural patterns around the conception and appreciation of silver. This chapter examines the imagining of silver itself in

Spain, especially at the trading centre of Seville and the Spanish court of Madrid, as well as for the Spanish Netherlands at the trading centre of Antwerp (Martínez Ruiz 2018). These cities were closely entangled with Spanish America and Asia via Manila and Acapulco. Representations of silver in the Dutch Republic are considered as well, because the northern Netherlands established their own cultural and economic connections with Asia and maintained close commercial contacts with Spain despite serious military confrontations. Finally, this chapter explores the representation of silver in Frankfurt, one of the major trading centres of Central Europe with well-known biannual fairs linked to both Amsterdam and Antwerp. All of the above locations were important hubs in the European cultural and informational network. Consequently, this chapter analyses the relationships between the construction of silver as money and as artefacts in paintings and news at Seville, Madrid, Antwerp, Amsterdam and Frankfurt across the first period of plenitude of the American silver streams in the first half of the seventeenth century, called the Golden Age of arts and culture in Spain (Jordan 1985) and in the Low Countries (North 1997).

Historiography has dealt extensively with luxury items made from Spanish American silver, especially their artistic expression in Spain and Spanish America and the influence of Native American forms and techniques (Philipps et al. 2004). The ambiguous representation of silver objects in paintings has received less attention. Although silver was an important subject in the paintings of the Dutch painter Pieter Claesz in the first half of the seventeenth century, there has been little research focusing on the relationship between silver as money and as artefact (Biesboer et al. 2004, p. 53). For the second half of the seventeenth century, an article by Byron Ellsworth Hamann analyses the presence of American objects in general in the famous painting *Las Meninas* by the Spanish court painter Diego de Velázquez. Hamann mentions that American silver appears in this oil painting only as a little tray (Hamann 2010). A specific analysis concerning the complex and divided imaginary of American silver as specie and as bullion in the first half of the seventeenth century is still lacking.

Beyond paintings, the imaginary associated with silver is evident in news media. Michel Morineau's studies find that north-western

European printed newspapers regarding the Spanish American treasure fleets generally dealt with the cargo of the ships and the quantities of silver delivered each year, but he did not further analyse the imagery associated with silver (Morineau 1984–1985). Studies by Arndt Brendecke and Horst Pietschmann sustain that the large number of vessels crossing the Atlantic in convoy underlined the political importance of the Spanish Crown and served the purpose of royal representation and communication (Brendecke 2016; Pietschmann 2013), as exemplified in Sánchez Coello's painting. Even though this is an important observation, it refers to the *fleets* and not to their *cargo*—the silver itself. The representation of silver and the imagery associated with silver in news and paintings still bears analysis.

At first, we will examine paintings, as the representation of silver objects is most obvious in this medium. Having established a grounding in visual art, we will move on to scrutinise written communication and its imaginary, showing the extent to which the estimation of silver as bullion and specie depended on the media or religious and political positions and whether the backing of money through otherwise highly valued precious metals contributed to the appreciation of specie.

Painting Silver

When, in the 1540s, the Dutch painter Marinus van Reymerswale satirised bankers, lawyers and tax collectors, he depicted them handling or receiving gold and silver coins, laid on a table in the foreground very nearly at the centre of the image (Mathes 2006, pp. 85–117, especially pp. 99–100). Silver specie was directly connected with avarice and immorality, related to or contrasted with paperwork and books. Besides the coins, no further silver objects appeared in Reymerswale's satirical paintings. Almost a century later one of the most famous and prolific Dutch artists, Pieter Claesz of Haarlem (Biesboer et al. 2004), painted a series of still lives including silver items: in 1627, he designed *Still Life with a Turkey Pie*, depicting a nautilus shell on a gilded silver stand (Claesz 1627). In Claesz's still life, the gilded silver was not associated with avarice but with conspicuous consumption and vanity;

the painting presented an ongoing, unfinished meal. The gilded stand, made of American silver, the turkey, stemming from the Americas, and the nautilus shell, imported from Asia, decorated a lavishly laden table and accompanied a pie containing Asian spices, local oysters, Mediterranean lemons, and Chinese porcelain. In contrast to the caricatures of the early sixteenth century, this early seventeenth-century Dutch still life made no reference to money or coinage.

This brief comparison of oil paintings already demonstrates a variable imagery of silver and the different conceptual functions of coins and artistic objects. The shifting meanings of silver in early modern paintings merit a more detailed analysis, to uncover the scope and interdependence of its cultural functions and the esteem in which it was held. Art historians have largely associated the presentation of silver objects with the representation of wealth (Merriam 2009, p. 205) but have not studied the relationship between imagery of *specie* and of *artefacts*. Because their primary purpose was the presentation of wealth and its shortcomings, this chapter uses a selection of still lives and vanity allegories to study the interdependence of money and art. Still lives and vanity allegories became commonplace in the northern and southern Low Countries during the late sixteenth century. They influenced paintings throughout Western Europe during the Golden Age of arts and the heyday of the "treasure fleets".

In Antwerp, in 1611, the artist Clara Peeters (Vergara 2016) painted a still life with flowers, a gilded silver cup, almonds, dried fruits, sweets on a tin or silver plate, a glass of wine and a tin jar. Some flowers and sugar sweets laid disorderly on the table pointing to the vanity of riches. The almonds and dried figs were imports from the Mediterranean, whereas the silver and gold for the cup and the sugar for the candies had come from Spanish and Portuguese America. The gilded silver cup was placed at the front and centre of the painting and dominated the scene (Peeters 1611). In this still life, the central position and the representation of gilded silver added to the sense of wealth being depicted, as well as demonstrating the far-reaching European and overseas connections of the southern Low Countries.

The centrality of silver was apparent in contemporary Dutch paintings as well. Pieter Claesz painted a still life with flowers in 1633 depicting a rummer glass of wine, a silver cup laying on the table, and oysters and a lemon on a silver or tin plate (Claesz 1633). As in the aforementioned composition with the turkey pie, the silver objects in the 1633 still life were accompanied by foreign and local luxuries: lemons, oysters, and the rummer glass, an expensive item at that time. In contrast to the still life from 1627, there is no reference to Asia, such as porcelain. Candies made of Brazilian sugar were absent as well; the only overseas luxury in this Dutch work of art was silver from Spanish America. Still, in Claesz's composition silver items were clearly associated with wealth and opulence, overlaid with connotations of vanity expressed by the position of the silver cup. The painting demonstrated that all sorts of "treasures", especially silver, were available even in times of intense warfare against Spain.

The influence of the Low Countries and their methods of representing Spanish American silver was also evident in Frankfurt. In the 1620s, the Moravian painter Georg Flegel (Ketelsen-Volkhardt 2003), initially a collaborator of the Flemish artist Lucas van Valckenborch at Frankfurt, designed one of his famously meticulous still lives. This impressive work, of relatively large dimensions (78 × 67 cm), presented a sumptuous meal with a parrot (Flegel 1620). As in the aforementioned still lives from the northern and southern Low Countries, Georg Flegel depicted a lavishly decorated table and two silver cups, a gilded silver vessel and a silver vase with gilded ornaments. As in Claesz's and Peters's paintings, the silver cups were accompanied by glass vessels, illustrating that silver and glass objects had an outstanding value. Flegel positioned one of the silver cups at the very centre of the painting alongside a stand filled with sugar candies. Walnuts, figs, peaches, pomegranates, grapes, and a melon laid on tin plates, and a single lemon lay directly on the table. A green parrot clinging to an earthenware jar dominated the scene. In contrast to Pieter Claesz's painting, Flegel's contained Mediterranean fruits and items from Spanish and Portuguese America, but not from Asia. The Brazilian parrot and sugar as well as the Spanish American silver and the melon alluded to Frankfurt's American connections and the networks of its mercantile elites. As in the still lives of Dutch and Flemish artists, silver artefacts in Flegel's Frankfurt paintings connoted wealth and represented connections with distant lands. However in contrast to Claesz's and Peeters's paintings, Flegel generally avoided critical vanity associations. As in the other examples, silver coins had no place in this still life.

The early seventeenth-century still lives we have examined represented silver in the form of ornamented and often gilded silver cups, accompanied by highly polished tin plates and pots, often made to resemble silver items. Fresh or dried fruits from the Mediterranean, sugar from Brazil, and Asian imports completed the scenes. The silver cups were often large and placed at the centre of the composition. In cases where still lives showed clear connotations of vanity, they tended to affect organic materials or glass vessels rather than the silver. At most, a silver cup lay on a table instead of standing on it but showed no damage. From the analysis of the aforementioned still lives, we can conclude that, in north-western Europe in the early seventeenth century, the presentation of silver artefacts in paintings was an important method of representing far-reaching global connections as well as worldly wealth and splendour in a time of intense warfare. Money had no place in this imagery.

In Southern Europe, the presentation of silver in still lives differed from that in north-western Europe because Mediterranean fruits lacked any foreign connotations. Peaches, pomegranates and so on were not considered special luxuries and conveyed no connection to the global market. In contrast to north-western Europe, Spanish still lives were called *bodegones*, literally *taverns* (Jordan 1985; Burke and Cherry 1997). Influenced by Italian artists like Michelangelo Caravaggio, Spanish *bodegones* usually represented ordinary scenes populated with baskets, clay receptacles, and typical fruits of the area such as olives, grapes, lemons, oranges and apples. In the early seventeenth century, plates were a luxury that appeared frequently in Spanish still lives, but which could have been made from either silver or tin. A well-known example from 1650 is a still life from Seville by Francisco de Zurbarán (Zurbarán 1650). In the painting, a relatively small gilded silver cup and two clay jugs stand in a line, the gilded silver bowl at the left and the fourth pitcher on the right, each placed on a polished tin plate. The second clay carafe was a *búcaro*, presumably produced in Tonalá, in modern Mexico. Thus, Zurbarán positioned gilded silver objects alongside earthenware and tin items without any special suggestions of wealth and well-being. The American origins of the silver were obvious but unremarkable; the same applied to the *búcaro*.

Besides Italian artists, Spanish still life paintings were also influenced by the Low Countries in the "Siglo de Oro", the golden seventeenth century of Spanish arts and culture. The best-known example is the Artist Juan van der Hamen (Jordan 2005). He was the son of a Flemish courtier who moved to Madrid from Brussels, and a half-Flemish mother, Dorotea Witman Gómez de León. Juan van der Hamen became a court painter. In the 1620s, he painted a still life with two monkeys sitting beside a chest filled with grapes and pomegranates under a branch of an orange tree (Jordan 2005, p. 107). In the background, a silver cup filled with candied fruits recalls the Dutch use of silver to communicate opulence and global connections. On both sides of the foreground are two vessels of red ceramics, probably from Mexican Tonalá. In this painting, silver is afforded less importance than ceramics from Tonalá and American monkeys. Nevertheless, in contrast to Zurbarán, van der Hamen connected silver with sugar for candies, whereas he presented the Mediterranean fruits in local chests. Even if van der Hamen echoed Flemish paintings in his presentation of silver objects, they were seen in a different light at the court of Madrid. Van Hamen linked silver to monkeys, sugar, earthenware jars and local fruits, whereas the still lives designed in north-western Europe placed silver objects alongside nautilus shells, Chinese porcelain, oysters and glassware. The objects presented in Spanish *bodegones* were far more ordinary than those designed by north-west European artists. Even references to overseas—monkeys and sugar from Brazil, pottery from Mexico—were considered less luxurious than nautilus shells and Chinese porcelain.

The difference in the treatment of silver in north-western European compared to Spanish still lives suggests that silver was not regarded so highly in the south. This impression is corroborated by an inventory set up for the passport of a young Flemish-German nobleman, the Duke of Aarschoot and prince of Arenberg. In the 1640s, after staying for several years in Madrid and marrying a high-ranking and much wealthier aristocrat from Valencia, the young couple prepared for the voyage to Flanders. An inventory of silver objects from the kitchen to be transported over the long distance between Madrid and Brussels contained a noteworthy collection of large silver plates, cups and other objects. In contrast, the post-mortem inventories of the Flemish relatives of the duke mention silver mostly as jewellery (Pieper 2019). The comparison shows that the amount of silver used at the Spanish court far surpassed what was available in Flanders. This had a predictable impact on the presentation of silver in still lives: whereas in north-western Europe decorated silver laid on tables and held vanity connotations, silver in Spain appeared alongside "ordinary-exotic" items, without any allusion to vanity. Thus, in Spain, even though artefacts of American silver still had connotations of global connectivity, they were well integrated into the globalised material culture that had already integrated a range of exotic goods—including monkeys and Mexican pottery—and therefore did not function as references to outstanding wealth.

The arrangements of silver objects in the aforementioned still lives did not contain silver coins. Specie had its place in allegories with human figures representing vanity. In line with the concepts employed by Marinus van Reymerswale in the first half of the sixteenth century. Reymerswale personified vanity allegories and used silver coins for moral critique. In contrast to the presentation of silver objects in still lives—associated with wealth and conspicuous consumption, which *might* be considered harmful, the depiction of silver coins in personified vanity paintings left no room for interpretation: in Flanders as well as Spain silver coins were money, and money was pernicious. By the early seventeenth century—one hundred years after Reymerswale—allegorical works of art combined silver objects with silver coins.

In the second decade of the seventeenth century, in Antwerp, Clara Peeters painted a personified vanity allegory (Vergara 2016, p. 64, Fig. 33). It presented twelve gold and three Flemish silver coins beside a large gilded silver cup which laid on a table, a pair of dice and another gilded silver vessel standing in the rear. A necklace of pearls and a golden hairband with pearls and rubies, like most items in this allegory, pointed to Spanish America. Whereas elaborate silver objects—the silver cups—clearly evoked wealth and global connections, the coins and dice introduced an element of criticism. The relationship between silver objects and silver coins is unequivocal: silver cups, standing or laying down, represented a tangible, intrinsic worth transferable to one's heirs even if it might vanish when the individual owner was dead. On the other hand, coins could only be used for gambling, a pernicious pleasure with no lasting value.

Silver coins are employed for a similar symbolic function in a vanity allegory, painted by Antonio de Pereda y Salgado in Madrid around 1634 (*The Spanish Golden Age* 2016; Pereda y Salgado 1634). In this image, 15 silver and eight Spanish American gold coins lay among playing cards and a disorderly collection of pearls and golden filigree on the edge of a table in front of a gilded clock. A gilded armoury and candlestick are the other metal objects in this allegory. A genius with ails pointing at a globe suggests global connection as a particular form of vanity. Once again, coins are associated with gambling, whereas the gilded silver candlestick represents long-lasting value, even though it would lose its personal worth in the face of mortality.

It is noteworthy that coins were not depicted in the still lives we have examined from the first half of the seventeenth century. Despite the contemporary use of coins as medals for collars, to commemorate baptisms, and as silver amulets to provide good luck and wealth; in vanitas oil paintings silver coins were presented critically, associated with gambling. On the other hand, the same vanity paintings employed elaborate silver objects to evoke wealth, even if this was considered vainglorious. This applied equally to still lives painted in Spain, which attributed less value and exoticism to silver artefacts than the still lives designed in north-western Europe. In Catholic Northern Europe as well as in the south, personified vanity paintings combined both forms of silver representation: valuable silver items and nefarious silver coins. In Protestant Europe, the still lives and their vanity allusions incorporated no personified vanity, made no reference to gambling and, therefore, did not depict money.

In the sample from Western Europe, still lives and paintings with a personified vanity associated silver artefacts with durable wealth and global connections. This representation of silver objects reflected a common attitude and experience irrespective of political or religious positions. Silver artefacts were used by the general public as well as by religious and governmental officials and the nobility as a secure and reliable item of value. In contrast, coins were not linked to any lasting, intrinsic value. Besides religiously inspired suspicions with respect to money, Europeans' low regard for silver coins, as evident in the early seventeenth century paintings we have examined, might have been the result of frequent coin debasements and fraud (Lane 2017). Despite the annual deliveries of American silver, debasement was necessary all over Europe (Quinn and Roberds 2007) to cope with financial shortages in a time of unceasing warfare.

Counting Silver

A varied and complex social imaginary of silver is also evident in contemporary accounts regarding the Spanish American "treasure fleets". Beginning in the mid-sixteenth century, convoys of several dozen ships crossed the Atlantic between Spain and Spanish America twice a year, following the example of the Italian merchant republics in the Mediterranean (García-Baquero and Shaw 2002). Whereas the departure of the transatlantic fleets from Europe went more-or-less unnoticed, their safe arrival overseas was impressively represented. In the Caribbean, large fairs celebrated the arrival of goods from Europe, which included commodities from Africa and even Asia up until 1600 (Pieper 2018). The fairs at Portobelo were famous even beyond the Iberian world and attracted attacks from Dutch and English privateers. The eyewitness account of Thomas Gage in his *New survey of the West-Indies*, first published in London in 1648, offered even more publicity. Describing the riches of the Portobelo fair, as Gage experienced them in 1637, the author deployed the following imagery:

> what most I wondred at was to see the requa's of Mules which came thither from Panama, laden with wedges of silver, [...] the heapes of silver wedges lay like heaps of stones in the street. [...] Within ten daies the fleet came, consisting of eight Galeons and ten Merchant ships [...] Then began the price of all things to rise, a fowl to be worth twelve Rials, which in the main land within I had often bought for one [...] It was worth seeing how Merchants sold their commodities, not by Ell or yard, but by the

piece and weight, not paying in coined pieces of money, but in wedges which were weighed and taken for commodities. This lasted but fifteen days, whilst the Galeons were lading with wedges of silver and nothing else; (Gage 1648, p. 196)

Gage made a clear distinction between bullion, which was brought by mules, exchanged for goods from the Old World and loaded on the ships as a commodity, and specie—the coins and money he had to pay for his subsistence. The silver wedges symbolised overwhelming riches; silver coins meant depletion and overcharge.

Besides this spectacular description of a "treasure fleet" loaded in Portobelo in 1637, consistent (if less sensational) accounts of the fleets' arrival from the Caribbean spread across Europe during the sixteenth century. News about the fate of "treasure fleets" from overseas was transmitted through several channels and for a range of purposes in early modern Europe: merchants were interested in the commercial value of the fleets' cargo, whereas governments needed precious metals to pay their debts, finance warfare and cover public expenditure. Both merchants and governments got their information, at least in part, through weekly newsletters and newspapers. Up until now, studies have centred on the sources and the audience of such media (Raymond and Moxham 2016). The accuracy of notices about American treasures in printed newspapers has been discussed, but the specifics of the imaginary used in discussion of American silver in newsletters and newspapers still lack a thorough analysis. After a short description of early modern newsletters and newspapers, this contribution will focus on the imaginary associated with bullion and specie in both manuscripts and imprints.

Beginning in the fifteenth century, handwritten newsletters appeared disseminating information about the latest events in Italy (Infelise 2017; Keller and Molino 2015). From the late sixteenth century onwards, specialised agencies collected notices for weekly newsletters and circulated the anonymous two folio letters all over Europe to inform the elites and persons related to them. Venice, Rome, Antwerp and Cologne were centres for the compilation and further distribution of newsletters concerning overseas events. In many smaller cities, professional bureaus re-copied the newsletters from information hubs and added local news, as did private individuals for their friends and relatives. Sometimes the subscribers of newsletters collected them. One famous collection of newsletters is the so-called *Fuggerzeitungen,* obtained by the Fugger merchants at Augsburg and still preserved at the National Library of Vienna (*Fugger-Zeitungen*). The series starts in 1569 and ends in 1604. For 1568 and 1605, very few newsletters still exist.

The first weekly printed newspapers emerged on the eve of the Thirty Years' War. Starting in 1618 a one-folio newspaper appeared in Amsterdam, the *Courante uyt Italien, Duitsland,* etc. (de Weduwen 2017; *Courante*). Soon afterwards, publishers in Antwerp began printing newspapers as well (Arblaster 2014). Paris, London and many other cities followed the Dutch example. Unlike the system of handwritten newsletters, which had hubs in Venice, Rome, Cologne and Antwerp, and disseminated information across Europe, printed newspapers had local editors who edited them for a local public, although their readers might forward already printed copies to friends and relatives in more distant places.

At first, printed newsletters adopted the external form of handwritten newsletters, using the location and date where the news was collected as headings for the paragraphs of each letter. From 1621 onwards— the beginning of the Thirty Years' War in the Dutch Republic— Amsterdam-printed newspapers introduced a change in their external format: after the notices from Europe, copied from handwritten newsletters and mentioning the places of origin in the heading, the printed newspapers omitted a heading for local news and simply drew a line. Beneath that line, hurried readers could find news from Amsterdam.

Notices about the American silver fleets were published in both handwritten and printed news. In the late sixteenth century, the handwritten newsletters collected by the Fugger merchants had usually obtained their information about the silver fleets by way of Venice or Antwerp, originally from Madrid or Seville. Notices regarding the silver fleets relied heavily on the data of the Spanish tax administration, rented by the merchants' guild of traders with Spanish America (*Consulado de Indias*). Information concerning the fate of the ships and the silver from overseas circulated first in Seville and Madrid, was collected and copied by merchants, ambassadors and authors of newsletters and then sent on to their correspondents in other parts of Europe. The central hubs of newsletter compilation collected these messages and copied them for the subscribers of newsletters such as the dukes of Bavaria or the Fugger merchants in Augsburg (Pieper 2016). From 1618, the information began to be published in weekly printed newspapers as well.

Newsletters and newspapers adhered to a compressed style. Although newsletters circulated across much of Europe and should perhaps have avoided politically biased language, they did not always do so. Especially in descriptions of warfare and overseas piracy, the wording differed depending on the origin of the information and the locale where the notices had been compiled. This was especially true in the case of printed newspapers, as they had a close relationship to their local public sphere. As a result, political points of view influenced the messages in printed news more heavily than in handwritten ones. Nonetheless, the previous analysis of paintings from Spain, the northern and the southern Netherlands revealed that attitudes towards silver did not vary according to political positions but was related to the *form* of the silver being presented, as either bullion or specie. The analysis of Thomas Gage's polemic regarding the Portobelo fair corroborated that the form of silver as either bars—"wedges", as he called them—or as silver coins influenced the image of silver more than political positions. These findings need to be tested for handwritten and printed regular news, which were not as contentious as Gage's text.

Gage described the loading of a "treasure fleet" in the Caribbean in 1637. On 27 November 1604, correspondents in Rome compiled a handwritten newsletter describing the arrival of another such fleet in Spain. The newsletter reported that letters from Spain said the arrival of the fleet from Peru was expected soon, loaded with silver and gold worth around "six million".[1] On 5 December 1621, the eve of the Thirty Years' War in the Netherlands, the Amsterdam newspaper *Courante uyt Italien* published news compiled in Venice

---

[1] *Fugger-Zeitung*, Rome, 27 November 1604: "Verschine Mittwoch ist die Ordinari Post mit Briefen aus Spania von ultimo pasato angelangt, die berichten, daß […] man der Flota auß Peru, mit Silber und Goldt, uf 6 Million reich täglich gerwertig".

on 12 November 1621, confirming that the fleet from New Spain had arrived with a rich cargo worth "5.5 million".[2] Like Thomas Gage's descriptions of silver wedges laying in the streets, manuscript and printed news described the arrival of silver in Europe as rich and focused on lump sums of millions, omitting the monetary unit. This would have been the peso, a coin and accounting unit of Spanish America with a silver content comparable to the thaler.

American silver, which was mainly shipped in bars, retained its image as a rich commodity in handwritten and printed news. There was no marked difference between the manuscript newsletter from Rome and the newspaper published in Amsterdam referring to information from Venice. In contrast to Gage's account, which characterised the form of the silver as wedges, European news did not mention the form of the silver arriving from overseas but instead its rough value—its *richness*. Even when notices about the silver fleet appeared in the news-from-Amsterdam section beneath the line in the newspaper, the tone was not critical, and it stressed the anticipated wealth or richness of the cargo. In December 1626 reports reached Amsterdam from Lisbon that the silver fleet had reached the Old Bahama Channel with a very rich cargo.[3] Spanish American silver was consistently represented as bullion and thought of as *wealth* upon its actual or expected arrival in Europe.

The idea of silver as highly valuable merchandise was augmented by the inclusion of silver in the enumeration of the most important goods arriving in the Iberian Peninsula. In 1598, a newsletter from Antwerp reported that the fleet from the Indies (Spanish America) had arrived in Seville with 7 million gold in cash and 8000 arrobas of cochineal. The term used for silver was "gold inn content", gold as a synonym of the German term for money, i.e. "Geld", and the expression "content" being a translation of the Spanish "*contante*" or cash as opposed to credit, without any distinction between coins and bars. The cochineal

---

[2] *Courante uyt Italien en Duytslandt [...]*, 5 December 1621, "Ut Spanien wert geconfirmeret dat die Vlote upt niew Spanjen vyf ent een half Millionen rijck aencomen was".

[3] *Courante uyt Italien en Duytslandt [...]*, 12 December 1626, "Briven vanden 17 ende 25 Octob. uyt Lisbona adviseren dat aldaer een Carveel vande Silver Vloote gecomen was met tydinghe dat deselve volghde ende inde Canael Baham lach seer rijekelijch gheladen".

mentioned in the news was a highly valued red dyestuff. In this case, it amounted to about 90 metric tons. As a point of comparison, using the price of cochineal in Amsterdam in 1609, the approximate value of the red dyes might have been 1.6 million pesos.[4] A similar construction of silver as merchandise included in a list of cargo appeared in the *Courante uyt Italien Duytsland, etc.* on 18 November 1628. The report announced the incredible riches captured by Piet Heyn, the famous Dutch privateer: 200 "tons of gold", with the explanation that it was "gold silver", i.e. probably 200,000 silver pesos, the cargo of one ship. In addition, the newspaper mentioned 2400 chests of indigo and cochineal, 37,000 Mexican skins, and more exquisite merchandise.[5] Both reports, one about the arrival of the silver fleet in Spain and the other about the seizure of a Spanish ship by a Dutch privateer, relied on the data and documents of the Spanish American tax administration, which accompanied the cargo on board the ships. Silver was represented as a very valuable commodity, but not markedly different from other merchandise.

The transformation of American silver from bullion to specie occurred within Europe, in the mints as well as in the news. Information concerning the movement of Spanish American silver within Europe employed a different terminology than what was used to describe transatlantic deliveries. On 25 January 1596, handwritten newsletters from Amsterdam informed the Fugger, via Antwerp, that a ship from San Lúcar de Barrameda had arrived at Rotterdam. The cargo included wine, a small amount of cochineal and olive oil, and 5000 Flemish pounds in cash. The amount was approximately equivalent to

---

[4] *Fugger-Zeitung*, Antwerp, 4 April 1598: "Mit gedachte Cuerier wirdt auch auß Madrill geschrieben wieder die flotta auß Indya Inn Seviglia mit 7 Million gold inn content auch so inn 8000 Arrobas Cutzeniglia angelangt", anno.onb.ac.at. Accessed on January 10, 2019. In 1609, the price of a pound of cochineal in Amsterdam was about 21.75 guilders (Posthumus 1943).

[5] *Courante uyt Italien en Duytslandt [...]*, 18 November 1628: "[...] die gheschat wordt op ontrent de 200.tonnen Gouts soo aen Gout Silver, 2400. Cassen Indigo ende Consinillie, 37000. Mestcanische Huyden, ende meer andere costelijcke Waren, diemen alle daghen hier te Lande verwachtende is". The amount of the cargo of the whole fleet was valued afterwards at 12 million guilders or 4.8 million pesos, equivalent to approximately 120 tons of silver.

12,500 pesos[6] The description of the cargo as well as its Spanish origin, the same Andalusian port where the American silver fleets arrived, meant that American silver had reached Amsterdam. The wording with which this information was presented, however, stands in contrast to the notices referring to the arrival of American goods in Europe, which reported quantities without any units. In the Antwerp newsletter, the amount was expressed in a Flemish monetary unit, even though the notice itself came from Amsterdam. The representation of Spanish American silver still maintained a certain character as bullion—it was listed alongside other commodities—but the description of the silver as cash and the use of a monetary unit transformed the representation of the silver from bullion into specie.

The same applied to silver deliveries mentioned in printed Amsterdam newspapers. On 15 May 1619, the *Courante uyt Italien, Duytsland,* etc. notified its readers that, according to news from Venice, 150,000 crowns from Spain had reached Genoa. The money was to be further remitted to Germany.[7] In this newspaper, the specific monetary unit, a "crown", was used to inform readers about the transfer of silver as specie within Europe. The crown was an English coin almost equivalent to the peso, indicating that the information from Venice had reached Amsterdam via London. In contrast to the previous notice about precious metals arriving at Rotterdam, the silver delivered to Genoa did not form part of a commercial venture as no further commodities were mentioned. It could be inferred that the specie had been sent from Spain to Genoa to pay for troops in Germany in the early phase of the Thirty Years' War. Importantly, no hint appeared in the Dutch newspaper that the Spanish remittances should be used for a war, which was still far from Amsterdam at a time when the Twelve Years' Truce between Spain and the Dutch Republic was in force.

---

[6]*Fugger-Zeitung,* Antwerp, 25 January 1596: "Aus Amsterdam di 20 ditto, Auf 17 dis ist zue Rotterdam ein Schüff von San Lúcar angelanngt, bringt Wein, was weniges cremse, und Oell, auch 5000 Pfundt Flemisch bar Gelt".

[7]*Courante uyt Italien en Duytslandt [...],* 15 May 1619: "Uut Venetien, den 26. April 1619: [...] tot Genua zyn wederomme 150 Duysent Croonen uyt Spaignen aenghecomen/ die on Duytslant souden gheremitteert werden".

The arrival of Spanish American silver in Europe was always represented in terms of wealth and richness, irrespective of the origin or the intended audience of the message. Political considerations had no influence. Precious metals were imagined as bullion and associated with other exotic merchandise. In contrast, when silver was shipped from one European port to the other, its presentation was not as a commodity but as specie. This transformation was evident both in descriptions of commercial transactions and reports about governmental expenditure for military purposes. Unlike the imagery employed to describe silver crossing the Atlantic, the transfer of silver within Europe was represented not as the transmission of richness but as the matter-of-fact movement of money. The description of specie in news media was neutral in tone, a contrast to the clearly negative connotations of silver-as-money in Thomas Gage's account.

Conclusion

Silver as bullion and as specie received different representations in early modern Europe. Silver bullion in the form of artefacts and unminted bars had very positive connotations. Silver artefacts like jars and cups in paintings symbolised wealth and were depicted in combination with goods from overseas like sugar and nautilus shells and exotic animals like monkeys. Silver in bars was mentioned in handwritten and printed news and books alongside explicit references to wealth. Thus, silver in its raw form and as artefact was understood as a precious metal, wealth in its most obvious form. This imagery did not depend on the particular medium—painting, manuscript or imprint—nor on the place where the images appeared, nor was there any influence of political or cultural viewpoints.

Silver as specie, on the other hand, received dramatically less enthusiastic descriptions. In paintings, it was associated with avarice, treachery, exploitation and gambling. In texts, the presentation of silver money was more nuanced. Whereas Thomas Gage's polemical book about the Spanish Caribbean linked the abundance of silver money to high prices and poverty, specie appeared in newsletters and newspapers with neither positive nor negative connotations. When silver appeared in news media as a shipment *en route* to Europe, it was described as bullion, but once it was moving *within* Europe, even without changing its external form and without coinage, it was regarded as specie.

The image of silver in the early modern imaginary varied according to the use of the metal. Silver-as-artefact was a precious metal; silver-as-coin was money, regarded at best with a neutral attitude in "technical" information, and in less nuanced descriptions and vanity paintings as something explicitly negative. While silver bullion was appreciated according to its metallic value, silver specie was the domain of Mammon.

Bibliography

Primary Sources

Manuscripts
Courante uyt Italien en Duytslandt [...]. Online: *delpher.nl*. Accessed June 15, 2018.
Fugger-Zeitungen. Online: *anno.onb.ac.at*. Accessed January 10, 2019.
"Inventory of Antonio González de Barreda, 20 April 1741, on the ship Nuestra Señora de Covadonga, on the way from Acapulco to Manila": Archivo General de Indias, Sevilla, Contratación 581 B, N5: 9r–17r.

Old Prints
Gage, Thomas, 1648. *The English-American His Travail by Sea and Land: or, A New survey of the West-Indias*. London: R. Cotes 1648.

Paintings
Claesz, Pieter, 1627. *Still Life with a Turkey Pie*, Haarlem. Rijksmuseum, Amsterdam, Inventory SK-A-4646.
Claesz, Pieter, 1633. *Breakfast with oysters*, Haarlem. Museumslandschaft Hessen-Kassel Gemäldegalerie Alte Meister, Kassel, Inventory GK 437.
Flegel, Georg, 1620 (ca.). *Grosses Schauessen mit Papagei*, Frankfurt. Alte Pinakothek München, Inventory 1622.
Peeters, Clara, 1611. *Still life with flowers, guilded silver cup, almonds, dried fruits, sweets, bread, vine and a tin jar*, Antwerp. Museo del Prado, Madrid, Inventory P 001620.

Pereda y Salgado, Antonio de, 1634. *Allegorie der Vergänglichkeit*, Madrid. Kunsthistorisches Museum Wien, Gemäldegalerie, Wien, Inventory 711.

Sánchez Coello, Alonso, 1576 (ca.). *Vista de Sevilla*, Sevilla. Museo de América, Madrid, Inventario 00016.

Zurbarán, Francisco de, 1650 (ca.). *Bodegón con cacharros*, Seville. Museo del Prado, Madrid, Inventory P002803.

Secondary Sources

Arblaster, Paul, 2014. *From Ghent to Aix: How they brought the News in the Habsburg Netherlands, 1550–1700*. Leiden/Boston: Brill.

Biesboer, Pieter; Brunner-Bulst, Martina; Gregory, Henry D., 2004. *Pieter Claesz: Master of Haarlem Still Life*. Zwolle: Waanders.

Brendecke, Arndt, 2016. *The Empirical Empire: Spanish Colonial Rule and the Politics of Knowledge*. Berlin and Boston: De Gruyter Oldenburg.

Brook, Timothy, 2007. *Vermeer's Hat: The Seventeenth Century and the Dawn of the Global World*. London: Bloomsbury Press.

Burke, Marcus B.; Cherry, Peter, 1997. *The Collections of Painting in Madrid, 1601–1755*, edited by Maria L. Gilbert, 2 vols, Part 1. Los Angeles: Provenance Index of the Getty Information Institute: 716–720.

De Weduwen, Arthur, 2017. *Dutch and Flemish Newspapers of the Seventeenth Century*, 2 vols. Leiden and Boston: Brill.

García-Baquero González, Antonio; Martínez Shaw, Carlos, 2002. *La Carrera de Indias: suma de la contratación y océano de negocios*. Granada: Universidad de Granada.

Infelise, Mario, 2017. *Gazzetta. Storia di una parola*. Venice: Marsilio.

Jordan, William B., 1985. *Spanish Still Life in the Golden Age 1600–1650*. Fort Worth: Kimbell Art Museum.

Jordan, William B., 2005. *Juan van der Hamen y León and the Court of Madrid*. New Haven and London: Yale University Press.

Hamann, Byron Ellsworth, 2010. "The Mirrors of Las Meninas. Cochineal, Silver and Clay." In: *The Art Bulletin* 92, 1–2: 6–35.

Keller, Katrin; Molino, Paola, 2015. *Die Fuggerzeitungen im Kontext: Zeitungssammlungen im Alten Reich und in Italien*. Wien and Köln: Böhlau.

Ketelsen-Volkhardt, Anne-Dore, 2003. *Georg Flegel, 1566–1638*. München and Berlin: Deutscher Kunstverlag.

Lacueva Muñoz, Jaime J., 2010. *La plata del rey y sus vasallos. Minería y metalúrgia en México (siglos XVI y XVII)*. Sevilla: CSIC/Universidad de Sevilla/Diputación de Sevilla.

Lane, Kris, 2017. "From Corrupt to Criminal: Reflections on the Great Potosí Mint Fraud of 1649." In: *Corruption in the Iberian Empires: Greed, Custom, and Colonial Networks*, edited by Christoph Rosenmüller. Albuquerque: University of New Mexico Press: 33–61.

Martínez Ruiz, José Ignacio, ed., 2018. *A Global Trading Network: The Spanish Empire in the World Economy (1580–1820)*. Sevilla: Editorial Universidad de Sevilla.

Mathes, Bettina, 2006. *Under Cover: Das Geschlecht in den Medien*. Bielefeld: Transcript.

Merriam, Susan, 2009. "The Reception of Garland Pictures in Seventeenth-Century Flanders and Italy." In: *Domestic Institutional Interiors in Early Modern Europe*, edited by Sandra Cavallo and Silvia Evangelisti. London and New York: Routledge: 201–220.

Morineau, Michel, 1984–1985. *Incroyables Gazettes et fabuleux métaux. Les rétours des trésors américains dans les gazettes hollandaises XVIe–XVIIIe siècles*. Paris: Maison des Sciences de l'Homme and Cambridge University Press.

North, Michael, 1997. *Art and Commerce in the Dutch Golden Age*. New Haven and London: Yale University Press.

Philipps, Elena; Hecht, Johanna; Esteras Martín, Cristina, eds., 2004. *The Colonial Andes: Tapestries and Silverwork, 1530–1830*. New York: The Metropolitan Museum of Art.

Pieper, Renate, 1995. "Amerikanische Edelmetalle in Europa (1492–1621). Ihr Einfluß auf die Verwendung von Gold und Silber." In: *Jahrbuch für Geschichte Lateinamerikas* 12: 163–192.

Pieper, Renate, 2016. "News from the New World: Spain's Monopoly in the European Network of Handwritten Newsletters during the Sixteenth Century." In: *New Networks in Early Modern Europe edited by Joad Raymond and Noah Moxham*. Leiden and Boston: Brill: 495–511.

Pieper, Renate, 2018. "Die Messen der Karibik im hispanoamerikanischen Handelsgefüge, 16.–18. Jahrhundert." In: *Europäische Messegeschichte 9.–19. Jahrhundert*, edited by Markus A. Denzel. Köln, Weimar, and Wien: Böhlau: 353–367.

Pieper, Renate, 2019. "Das Haus Arenberg im Spanischen Imperium (16. und 17. Jahrhundert)." In: *Das Haus Arenberg und die Habsburgermonarchie. Eine transterritoriale Adelsfamilie zwischen Fürstendienst und Eigenständigkeit*

(16.–20. Jahrhundert), edited by William D. Godsey and Veronika Hyden-Hanscho. Regensburg: Schnell & Steiner (in Preparation).

Pietschmann, Horst, 2013. "Imperio y comercio en la formacón del Atlántico español." In: *El sistema comercial español en la economía mundial (siglos XVII–XVIII). Homenaje a Jesús Aguado de los Reyes*. Huelva: Universidad de Huelva: 71–95.

Posthumus, N. W., 1943. "Prices from *Nederlandsche Prijsgeschiedenis* (Leiden 1943)." In: *The Medieval and Early Modern Data Bank*, edited by Rudolph M. Bell and Martha Howell. http://www2.scc.rutgers.edu/memdb/. Accessed January 15, 2019.

Quinn, Stephen; Roberds, William, 2007. "The Bank of Amsterdam and the Leap to Central Bank Money." In: *The American Economic Review* 97, 2: 262–265.

Raymond, Joad; Moxham, Noah, eds., 2016. *News Networks in Early Modern Europe*. Leiden and Boston: Brill.

Siebenhüner, Kim, 2018. *Die Spur der Juwelen. Materielle Kultur und transkontinentale Verbindungen zwischen Indien und Europa in der Frühen Neuzeit*. Köln and Weimar: Böhlau.

The Spanish Golden Age: Painting and Sculpture in the Time of Velázquez, 2016, edited by Staatliche Museen zu Berlin and Kunsthalle München. München: Hirmer.

TePaske, John J., 2010. *A New World of Gold and Silver*, edited by Kendall W. Brown. Leiden: Brill.

Vergara, Alexander, 2016, *The Art of Clara Peeters*. Antwerp: Koninklijk Museum voor Schone Kunsten and Madrid: Museo del Prado.

Young, Amanda Ruth, 2013. *Making Reflections/Reflecting Making: Clara Peeters and the Representation of Early Modern Authorship*. Santa Barbara: University of California.

Part III

Mining

5

A Matter of Scales: Understanding Spatial Patterns of Colonial Spanish America's Silver Mining in the Digital Age

Werner Stangl

Introduction

Silver mining in New Spain and Upper Peru was the primary source of wealth for the Spanish colonial economy. Around that central source, there was a network of related activities and effects: secondary supply economies commercially bound to the mining centres; the organisation of labour forces; the administrative and fiscal territorial organisations which controlled and channelled mining-related revenues; and many demographic transformations. All of these related processes followed spatial patterns subject to temporal change. There is abundant literature on the subject of colonial mining, which usually does encompass some sort of geographical information or organisation of contribution by sub-spaces, but spatial relations themselves—both as object of study and as

This article is part of the project, "Reconstructing Colonial Rule: A Historical Web-*GIS* for Spanish America", funded by the Austrian Science Fund (FWF), P 26379 G-18.

W. Stangl (✉)
Institute of History, University of Graz, Graz, Austria

© The Author(s) 2019
R. Pieper et al. (eds.), *Mining, Money and Markets in the Early Modern Atlantic*, Palgrave Studies in Economic History,
https://doi.org/10.1007/978-3-030-23894-0_5

explicans of phenomena—have received comparatively little attention in this context.

As with all studies with a strong geographical component, epistemologies and sources are bound by the scale at which they are aimed or within which they were produced (cf. Watson 1978). Given the local nature of extraction sites and fluctuations in mining cycles in Spanish America, most studies have a local or regional focus, while overarching publications consist in edited volumes that unite different case studies and are frequently focused on either "New Spanish" (Brading 1971) or "Peruvian" (Brown 2015) mining. Both approaches make sense given the fundamental differences between the two main spheres. Bernd Hausberger has most clearly sketched the differences between the two regimes: New Spain was characterised by unstable mining cycles, shifting regional distribution and higher dynamism, resulting in a better risk diversification (Hausberger 1995). Studies tackling colonial mining hemispherically are logically and thankfully scarcer, given the nature of the subject. In order to be meaningful—beyond general summaries like Bakewell's (1990)—they need to take a strongly comparative approach, contrasting different regional realities and case studies (Bakewell 1997; Povea and Zagalsky 2017), or to choose a topic that is geographically more uniform, such as pure numbers of production or fiscal data (Klein and TePaske 1982–1990).

However, practically nothing has been written on how different modes of production, forms of civil and ecclesiastical administration, or settlement topography framed the perception and representation of mining geography by travellers, observers, administrators and cartographers who sought to describe the totality of America. Bridging the gap between the different scales by synthesising them without disregarding or misunderstanding local or regional idiosyncrasies is an ideal that cannot be fully achieved. Alexander von Humboldt was outstanding in this respect because of his ability to use sources produced at all scales—maps from general charts to sketches of local circumstances, written texts from general descriptions to detailed reports, oral information and personal in situ reconnaissance—to pull together the disparate pieces of information, forming a spatially all-encompassing vision.

In the past few years, the FWF-funded project HGIS de las Indias (Stangl 2015a, b) has compiled a spatio-temporal database tracking the administrative organisation and development of settlement patterns in eighteenth-century and early nineteenth-century Spanish America. HGIS de las Indias's digital infrastructure allows researchers to combine different datasets within a common framework. It mirrors the "spatial containers" (provinces, districts, places...) which colonial administrators and authors used to organise information. In doing so, it juxtaposes the various overlapping datasets within a consistent frame and makes them more easily comparable. HGIS de las Indias further allows third parties to insert their own tabular data in the system, expanding its database. In this article, I show exactly how such a framework can contribute to a better understanding of the developments in the central economic activity of Spanish imperial silver mining and how it can assist in interpreting information gathered at different scales.

Unsurprisingly, given the centrality of the mining economy, the current HGIS de las Indias database already reflects settlement and economic patterns, which will be discussed in the first chapter. A separate section focused on a specific local context (Guantajaya in Tarapacá) will demonstrate how the description of local realities, such as the distribution of population and "settlements" (defined as a group of built structures and institutions) relate to the database's general vision of settlements compiled from data at a macro-scale. The second chapter will return to the problem of scale and analyse the spatio-temporal patterns in the well-known silver registry data compiled by Herbert Klein and John Jay TePaske (1982–1990), a dataset employed by other contributors in this volume (Hofmann; Jeffries; Pieper). The third and fourth chapters will discuss, in the contexts of New Spain and Peru, respectively, the effects of mining on the development of colonial institutions like territorial organisation and the postal system—reforms which also affected the development of mining. Another regional example (New Galicia in New Spain) helps to explicate the macro-level with the chorographic realities of mining geography. Unlike the Guantajaya case, which to some extent provides a warning against overemphasising the contribution of GIS infrastructure to historical understanding,

New Galician mining offers an example of how GIS can contribute to a better understanding of specific regional settings. Finally, the GIS tools will be used to assess Alexander von Humboldt's efforts at understanding the spatial patterns of colonial silver mining. By using GIS to analyse Humboldt, I will reveal the value of digital tools without overemphasising the explanatory power of spatial calculations and measurements. Ultimately, I argue that using GIS in building a historical narrative offers a means to direct attention towards otherwise undetected patterns, and that—for most early modern data—it should not be understood as a cliometric tool that produces "hard" data or used to explain causation on its own.

The Mining Town as a Category of Settlement in Spanish America

Geographic dictionaries and descriptions, as well as maps, represent settlements as fitting into a set schema of categories: for example as city, town or village. In the typical colonial text or geographic description, settlements were categorised by a mix of religious, ethnic, military and legal characteristics, as well as the settlement's layout or economic activity. Occasionally, when a place had two "strong" identities, the author would include a secondary identifier, like the "*city* and *port* of Acapulco". It is not surprising that mining, and especially silver mining, was one such identity marker for observers throughout the colonial period, and that settlements related to the mining industry fell into various categorisations of "mining town". Working through hundreds of geographical descriptions and thousands of archival records, HGIS de las Indias's thesaurus of settlement categories integrates the various terms used in a variety of contemporary sources to refer to places of mining activity, depending on the regional context and the prevailing spatial organisation of the settlements.

In New Spain and its northern provinces, the most common term for mining towns was the "real de minas" or simply "real". Legally, a "real de minas" was a mining district, but it was very common to identify the "real de minas" with its central settlement. And, legally speaking, not every town called "real" necessarily was one. The famous "real de Arizona", where huge silver nuggets were found in the 1730s, was termed a "real" for a long time even though there were never any mining institutions. Other small or informal isolated mining settlements in the Sonora region appear as "realito" in documents. In other parts of New Spain, there were dozens of minor mining settlements that did not constitute their own district, termed "mineral de XY". In the case of the minerales "del Monte", "del Chico" and "de Omitlán" near Pachuca, these places gained importance over time, turning into "reales de minas". By then, the prefix "mineral" had become part of the toponymy (and continues to be so today). In the mining region of Honduras, one finds the same divisions, with one central "real de minas" (Tegucigalpa) and several smaller "minerales" dotting the province.

In the gold-panning areas on the Pacific coast of the Nuevo Reino de Granada (in modern-day Colombia), historical texts sometimes term populational units with mining activity simply as "river of XY" or "beach of XY". In that area, very few formal settlements existed, but mining operations based on working groups, mostly of slaves (cuadrillas), operated along riverbanks, often with a gold processing unit located at the confluence of key rivers. A manuscript description of one province in the region (Iscuandé) reports: "On these beaches and rivers there are washing plants for placer gold which is extracted by one or the other citizen without formality nor method, only if they are in need of the metal" (Relación 1804, f. 193r).[1] More compact informal settlements in the Nuevo Reino de Granada might again appear under the category "mineral", while formal mining towns are also "real de minas".

In the rest of South America, mining towns are typically not referred to as "reales" but as "asiento de minas" ("asiento" being a general term for non-Indian settlements). But there too, we find the more informal mining camp being called "mineral", a term which puts less emphasis on the settlement itself. Even if it was temporarily or permanently abandoned, a place with silver veins could continue to be referred to as

---

[1]All direct quotes in this article are translations by the author. The Spanish original text can be consulted in the original source.

"mineral". Those "minerales" were ignored by the geographic literature and census tables of Peru, which tended to focus on Indian pueblos and formal Spanish settlements, ignoring haciendas and mining sites. Most of the settlements with mining activities were related to silver or gold mining. In Chile, on the other hand, "asientos de minas" were often related to lesser ores as well, and there were a handful of mining towns scattered across the whole hemisphere that derived their identity from copper mining. Map 5.1 shows all inhabited places with mining identities in the HGIS de las Indias database which existed at some point between 1701 and 1808.

This map is a static representation of the HGIS database. However, the programme is also capable of representing changes over time. One can use the database's stored dates of establishment and abandonment of settlements, as well as the presence or absence of a resident priest as a rough proxy for the centrality and importance of a mining site, to reveal the spatio-temporal developments in mining patterns across Spanish America.

Map 5.2a–d show the temporal change in the densities of mining towns in 25-year periods, giving weight to parish centres over other settlements. There is very little dynamism in the first quarter of the century; the founding of Chihuahua is the only significant event. The period between 1725 and 1750 sees a rise of mining in the southwest in the Nayarit area and Sierra Tepeque, including the boom mining town of Bolaños, which swiftly came to rival large mining centres like Zacatecas, well documented in the literature (Carbajal López 2002). Also, in the north, Sonoran mining sites began to disappear while neighbouring areas of Ostimuri, Sinaloa and to the east, in the Tarahumara country including the newly founded Chihuahua, saw them increase. The third quarter gives the most mixed impression. The decline in Sonora continued, and in contrast to the previous 25 years, we know of direct external factors: the second half of the century saw the rebellion of the Pima Indians in 1751 and a generally weakening regional stability due to increased Apache raids and the expulsion of the Jesuits on top of the traditionally high fluctuation of mining in areas related to small or superficial deposits. While old sites were abandoned as usual, few to no new ones were founded. In the rest of the

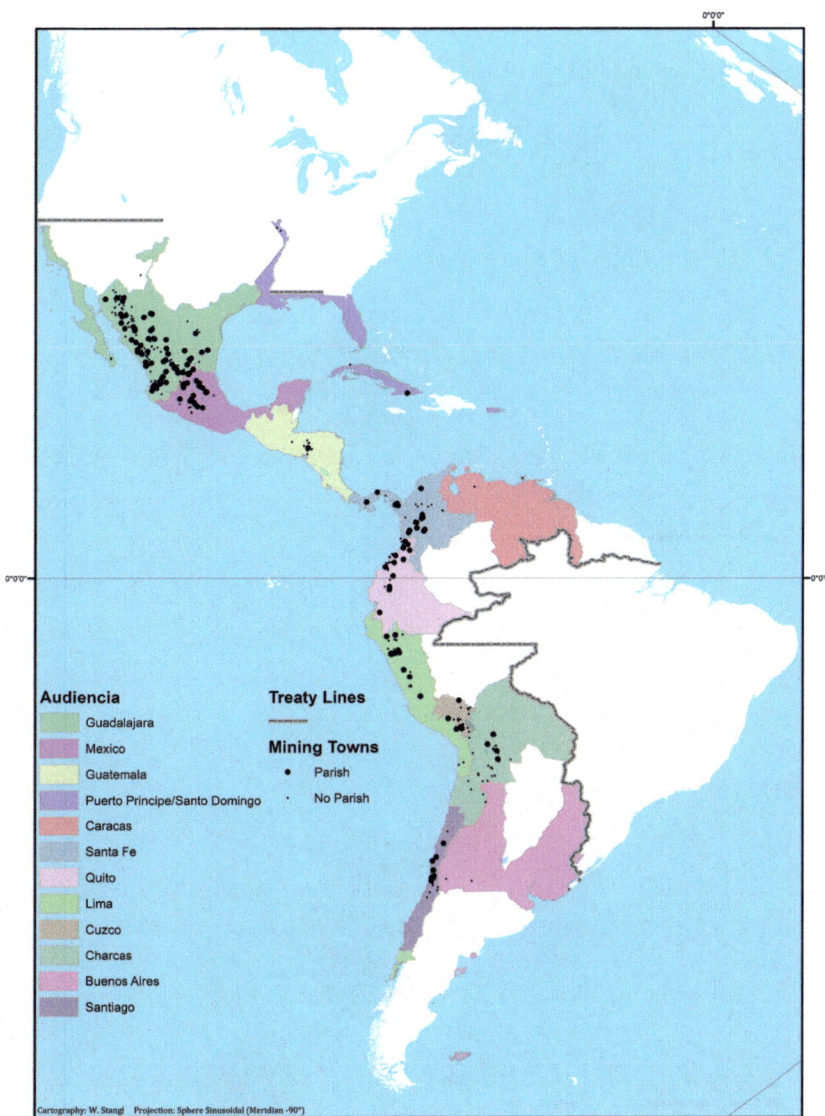

Map 5.1 Eighteenth-century mining sites in HGIS de las Indias (large dots: parish centres; small dots: no parish centres)

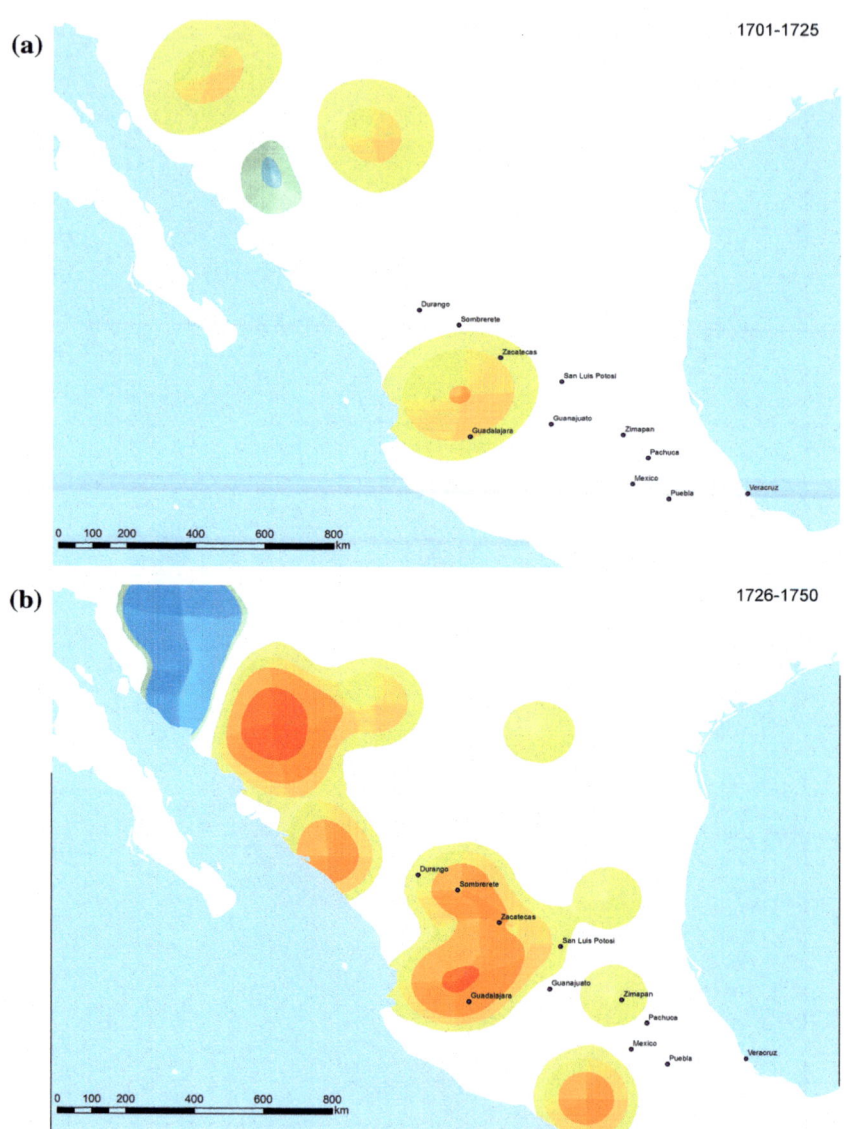

Map 5.2   a–d Increase (yellow/red) and decrease (green/blue) in active mining sites in New Spain during the eighteenth century (*Source* Stangl 2015a)

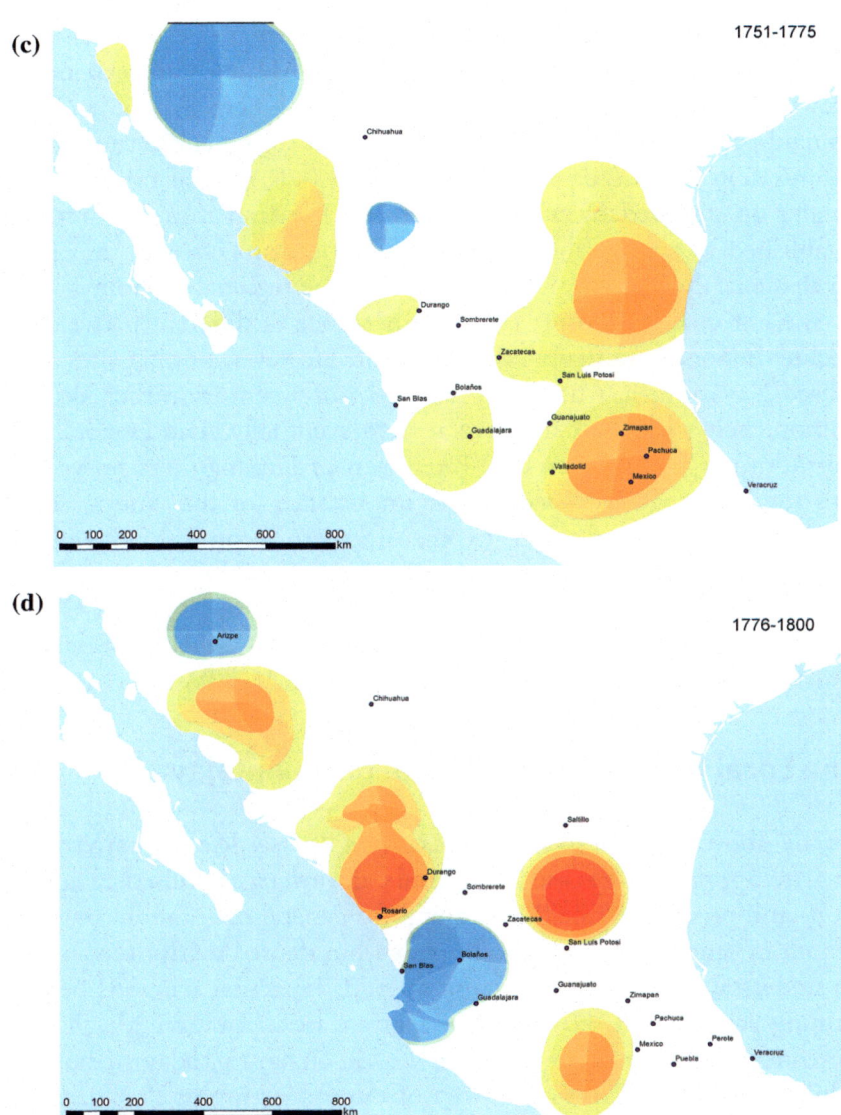

Map 5.2 (continued)

territory, dynamism seems to have slowed, with two exceptions: first, the more traditional mining areas north of Mexico City experienced growth. Second, in the Sierra Madre Oriental (Nuevo Santander, today: Tamaulipas), a formerly uncontrolled frontier region subject to planned colonisation by José de Escandón in the 1740s, several mining towns sprang up. Ignored by most literature on Mexican mining, they were established as part of the general colonisation process and constituted an element, though not the centrepiece, of the regional economy.

The last quarter century saw the emergence of different patterns: the decline in Sonora virtually came to a halt. However, mining sites in the Jalisco/Nayarit area, which had boomed half a century earlier, declined without being replaced and several were eventually abandoned. To the north, mining in the Sierra de Topia west of Durango was revived and this southern region gained greater importance for the Nueva Vizcaya mining industry compared to the northern centres of Parral and Chihuahua. The most prominent development occurred in the east, where mining boomed in a compact area around the reales of Charcas and Catorce which, at that time, were the effective centre of what we usually subsume under the label "mining in San Luis Potosí".

The Local Context: The Mineral de Guantajaya

In the above examples, we have examined at settlement patterns and their temporal changes on a large scale and over generalised categorisations and proxies. While this is useful for general orientation, one must "zoom in" and situate these concepts within dense descriptions in order to understand their regional contexts and the actual relations between mining activity and settlement in a given area. The example that will be considered here is the "mineral de Guantajaya", today in northern Chile, then in the Tarapacá district of the southernmost Peruvian jurisdiction of Arica.

Guantajaya was well known from the sixteenth century for its fine ore, but exploitation was impeded by a lack of confidence in sustainable veins, a lack of funds, the availability of workers and a problematic water supply stemming from its desert climate and soil. In 1758, a local miner received an allowance of 50 Indian mita workers for his mines, and the Crown marvelled at how such a famous silver mine could contribute so little to the treasury. This encouraged reforms, and Antonio O'Brien was dispatched by the Crown as a representative to assess the situation. In 1767, O'Brien was named governor of Tarapacá, which thus became a separate province from Arica—a rare decision, motivated by the hope of making silver mining at Guantajaya more profitable (Hidalgo Lehuedé 2004, pp. 249–255). Nonetheless, Guantajaya as a place even in later documents is mostly presented as a nondescript "mineral". It never had a parish priest assigned to it, and it is absent as an entity from the 1791/1792 census of the Province of Tarapacá (Estado de la Población 1792, f. 14r). From the perspective of the HGIS de las Indias database, it is a place of low profile.

This discrepancy demands that we try to understand where the labour force of the mines resided and what other infrastructure existed at the site. The closest formal settlement to Guantajaya was the port of Iquique—some ten kilometres away—which is, at least sometimes, mentioned as a dependent pueblo within the Tarapacá parish (Estado de la Población 1792). O'Brien produced a detailed report and map of the area (Plano de la mina 1764). It shows a number of houses at the port, while on the hill of Guantajaya there is only a single house and a few heaps marking individual claims. That map still seemed to hint that the workforce mainly lived in Iquique, even if walking ten kilometres from Iquique seems like a long commute.

Antonio O'Brien's textual report on his visit gives us some clarity about the distribution of population in the area (Hidalgo Lehuedé 2009). There he reports on the population of the port of Iquique in 1764. Its inhabitants included the tenant of the port, two Indian authorities, 25 encomienda Indians and a few mestizos—hardly a suitable workforce for the mines. But he also provides information about the population at the mineral itself: one civilian commissioned judge, one deputy priest, 4 mining administrators with 4 adjutants, some 350 workers with their families, an unspecified number of

> [...] men and women of all races, villages and provinces who are busy in the rescate de plata and searching the dump heaps for good scraps.

The number of people cannot be determined because it uses to be high one week and low the next. It can be estimated that by average it may be 200 people. Beyond that citizenry, some European and criollo merchants with shops [...] but all in the business of rescate de plata. And in the same way, there are other merchants from Chile, Buenos Aires, Tucuman and the interior provinces. (Hidalgo Lehuedé 2009, pp. 28–29)

The *rescate* was the economic activity of exchanging untaxed raw silver as payment instead of coins. In principle, it was a mechanism to collect produce at mining sites, with a commission, before transporting it to the treasury. Until 1755, merchants at Guantajaya were explicitly allowed to participate in the *rescate* (Gavira Márquez 2005). But, quite apart from the question of legality, the *rescate* was a frequent mechanism of informal exchange in the under-monetarised periphery, and as this report shows, it was the preferred method for economic transactions at Guantajaya. O'Brien ends his description of Guantajaya giving with a total population number for the whole site of "close to 700 inhabitants". The composition of the population according to him was mixed: the workers were mostly "zambos, mulatos, cholos and mestizos" from the "pueblo" of Tarapacá itself and its subject "asientos" of San José de Guarasiña and Tilivilca, while the Indians would not lower themselves to work under any terms not imposed by themselves. The mita Indians were employed in the processing of silver but not in the mines because "they go to the hill when they want, work the time they see fit and go back home when they like" (Hidalgo Lehuedé 2009, p. 28). O'Brien describes realities inverse to the depiction in his own map and the official church hierarchy, which both highlight Iquique as the centre: the deputy priest of Iquique resided at Guantajaya, as we have seen.

Twenty years later, in 1786, Guantajaya became seat of its own mining deputation, a representative institution of miners within the Royal Mining Tribunal of Lima, which explains the occasional use of more ostentatious titles like "Real Asiento de minas de San Agustín de Guantajaya" beginning in that year. For the most part, though, Guantajaya remained invisible. Antonio de Alcedo in his geographical dictionary of America, with five volumes and over 19,000 entries, describes it as: "a mountain of the province of Arica in Peru, at two leagues distance from the sea, where there are some rich silver mines, which yield but little through a scarcity of water" (Alcedo 1786–1789, vol. 2, p. 384).[2] The full population of the Tarapacá parish in 1791 is given as 2641 individuals in five settlements: Tarapacá, Guaviña, Camsana, Mañina and Iquique (Estado de la Población 1792, f. 14r). According to census logic, Guantajaya still did not even exist, while in actuality it was a buzzing town (for the standards of the region), accounting for over 25% of the parish population and most of its mixed-race population. Its residents, most probably, were registered at places like Tarapacá, Guarasiña and Tilivilca or at Iquique. A panoramic drawing of Guantajaya from 1807 represents the camp as a formal town, with some 30 buildings, including a wooden church (Vista del célebre Mineral 1807), giving an impression of Guantajaya at the end of the colonial period.

Large-scale cartographic representations and geographical descriptions, as well as modern reference databases based upon these works, are useful for giving a broad sense of "spatial patterns" in mining, but due to the notorious lack of consistency of method in colonial sources they have obvious limitations in grasping and representing local configurations. Only the reading of a close-up description and a panorama image provide us with a strong sense of how mining at Guantajaya was spatially organised. We will have to turn to other examples to probe the ways in which GIS infrastructure may really help to improve our knowledge on local or regional contexts.

Silver Registration and Silver Production

Another way to look at large-scale mining geography is by assessing the quantities of production as registered by the Spanish authorities at the royal treasuries (*cajas reales*), although one quickly runs into problems with this approach. It will be apparent to the informed reader

---

[2]This translation is not the author's but that of the English edition of Alcedo by Thompson from 1812.

that focusing on the treasuries themselves is not a good representation of mining geography, particularly as a basis for spatial analysis. Expert knowledge on local context and a specific timeframe is necessary to produce meaningful knowledge based on an analysis of geographic data. Johan García Zaldúa and Amélia Polonia's contribution to this volume is a great example of how GIS can be employed to assess the ecological impact of copper mining-related charcoal production on the forests of Michoacán.

Without an adequate spatial and temporal focus basing analysis on the well-known data on treasury registries (Klein and TePaske 1982–1990)—as some historians have (Studnicki-Gizbert and Schecter 2010)—is a futile exercise. Studnicki-Gizbert and Schecter studied mining-related deforestation in colonial Mexico and produced a GIS map which visualised areas they calculated would have been needed for fuel-wood production if all the colonial silver production had happened at once (ibid., p. 99). The authors acknowledge that this map does not consider differences in vegetation density, therefore only being a "rough" representation of the affected area over time (and, to be fair, their study is not limited to spatial representation, so the map is only a small component of their findings). However, the extremely long period represented in a single spatial analysis, as well as the problematic association between the administrative centres of mining-related fiscal authorities and mining activity itself, is sufficient to cast considerable doubt on whether such a map is of any use at.

Mexico City, around which Studnicki-Gizbert/Schecter placed the largest deforestation circle, was certainly not a centre of mining. The same is true for Durango, capital of Nueva Vizcaya, a mining province par excellence. The city of Durango itself was an administrative centre and episcopal see with the nearest silver mines (Avinito) some 60 kilometres away as the crow flies. Bernd Hausberger has elaborated on the diverse regional mining landscape (Hausberger 1995). The treasury of Durango registered the silver of the whole Province of Nueva Vizcaya (and until the founding of the treasury at Alamos/Rosario, also the silver from Sonora), an area approximately the size of France (540,000 km^2) which has been described as a continuous mining area. Also in San Luis Potosí, most definitely a "mining town", the seemingly stable production numbers reported from the royal treasury hide the booms and busts at San Luis Potosí itself, at Charcas and later at the Real de Catorce, giving it stability as a treasury while production changed significantly. Even in the case of the "Peruvian" Potosí, Intendant Pino Manrique estimated the silver output of the Chichas Province by 1787 at 400,000 pesos ($) and that of Chayanta at $850,000, most of them from the "asiento de Aullagas". Even if he calculates some $2,250,000 at Potosí itself, at least a third of Potosí's nominal output was actually produced in the provinces (Pino Manrique 1787). In the seventeenth century, Peter Bakewell indicates a similar percentage of 20–40% of "provincial contribution" to the Potosí output by 1660–1685—the Province Lipez being the most significant contributor (Bakewell 1975). This complicates the persistent image of the dependence of South American mining on Potosí itself, when in fact, as in Mexico, it was the contributions from minor centres that enabled the empire's long-term stability in mining outputs.

Overall, using aggregated treasury-based data for analysis is somewhat superficial and perpetuates the bias of showing production at the centre of administration. However, adding a temporal axis to a spatial representation provides new possibilities of exploring production. Until now, that data has generally been presented either along the time axis in the form of a diagram, or spatially as a map even if the original data, organised in tables, includes both dimensions. Modern GIS software can visualise both the spatial and temporal information from such a table by using two columns as start and end markers for the validity of an entry. Like Studnicki-Gizbert/Schecter and many other scholars, I used the Excel spreadsheets created by Richard Garner, based on datasets by TePaske (Garner 2007). The widespread creative use of Garner's tables is a great example of the importance of sharing data freely among scholars: it helps in creating new knowledge, and the provision of such data does not go unrewarded as the provenance of the data is usually fully acknowledged. I transformed Garner's tables to a format the GIS software could read, related the table to the cities of the royal treasuries in my database and created a new GIS dataset (shapefile) that still holds largely the same information as the original tables, now associated with located places. My dataset is available under a creative commons license at the project's homepage (Stangl 2018). The data of this file can be visualised (using GIS software) in unlimited ways as maps or animations, the tabular data contained in the file can be queried and statistically analysed, combined with other data, etc.

As a simple visualisation, I created two videos from the data which illustrates the quantities of silver registered by the royal treasuries from 1545 to 1808 in 5-year averages (Stangl 2017a, b). The Mexico video shows, for example, how at the creation of a new treasury, the production numbers of another one nearby dropped sharply. This is logical: the creation of a new treasury would be preceded by a significant mining boom which seemed stable enough to justify the institutionalisation of registry in situ. In 1632, for instance, the treasury at Guanajuato surged and numbers at Mexico City fell by a third (Stangl 2017a, 0'47'–0'48'"); numbers for Zacatecas grew dramatically in the 1660s and then fell by half in 1687 (Stangl 2017a, 1'08"–1'18") when the treasury of Sombrerete was created. This reveals the earlier growth of "Zacatecas" as actually being the growth of Sombrerete, which led to the formation of its own treasury—something that could escape our attention if the data was presented on a graph, table or static map. Further, the growing spatial diversification of silver registration in New Spain compared to the dependency of Peru and Upper Peru on Potosí, Oruro and later on Pasca is evident if we compare the two videos.

Mining and Institutional Development in New Spain

The discrepancy between Mexico and Peru in terms of the dynamism of the institutional landscape is not limited to the realm of royal treasuries. Another potential use of HGIS de las Indias, complementary to the development of settlements, is to track administrative organisation, down to the district level (*partidos*), as it developed over the eighteenth century. Revenues from mining were, of course, the focus of intense power struggles among local elites and between royal administrators. Jurisdiction over mining sites was heavily contended between the courts (*Audiencias*), governors, alcaldes mayores (district magistrates), the

Jesuits and city councils. Also, the danger of tax evasion and fraud drew attention from metropolitan institutions in Mexico, who occasionally established special administrations. A prominent example of such a struggle is the mining district of Bolaños in the Sierra Tepeque, which experienced a sudden mining boom in the 1740s and sparked the whole region's territorial reconfiguration (Carbajal López 2002; Rojas Galván 2013; Stangl 2017a, b).

One example of the interaction between mining cycles and local government comes from the northern provinces of Nueva Vizcaya and Sinaloa. The provinces are divided in a north–south direction by the Sierra Madre Occidental, the site of a number of different mining areas. By the seventeenth century, the governor of Nueva Vizcaya was competing with the technically subordinate but largely independent military commander of Sinaloa for the mines of Sonora and, around the turn of the seventeenth century to the eighteenth century, also with the administrator of the Rosario mining district. The principal instrument of the Nueva Vizcaya governors was to immediately name a deputy or an alcalde mayor when a new real de minas emerged. The dynamism of creating and abandoning districts there is staggering in comparison with the rest of New Spain. Of over 200 alcaldías mayores in central New Spain, about ten fledging Indian districts were lumped into other districts and only two subdivided in the eighteenth century before the intendency reforms (Tepeaca-Tecali around 1709 and Ixtlahuaca-Metepec in 1762); in Nueva Galicia, among some 40 alcaldías mayores, only two new ones were created (Bolaños and Nieves). By contrast, Nueva Vizcaya including Sinaloa in 1701 was composed of 24 jurisdictions. The pre-intendency era saw the creation of districts at the reales de minas of Maloya (~1705), Chihuahua (1708), Guarisamey (1757), Bacís (1763), Tabahueto (1765), Topago, Real del Oro, San Juan Nepomuceno and San Joaquín de los Arrieros (all before or in 1777), some of which quickly disappeared again: Tabahueto by 1777; Bacís, San Joaquín and San Juan Nepomuceno by 1787. The difficult terrain of the Sierra Madre invited anarchy in the absence of administrators, and the complications of defining a border between Sinaloa and Nueva Vizcaya in the contested mountains played their part too, incentivising a flexible administrative structure. The administration of mining was shifted towards institutions of self-government and representation in 1777, with the creation of the Real Tribunal de Minería in Mexico, which from 1783 on was composed of a number of territorial diputations, each consisting of individual mining districts or reales de mina. Miners of the reales had representation in the territorial diputations, which in turn formed a central electorate. Over the years, the number of territorial diputations increased and the number of reales changed, depending on mining cycles. Furthermore, the organisation of the diputations depended not only on the territorial organisation of civil administration but also on the prevailing economic patterns.

One other important institution that was affected by mining towns, and which influenced their development especially during the last decades of the colonial period, was the postal system. In the time before the state assumed control over the post in 1768, mail routes were essentially limited to connections between the main centres of commerce and administration. By 1768, post offices were almost exclusively located in Spanish *ciudades* and *villas*, all of them administrative centres. The few mining towns among them (Guanajuato, San Luis Potosí, Zacatecas and Chihuahua) were at the same time important administrative centres with city status.

This landscape changed drastically over the next couple of decades. While the expansion of the postal network was not confined to accommodating mining business, mining towns and military installations were the clear focus (Map 5.3a and b). To better understand the postal system's bias towards mining centres, I counted the number of post offices established in Indian towns and mining towns within each intendency by 1808 and calculated the ratio between this number and the total number of both settlement types in that territory according to our database. The contrast could not be starker: for mining towns, the range of ratios lies between one 1:1.5 to 1:6.5 towns per post office, whereas for Indian towns it ranges between 1:8.5 (29 out of 249) in Guadalajara to 1:155 in Yucatan—and none of the 40 Indian towns in Zacatecas had its own office. In other words, even structurally weak mining regions benefitted from the new institution significantly more than the best integrated Indian populations. This emphasis seems

(a)   **Access to the postal network. New Spain, 1768**

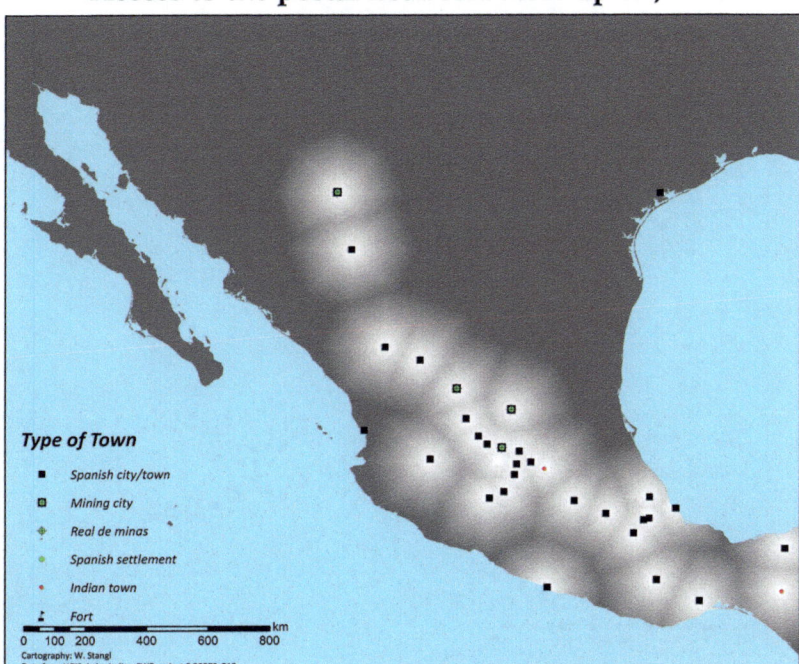

Map 5.3   a and **b** Post offices according to type of settlement (1768; 1808) (*Source* Stangl 2015a)

economically rational: it provided the volatile economy with access to market information, allocation of supplies and means of transactions. It allowed stable flows of communication and the shipping of registered parcels, but also made possible private bills of exchange and early postal money orders (*libranzas*) since some post offices upheld balances between each other (Harris 2009).

Even among mining towns, the post office was a central institution. For comparison, in New Spain and its northern provinces, there are roughly 5400 locales listed in the database, of which some 1100 were parishes in 1808, but only 360 post offices. As a consequence, when building the dataset of post offices, the last thing I expected was for one

(b)   **Access to the postal network. New Spain, 1808**

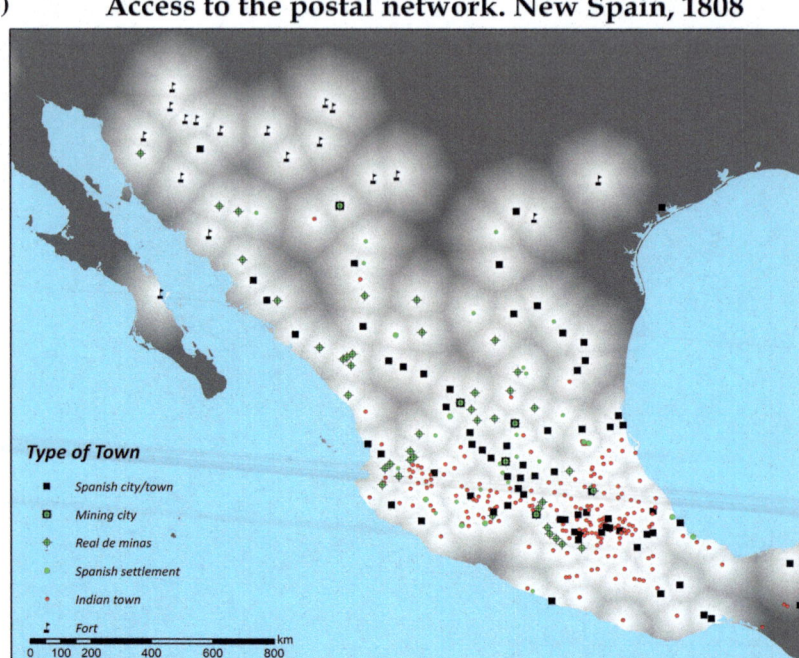

Type of Town

- ■ *Spanish city/town*
- ▣ *Mining city*
- ✛ *Real de minas*
- ● *Spanish settlement*
- · *Indian town*
- ⌇ *Fort*

0 100 200   400   600   800 km

Cartography: W. Stangl
Data from HGIS de las Indias, FWF project P 26379-G18.

Map 5.3 (continued)

of those places to be "missing". Nevertheless, I stumbled upon the clearing accounts of a post office called "Aranjuez" for the years 1801–1809. To my knowledge, not a single map or general geographic description mentions this place. Only after some additional research, I located a text by Antonio Ibarra, a specialist of colonial mining in Guadalajara, that does mention a "Real de San José de Aranjuez", established in 1801 and so short-lived that it escaped contemporary cartographic as well as textual representation (Ibarra 1993). This anecdotal episode may serve as additional evidence of how swift and decisive the late colonial state in New Spain was in its effort to reduce the distance of governance and to strengthen its grip on silver production.

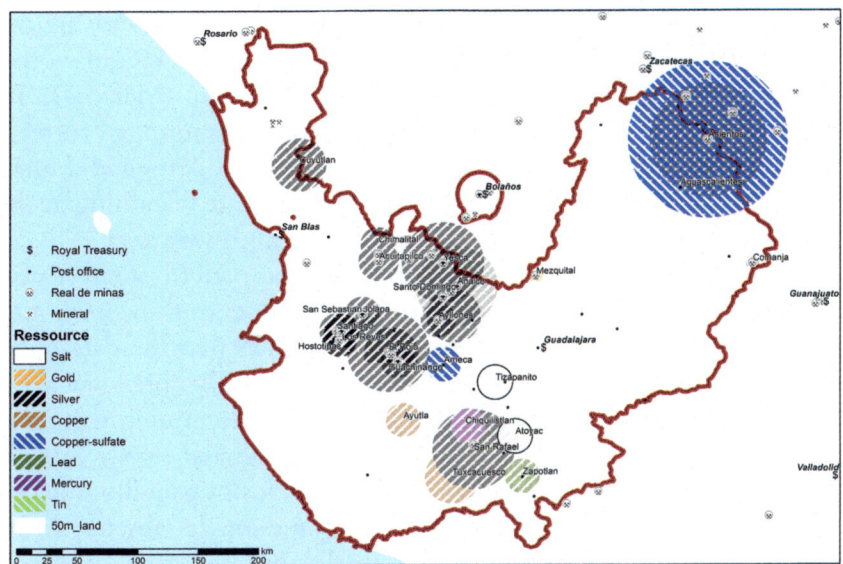

Map 5.4   Visualisation of the 1792 mining report by Menéndez in the context of HGIS de las Indias (*Sources* Ibarra 2000; Stangl 2015a)

The Regional Context: Mining in the Intendency of Guadalajara, 1792

As we have seen in the first two chapters on categories and silver registrations, understanding mining geography using locational or fiscal data may be of use but there are limits in how far they can advance our knowledge. It is important to complement and contrast the overall picture with an examination of regionally and locally focused data. For this purpose, I made use of the HGIS de las Indias database to visualise the geography of a report on mining by the intendant of Guadalajara in 1792, Menéndez Valdez, which had already been extracted and compiled in a table by Antonio Ibarra (2000, pp. 88–89).

The HGIS de las Indias website allows researchers to upload such tables when they meet a certain set of criteria and convert them into spatial data (shapefile). The original table was transformed into a standard format, the referenced localities were identified, and the corresponding IDs of the places included. I used this feature and worked with the resulting file in GIS software to create a map (Map 5.4). The size and transparency of the circles that represent the mining centres in the document represent the number of active mines (size) and a qualitative interpretation of the notes concerning the state of mining at that place (transparency; higher transparency indicating a worse state).

The resulting map reveals not only the spatial pattern of silver mining in the intendency at that time—clustered in the south-western-most part of the Sierra Madre Occidental. It also shows that the data of the table is incomplete, missing several mining areas of the intendency, and that identity markers are again at stake: the whole mining district of Bolaños is missing, possibly because it only integrated slowly into the intendency. More interesting is the absence of Hostotipaquillo, the centre of the province, seat of a territorial diputation. Its absence makes sense because, as a settlement, it frequently appeared as a "pueblo de indios" instead of "real de minas", even though it was totally surrounded by smaller mining sites and processing facilities—as will be discussed in the chapter on Humboldt. Arguably, this location is omitted in the document because it is considered an administrative centre of a mining region rather than a mining town itself. There is also an instance of the opposite occurring: a mention of three silver mines at Cuyutlán in the Acaponeta district, an Indian town nowhere else considered a mining town and without an economy centred on mining. The actual reales de minas in the jurisdiction (Motaje, Frontal and Tule) are absent from the table, so it is likely that Cuyutlán, being the administrative centre, made its way into the table in lieu of those places.

Apart from the unusual cases of Hostotipaquillo and Cuyutlán, the map confirms what has been stated earlier about mining town identity and its close relation to the mining of silver and gold. Mining of other materials—salt, copper, lead, copper sulphate and mercury—was not exploited under that regime. In the database, all places with such supplementary mining (as most of these materials were needed in silver mining) are registered as "pueblo de indios". Copper and other mining was considered a regular economic activity within the general economic sphere.

The map, as it is, is only one of the infinite potential ways to visualise the data of the report and arrange it with other layers of information or even compute results based on spatial analysis. The map shows royal treasuries, reales de minas and post offices as they existed in 1792, but the same could have been done with the silver registry figures, with parish centres or with the population of Guadalajara parishes as recorded in the census of 1778—all of which exist in the HGIS de las Indias database and are freely accessible. The region's census data of the Revillagigedo census (1790–1792), which would fit the moment of the report, is not (yet) in the database because I have not found a publication on this census where population numbers for Guadalajara are broken down to the parish level, although detailed original records do exist in the archives—whether they are complete or not would have to be scrutinised. Such information, as any other, may be contributed at a later stage to the project and thus add to the possibilities of further exploring relations between data.

Mining and Institutional Development in Peru and Charcas

In Peru and Upper Peru (Charcas), institutional development in the eighteenth century was completely different and much less dynamic. Fiscal revenue from mining had been a decisive factor when the Crown decided to create a new Viceroyalty in Buenos Aires. The reasons for this foundation were mostly geopolitical since the Southern Cone's Atlantic coast and the regions bordering Brazil were especially vulnerable to imperial competition. Funding for colonisation and defence in the area was to come from the revenues generated by the mines of Potosí and Oruro. Only one territorial reform was carried out for the better organisation of mining, when the Tarapacá Province was separated from Arica in 1768. Other than that, the provincial districts in 1800 were essentially the same as when they were established in the 1580s. And, while the New Spanish system of the Real Tribunal de Minería was also introduced in Peru (1785), it lacked the dynamism and self-organising development seen in Mexico. Disregarding geographical and economic dimensions and relations, the territorial diputations simply coincided with the areas of intendencies—a circumstance considered to have been at the core of institutional failure and stagnation (Molina 1986, p. 120)—and the diputations that existed in 1786 were the same as those at the end of the colonial period. The same consistency and lack of dynamism are evident in the development of the region's postal system. When the postal reform was put into effect in Peru and Upper Peru (1772), there were 36 post offices in the Audiencias of Lima and Charcas. Like in New Spain, the initial offices were established in administrative centres. By 1808, that number had more or less doubled to 75, while in New Spain it had expanded almost tenfold, from 41 to 360. Nevertheless, the centrality of mining to the postal organisation is also evident in Peru. Of the 75 post offices in Peru, 8 were located at mining towns and 38 in Indian pueblos. That corresponds to 11.25% of our registered mining settlements, but only 1.88% of the pueblos in the database.

Generally speaking, the perception of Peruvian mining by many observers is limited to the large centres of Potosí, Oruro and Pasco, occasionally including Hualgayoc (Chota). The Peruvian Viceroy Ambrosio O'Higgins commented in a treatise on the state of Peru in 1799, "This branch of industry [mining], which was at some point the credit and fame of Peru, today is reduced in this viceroyalty to the hills of Pasco in Tarma and that of Chota in Trujillo. There are many subaltern *minerales* which always give something, but the principal extraction is that of those [two]" (Sobre estado 1799, 6r–v). The viceroy judged that those sites could contribute much more, but that the indigenous population was busy with agriculture and other chores—instead he proposed to put "vagrants and idles" to use (ibid., 6v–7r).

Whether or not this assessment was correct, the organisation of the workforce in Peru also suffered from a lack of reform. The system of the supraregional mining *mitas* at Potosí and Huancavelica remained in full effect, as did the mechanisms of labour recruitment to other mining areas, which were indirectly dependent because the former *mitayos* were an important part of the regular labour market. Already, in the

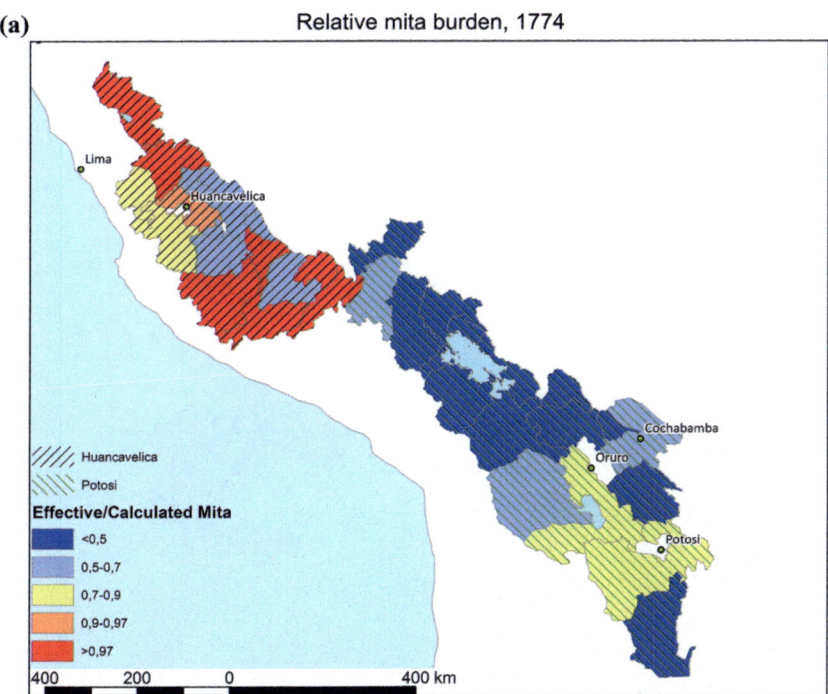

Map 5.5   a and **b** Relative Mita Burden in 1774 and 1780 (*Sources* Amat y Juniet 1776; Guirior 1780; Stangl 2015a)

first decades of its existence, the *mita* has been characterised as inflexible to demographic and economic developments (Zagalsky 2014), and very little changed in the seventeenth and eighteenth centuries. Map 5.5a and b show the provinces subject to the supraregional *mita* systems for the mercury mines of Huancavelica and for Potosí, a geography which hardly changed for over two centuries until the end of the colonial period (Dell 2008). The geographical coverage is slightly different from maps found elsewhere: the empty areas are the "Spanish" jurisdictions in the area, although Indians of those jurisdictions were likely subsumed in the repartimientos surrounding the "Indian provinces" of Angaraes (Huancavelica), Porco (Potosí) and Paria (Oruro). Also, the map excludes the jurisdiction of Tarija proper in the south-east.

(b)

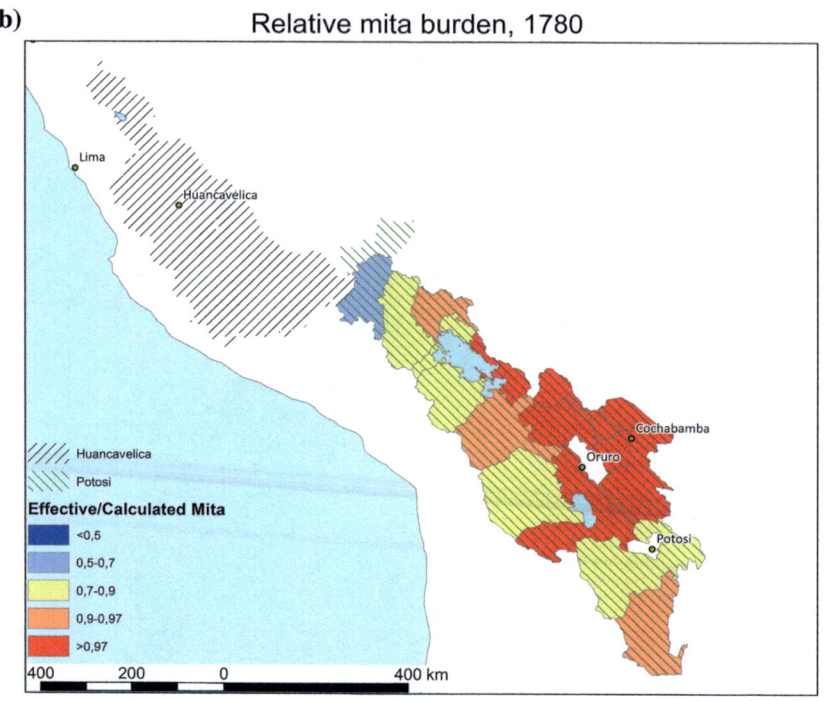

Map 5.5 (continued)

What did change was the reality underlying the *mita* structure, on both the local and provincial levels. Lamentably, knowledge of the individual repartimientos and how they changed is incomplete. Particularly in the eighteenth century, no complete listing of these units of labour organisation seems to exist in published form. Further, some references suggest that the Duque de Monclova, Viceroy between 1689 and 1705, issued an exemption of 18 pueblos, organised in 16 new parishes, in the original *mita* provinces (Ezguerra 1970, p. 495).

In Viceroy Manuel de Amat's instructions to his successor, from 1776, he gives the numbers of repartimientos in each province but not their names (Amat y Juniet 1776). Still, at least it is known that the number of 72 individual repartimientos in 1575 (Lohmann and Sarabia 1989, pp. 359–378) rose to 142 two centuries later (which does not

indicate a larger number of Indians, just a different method of organisation in smaller units). Pino Manrique, intendent of Potosí, gives more
information on some provinces for 1791 (Guía 1791), and the numbers coincide with the 1774 instructions. For the Tarija, Pino mentions Talina, Calcha and Santiago (Cotagaita) as repartimientos, towns
situated in "the province of Chichas". This clearly illustrates that Tarija
itself, which formed a combined jurisdiction with Chichas, was in fact
never subject to the *mita*, contrary to what maps typically show. And
even in "Chichas proper", the towns in the parishes of Tupiza and Tatasi
were apparently not subject to the supraregional *mita*, as—probably—
some of minor towns were subject to Talina, Calcha and Santiago.

Map 5.5a and b show the relative burden of the effectively enforced
mita at two points in time (by province): 1774, shortly before a series of
reforms that tightened the grip of authorities in fiscal and other matters,
and in 1780, the year of the Great Andean Rebellion, associated with
those reforms (Amat y Juniet 1776; Guirior 1780). For 1780, the only
available data is for the Potosí *mita* provinces, not for Huancavelica.
"Relative burden" means the portion of forced labourers who were actually at Huancavelica and Potosí (*de contínuo trabajo*) compared to the
calculated number. The theoretical *mita* was defined by starting from a
supposed Indian tributary population in the area (*gruesa*), occasionally
assessed by estimates, then dividing it by seven (*séptima*)—those values
are explicitly annotated in the original sources. For the calculus of the
calculated quota, I divided the *séptima* by three (since only one in three
should work at any given time while two would "rest") and then compared the number to that indicated in the manuscript.

If one looks at the maps, several things are immediately apparent:
first, provinces subject to the Huancavelica mining *mita* were fulfilling
their quota much more effectively than those sending *mitayos* to Potosí
in 1774. By 1780, the burden in most Potosí provinces was close to
the calculated quota, which is accounted for by the period's consistent
upward trend of silver mining in Potosí, illustrated in treasury records.
The exceptions are the northern regions of Tinta and Quispicanchis,
where there are low or no numbers, most probably as a direct result of
the Tupac Amaru rebellion which started in that area. While easily accessible data exists on the effective *mita* in earlier times (1724, 1736, 1740),

there are no numbers on the corresponding *tasas* to feed into the database. They most certainly exist, and my inability to access them for this purpose is an excellent example of the importance of a common spatio-temporal infrastructure in the field.

Understanding Spatial Patterns of Spanish American Mining with GIS: Alexander von Humboldt 2.0

In a way, Alexander von Humboldt did what this essay does: he tried to understand the spatial patterns of colonial Spanish American mining and its relations to development and political organisation, based on official papers, reports, maps and observations, and attempted to consolidate that information into a synoptic vision. Aside from his major publication on Mexico—the *Political Essay of the Kingdom of New Spain* (Humboldt 1811)—the medium Humboldt used to express his information was the map (General chart 1804). Humboldt's map is unusual because the "mining town" as a marker of distinction for settlement symbology is particularly prominent. Earlier large-scale maps of provinces or even larger areas instead use distinctions such as Indian town, Spanish town, hacienda, fortress, capital and parish centre as markers. If the real/asiento de minas is distinguished at all, it is typically one additional mark on a sign and not a central focus. Humboldt's map is entirely different; the manuscript draft of his map in particular makes it look as though it was designed to be a map of mining in a general context. Humboldt distinguishes three symbols for mining-related localities: city-mining town (city marker with two crossed hammers); major sites (crossed hammers); and minor sites (single hammer). The mining towns stand out among the regular settlements, and their number is much larger—more detailed—than on any other contemporary map, where it is usually only major sites that are represented. This cartographic representation of mining towns is important fresh information; it could not have been simply transferred from some older map.

In order to understand Humboldt's map on a deeper level, I first georeferenced it. Georeferencing is the process of aligning points on a map—such as prominent features of coastlines or capital cities—with their actual position over a satellite image or via a service like Open Streets Map. In the case of historical maps—if the process is feasible in the first place—this typically results in larger distortions which can be visualised by showing either the distortion on the grid or an unevenly stretched map-image (Jenny and Hurni 2011). Map 5.6a and b show both variants. The distortion grid reveals the main regions of error in Humboldt's work. First, a Sierra Madre Occidental in the north-western part with an excessive east–west extension, and second, two areas (Michoacán and southern Sinaloa) where settlements were apparently misplaced in relation to each other, warping the whole portion of the map. The warped map-image assisted with digitising the mining sites according to their symbol levels.

Analysing the places and comparing them to their categorisation in our database, it becomes apparent that Humboldt's definition of mining geography on the map is a purely administrative one. In fact, it apparently shows the same places that he mentions in the *Political Essay* when describing the organisation of the Real Tribunal de Minería, its territorial diputations and subject reales de minas (Humboldt 1811, vol. 3, pp. 119–128). For example, it shows the city of Oaxaca as a "mining city". To anyone minimally familiar with Oaxaca, this is more than odd, as its main economic activity was the cochineal trade. There were a few rather isolated, widespread and not particularly wealthy mining sites within its larger province, but until at least 1794 those communities did not even have a proper representation in the mining tribunal, as Fausto de Elhuyar—general director of New Spanish mines and major reformer of the industry—reported (Tamayo 1943). However, at some point between 1794 and 1803, Oaxaca must have got its proper territorial diputation, becoming—in Humboldt's eyes—a "mining city".

Humboldt's fixation on administrative designations produced more odd artefacts. For example, in his *Essay* and map, he registered two places with the name of San Juan Nepomuceno among the mining sites of Sinaloa and Nueva Vizcaya, respectively. On his map, they

(a)

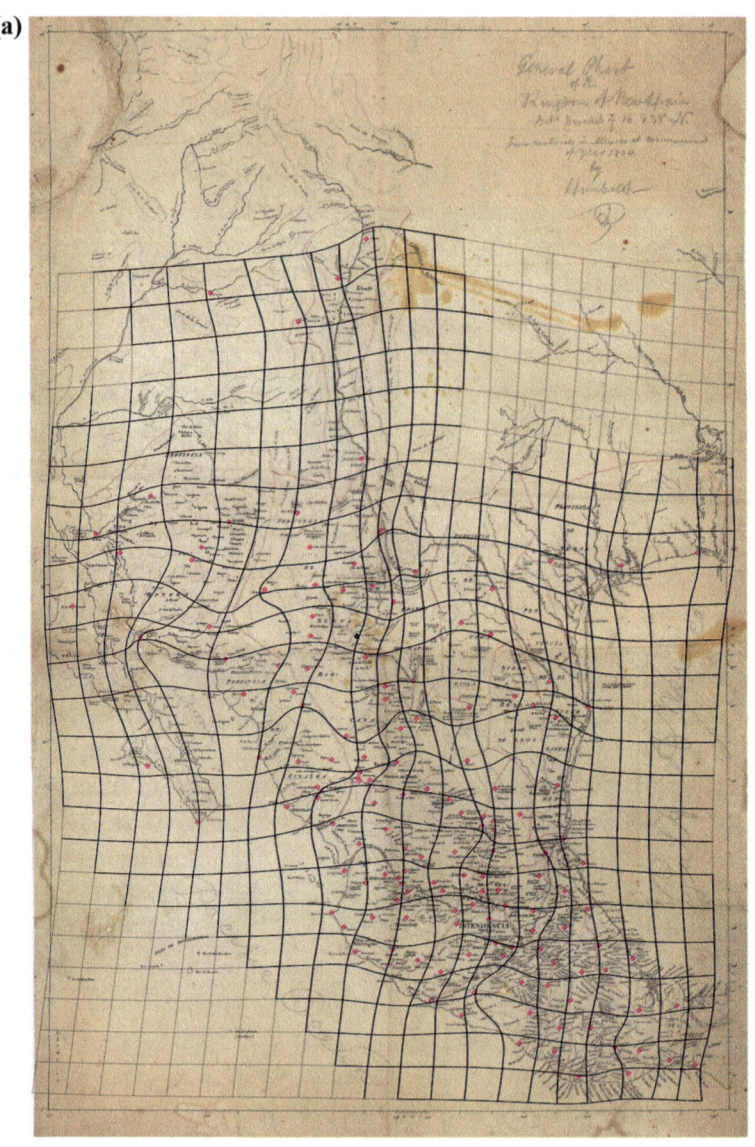

Map 5.6 Alignments of Humboldt's map with actual position of represented features: **a** Distorted grid and undistorted map; **b** Undistorted grid and distorted map (*Sources* General chart 1804. Software: **a** MapAnalyst; **b** ArcMap 10.6)

(b)

Map 5.6 (continued)

appear close to each other in the Sierra Madre Occidental. In fact, it is more than likely that both references regard to the same place. Situated deep in the mountain range of Tarahumara country, San Juan Nepomuceno and San Joaquín de los Arrieros were discovered in the mid-eighteenth century by Sinaloa-based miners and sparked one of the aforementioned jurisdictional conflicts between Sinaloa and Nueva Vizcaya (Gerhard 1982). Ultimately, the case ended with the official placement of the area under the authority of Nueva Vizcaya and the creation of two new minor jurisdictions for a few years, until the mines were subjected to the Batopilas district. However, we know that San Joaquín miners in 1790 petitioned to have this situation reversed (Molina Molina 1979, pp. 12–13), so it is a possibility that the two opposing communities of miners at San Juan Nepomuceno were eventually allowed to be represented in both territorial diputations; thus, San Juan Nepomuceno appears twice in Humboldt's list and map. This kind of information gathering aligns well with the aforementioned over-extension of the Sierra Madre itself: the location of interior places in

Sinaloa was probably established by determining their position relative to the coast, while the position of locales in Nueva Vizcaya was calculated from Durango or Parral. With the representation of a coastline that goes too far north-west instead of north, the resulting inaccuracy mostly affects the Sierra Madre, and the distance between the two "San Juan Nepomuceno" is a good indicator for the scale of the mismatch.

Comparing the overall density of registered mining places in our database with Humboldt's geography, it is apparent that Humboldt's numbers are much higher in most areas. Particularly in the Province of Michoacán, our database reports hardly any reales de minas, while in Humboldt's text they abound. This, again, is an effect of the "hidden nature" of copper mining in contemporary observation. Mining in Michoacán was predominantly of copper and, until the creation of the mining tribunal, nobody would refer to the sites of extraction as "reales de minas". However, as an exception to the rule, the mining tribunal did include four mining deputations in the province in 1803, three of which represented exclusively copper miners (Inguarán, Zitácuaro, Angangueo), bringing two dozen copper mining sites suddenly into the realm of reales de minas.

Even aside from this particular case, many of the places Humboldt listed are not mentioned or represented anywhere else in contemporary overviews. Between the mining report of Fausto de Elhuyar of 1794 and Humboldt's figures from 1803 (surely based on documents provided by Elhuyar), the number of reales de minas almost doubled from 175 to 347. Jorge Tamayo attributed this to the factors of "information refinement" and "progress in mining", judging the latter to be more important (Tamayo 1943, p. 316). However, as may already be clear from the arguments in this article, it could also be institutionalisation that drove the dramatic increase.

The conclusive regional example comes from the mining area of Nueva Galicia and the mysterious absence of Hostotipaquillo in the mining report of 1792 against the fact that for Humboldt the place is not only a simple "real de minas" but a major site. Looking into the textual description of the Guadalajara Province in 1792, the industry and commerce of Hostotipaquillo are given as related to the *rescate* (collection) and *beneficio* (processing) of silver, as well as *surtimiento* (provisioning) of miners (*Noticias varias* 1878, pp. 25–26). Apparently, the town had all the central functions of a small regional mining economy for a long time, but the intendent did not consider it a site of mining activity. According to Elhuyar in 1794, Hostotipaquillo was a real de minas subject to the deputation (and thus also royal treasury) of Bolaños, which also may explain its absence from listings of Guadalajara sites.

Furthermore, Humboldt's map shows several reales de minas around Hostotipaquillo, of which only two appear in my database: Santa María de Yesca and Santo Domingo. To understand Humboldt's information and the discrepancies between the sources, I first tried to locate Humboldt's places on the modern map. For some, I could find a correspondence using Open Street Maps or Geonames (Copala, Jocotán, Amaxac, Tecomatán), but two sites (Guilotitán and Aguacatancillo) could not be found using such tools. The Archivo General de Indias in Seville, however, houses two interesting maps of the area from late 1777—at least 20 years before either Menéndez, Elhuyar or Humboldt (Mapa Hostotipaquillo 1777; Mapa Cacaluta y Acatic 1777).

All the sites indicated by Humboldt as contiguous to Hostotipaquillo appear on those manuscript maps, and even though their positions are somewhat at odds with the data given by Open Street Map, their general distribution—mostly north-west of the central town—is accurate, contrary to the case of Humboldt.

Map 5.7 shows the distortions in Humboldt by indicating the corresponding spatial positions of outlaying places compared to the central place, Hostotipaquillo, with vectors. Places with red circles indicate places also shown on the mentioned manuscript maps. It appears that Humboldt's methodology for locating the mining sites on the map simply consisted in randomly placing them round a hierarchically superior place—in this case Hostotipaquillo.

The evidence in the archival maps shows not only that Humboldt's "novel" mining sites did in fact exist much earlier—which lends weight to the "refinement of information" hypothesis suggested by Tamayo. The map also attaches categories to these minor places, filing them as "ranchos", "haciendas de beneficio" ("silver-extracting haciendas") or "quadrillas", which translate into "working gang". In fact, some 50

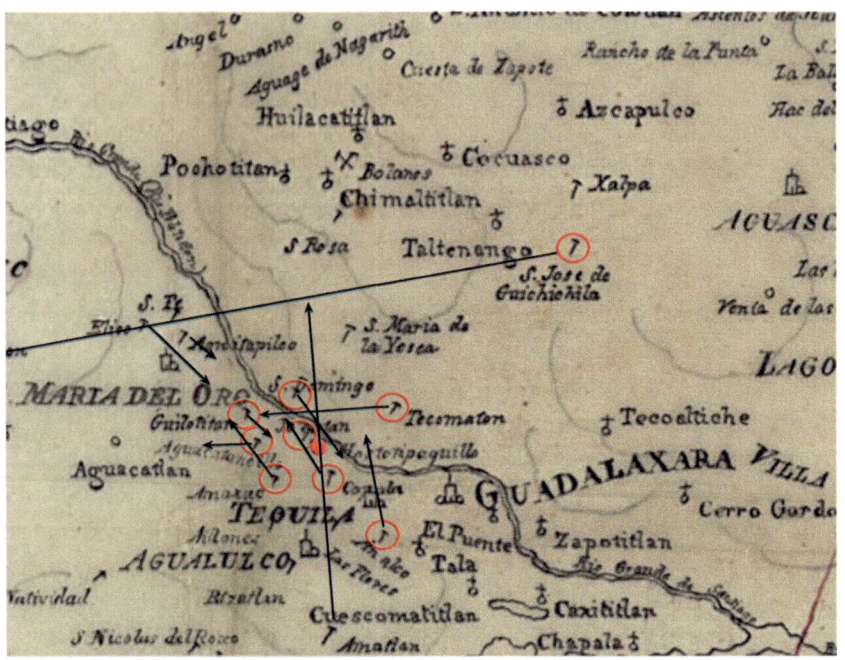

Map 5.7 Segment of the Hostotipaquillo area on Humboldt's map with distortion vectors (*Sources* Mapa Hostotipaquillo 1777; Mapa Cacaluta y Acatic 1777; General chart 1804, Open Street Map)

other rural extraction sites are prominently placed on the map—and nothing points to a conversion of the "emerging mining towns" into formal settlements or particular bonanza areas. In fact, the explosion in the numbers of reales de minas around 1800 may correspond to an arms race between groups of miners to gain influence in the territorial deputations and the Tribunal de Minería itself, an institutional development crystallised into visibility in Humboldt's vision of mining geography.

Looking deeper, Humboldt's random placement of mines around a centre repeats itself: he placed Analco south of Tequila, while in reality it is in the northernmost corner of its modern municipality, north-east of Hostotipaquillo. Humboldt's tacit practice explains the extreme warp-zones in the georeferenced map: when choosing the control points to make his map fit on actual geography, I had used an overproportioned number of randomly placed settlements in the affected areas. If I had controlled for all the settlements on the map, I would have exposed many more such distorted areas. The largest error concerns Guichichila, a mining town near Tepic, far in the western part of the intendency which on Humboldt's map appear in the opposite direction, east of Tlaltenango. It could be argued that Humboldt may have possessed some information on the distance from Bolaños but somehow misunderstood or badly guessed on the direction—in total, the error between the digitised point and its real position is over 200 kilometres! Humboldt's generally well-deserved fame as an authority, however, contributed to the errors of his map being copied and repeated over and over again by other encyclopaedic authors in the nineteenth century, who based their own articles on Humboldt's map. Both a Spanish *Diccionario universal* (1831, vol. 4, p. 977) and a British *Encyclopaedia Metropolitana* (Smedley et al. 1845, p. 775) repeat the false positioning of Guichichila.

Conclusions

Almost all historical information has an underlying spatial dimension, a dimension very often hinted to but concealed in letters and text. This is not sufficient to communicate spatial patterns for our minds; we need some sort of visualisation. We often transform the information into a sort of mental map, but we have limited capacities to do so and of course we are oblivious to the misinterpretations that can be caused by false preconceptions and lack of knowledge. Maps are a powerful medium of communication and allow for direct access to our modes of cognitive interpretation. In our culture, we are well conditioned to understand the Cartesian map and its symbols. Yet, a map is also piece of art, the production of which demands considerable effort. In modern cartography, the required amount of work can be significantly reduced with access to reusable geoinformation and data. Almost all existing data of this sort, however, is contemporary. The historian usually suffers from having to work from scratch. This is where a spatio-temporal infrastructure for any field in history comes in: it comes with geometries and basic data and thus allows us to spatially contextualise textual and tabular data, helping us complicate and improve upon the paltry mental maps we create when digesting spatial relations. Seeing information plotted on the map by the GIS software—even if it does not result in a proper cartographic artwork—triggers another mode of interpretation and helps us form questions and thoughts about the data which we would not otherwise have considered.

A common frame of reference for places and areas and their changes over time also helps to make different sets of information comparable. Feeding the database information on (for example) the extent of forced *mita* labour, demographic data, commercial relations, post-courses or, as in this article, mining generates even more possibilities for combining, contrasting and cross-analysing data. With a shared infrastructure like HGIS de las Indias, independent datasets can be brought into a spatially consistent framework and combined for analysis. Even if data only exists for a limited area or seems incomplete, it can be compared or aggregated with other data that includes an additional area, a different period, or other scopes and scales. It is possible to add structured datasets (Excel, CSV) via an interface on the website and thus contribute to a better picture not only of mining but also of other aspects of historical geography, whether they are economic, demographic or cultural.

Further, GIS systematisation of spatial information at the macro-level is not at odds with the analysis of local or regional circumstances. On the contrary, such a database is an ideal frame for creating a large-scale vision based on small-scale information, assembling them from the bottom up and filling (spatial as well as temporal) gaps which become visible when plotting the database on the map. What is crucial for the use of such an infrastructure is to maintain the direct relationship with non-aggregated sources of information instead of relying in the first place on already aggregated general dictionaries (Alcedo 1786–1789) or large-scale maps, even if they are as valuable as Humboldt's (General chart 1804). Making the inherent geography of a source (like the 1792 mining report) explicit with the help of the GIS database is a major contribution to a better understanding of what data it actually contains—it is a dialectic process.

One thing is certain: historical developments, complex systems and causal relationships cannot be understood by simply inputting data into a cybernetic machine like a GIS, hitting a button and looking at the output. Nevertheless, a common spatio-temporal infrastructure and tools such as georeferencing can help to build a better understanding and contextualisation of sources and data, as long as one examines at the resulting maps, calculations and visualisations with the eye of a historian.

Bibliography

Primary Sources

Archival and Manuscript Texts
"Estado de la Población del Perú," 1792. Archivo General de Indias, ESTADO,73,N.40.

"Plano de la Mina de Guantajaya y del Puerto de Yquique, cituado en la Costa del Perú en los 21 Grados 45 Minutos de Latitud," 1764. Archivo General de Indias, MP-PERU_CHILE,43.

"Relación que manifiesta por menor el nombre y número de pueblos y sitios comprehendidos en cada Partido de los dieziseis que componen la provincia y gobierno de Popayán," 1804. Archivo General de la Nación (Colombia), Virreyes,16: f. 191r-194v.

"Sobre estado del Virreinato de Perú," 1799. Archivo General de Indias, ESTADO,73,N.86.

Amat y Juniet, Manuel de, 1776 (ca). "Relación de gobierno que hace D. Manuel de Amat y Junyent, Virrey del Perú, a su sucesor, D. Manuel de Guirior." Biblioteca Nacional de España, MSS/3110V.1/V.2. http://bdh.bne.es. Accessed September 11, 2018.

Guirior, Manuel de, 1780 (ca). "Relación que hizo de su gobierno D. Manuel de Guirior, Virrey del Perú, a D. Agustín de Jáuregui y Aldecoa, su sucesor." MSS/3114. http://bdh.bne.es. Accessed September 11, 2018.

Manuscript Maps
"Vista del célebre Mineral de Huantajaya," 1807. Archivo General de Indias, MP-Peru_Chile,162.

"General chart of the kingdom of New Spain [...]," 1804. Library of Congress Geography and Map Division Washington, DC, G4410 1804.H8. http://hdl.loc.gov/loc.gmd/g4410.ct000554. Accessed September 11, 2018.

"Mapa del curato de Hostotipaquillo y su jurisdicción," 1777. Archivo General de Indias, MP-MEXICO,297.

"Mapa de los pueblos de Cacaluta y Acatic," 1777. Archivo General de Indias, MP-MEXICO,343.

Source Editions and Old Prints

Alcedo, Antonio, 1787. *Diccionario geográfico-histórico de las Indias Occidentales o América*, vol. 2. Madrid: Manuel González.

Cañete y Domínguez, Pedro Vicente, 1791/1939. *Guía histórica, geográfica, política, civil y legal del gobierno e intendencia de la provincia de Potosí*. La Paz: Artística.

Diccionario geográfico universal. Barcelona: José Torner, 1831.

Humboldt, Alexander von, 1811. *Political Essay on the Kingdom of New Spain*. London: Longman, Hurst, Rees, Orme, and Brown.

Molina Molina, Flavio, 1979. *Límites de Sonora, Sinaloa y Californias 1790*. Hermosillo: [s.d].

Pino Manrique, Juan del, 1787. "Descripción de la villa de Potosí y de los partidos sujetos a su intendencia." http://www.cervantesvirtual.com/obra-visor/descripcion-de-la-villa-de-potosi-y-de-los-partidos-sujetos-a-su-intendencia--0/html/ff8732ae-82b1-11df-acc7-002185ce6064_2.html. Accessed September 11, 2018.

Smedley, Edward, et al., 1845. *Encyclopaedia Metropolitana, or, Universal Dictionary of Knowledge*, vol. 19. London: B. Fellowes.

Tamayo, Jorge L., 1943. "La minería de Nueva España en 1794." In: *El Trimestre Económico*, 10, 38/2: 287–319.

Secondary Sources

Bakewell, Peter J., 1975. "Registered Silver Production in the Potosí District, 1550–1735." In: *Jahrbuch für Geschichte Lateinamerikas*, 12: 67–103.

Bakewell, Peter J., 1990. "La minería en la Hispanoamérica colonial." In: *Historia de América Latina*, edited by Leslie Bethell, vol. 3. Barcelona: Edición crítica: 49–81.

Bakewell, Peter J., ed., 1997, *Mines of Silver and Gold in the Americas*. Aldershot: Variorum.

Brading, David A., 1971, *Miners and Merchants in Bourbon México*. Cambridge: Cambridge University Press.

Brown, Kendall, 2015. *Minería e imperio en hispanoamérica colonial: producción, mercado y trabajo*. Lima: Instituto de Estudios Peruanos.

Carbajal López, David, 2002. *La minería en Bolaños, 1748–1810: ciclos productivos y actores económicos*. Zamora: El colegio de Michoacán.

Dell, Melissa, 2008. "The Mining Mita: Explaining Institutional Persistence (Conference Paper)." https://web.stanford.edu/group/peg/april_2008_conference/080330mita.pdf. Accessed September 11, 2018.

Ezquerra Abadía, Ramón, 1970. "Problemas de la mita de Potosí en el siglo XVIII." In: *VI Congreso Internacional de Minería*, vol. I. León: Cátedral de San Isidoro: 483–511.

Garner, Richard, 2007. http://www.insidemydesk.com/hdd.html. Accessed September 11, 2018.

Gavira Márquez, María Concepción, 2005. "Producción de plata en el mineral de San Agustin de Huantajaya (Chile)." In: *Chungará. Revista de Antropología Chilena*, 37, 1: 37–57.

Gavira Márquez, María Concepción, 2014. "El triunfo de la minería informal. Conflictos por el control de los recursos mineros en Carangas a fines del siglo XVIII." In: *Estudios atacameños*, 48: 71–84.

Gerhard, Peter, 1982. *The North Frontier of New Spain*. Princeton: Princeton University Press.

Harris, Leo J., 2009. "Libranzas in Colonial Latin America." In: *Postal History Journal*, 144: 4–10.

Hausberger, Bernd, 1995. "La minería novohispana a través de los 'libros de carga y data' de la Real Hacienda (1761–1767)." In: *Estudios de Historia Novohispana*, 15: 35–66.

Hidalgo Lehuedé, Jorge, 2004. *Historia andina en Chile*, vol. 1. Santiago de Chile: Editorial universitaria.

Hidalgo Lehuedé, Jorge, 2009. "Civilización y fomento: La 'descripción de Tarapacá' de Antonio O'Brien, 1765." In: *Chungará. Revista de Antropología Chilena*, 41, 1: 5–44.

Ibarra, Antonio, 1993. "La minería local y el comercio colonial: el Real de San José de Aranjuez, 1801–1803." In: *Estudios Jaliscienses*, 11: 4–27.

Ibarra, Antonio, 2000. *La organización regional del mercado interno colonial novohispano: la economía de Guadalajara, 1770–1804*. México: UNAM.

Jenny, B.; Hurni, L., 2011. "Studying Cartographic Heritage: Analysis and Visualization of Geometric Distortions." In: *Computers & Graphics*, 35, 2: 402–411.

Klein, Herbert; TePaske, John Jay, 1982–1990. *The Royal Treasuries of the Spanish Empire in America*, 4 vols. Durham and London: Duke University Press.

Lohmann Villena, Guillermo; Sarabia Viejo, María Justina, 1989. *Francisco de Toledo. Disposiciones gubernativas para el Virreinato del Perú, 1575–1580.* Sevilla: Escuela de Estudios Hispano-Americanos.

Molina Martínez, Miguel, 1986. *El Real Tribunal de Minería de Lima (1785–1821).* Sevilla: Diputación provincial de Sevilla.

Povea Moreno, Isabel; Zagalsky, Paula, 2017. "Presentación del Dossier 'Conflictos y violencia en los distritos mineros de la América española (siglos XVI–XVIII)'." In: *Revista Historia y Justicia*, 9: 6–10.

Rojas Galván, José, 2013. "La participación de los grupos de poder en la historia del gobierno de las Fronteras de San Luis de Colotlán." In: *Letras históricas*, 7: 71–94.

Stangl, Werner, 2015a. "HGIS de las Indias." https://www.hgis-indias.net. Accessed September 11, 2018.

Stangl, Werner, 2015b. "Scylla and Charybdis 2.0: Reconstructing Colonial SPANISH American Territories Between Metropolitan Dream and Effective Control, Historical Ambiguities and Cybernetic Determinism." In: *Culture & History Digital Journal*, 4, 1. http://dx.doi.org/10.3989/chdj.2015.008. Accessed September 11, 2018.

Stangl, Werner, 2017a. "Plata registrada en las cajas reales de Nueva España, 1565–1808" (Video). https://youtu.be/MUGQyo7M0rY. Accessed August 08, 2019.

Stangl, Werner, 2017b. "Plata registrada en las cajas reales del Perú colonial" (Video). https://youtu.be/L4QR6mbxqmo. Accessed August 08, 2019.

Stangl, Werner, 2018. "Producción anual de plata por caja (México y Perú, 1545–1808), elaborado sobre Garner/TePaske." https://www.hgis-indias.net/downloads/Datos%20agregados/plata_cajas_total_garner.zip. Accessed September 11, 2018.

Studnicki-Gizbert, Daviken; Schecter, David, 2010. "The Environmental Dynamics of a Colonial Fuel-Rush: Silver Mining and Deforestation in New Spain, 1522 to 1810." In: *Environmental History*, 15, 1: 94–119.

Watson, Mary K., 1978. "The Scale Problem in Human Geography." In: *Geografiska Annaler. Series B, Human Geography*, 60, 1: 36–47.

Zagalsky, Paula C., 2014. "La mita de Potosí: una imposición colonial invariable en un contexto de múltiples transformaciones (siglos XVI–XVII; Charcas, Virreinato del Perú)." *Chungará. Revista de Antropología Chilena*, 46, 3: 375–395.

6

Manufacturing Landscapes in Spanish America: The Case Study of Copper Exploitation in Mexico (Sixteenth–Eighteenth Centuries)

Amélia Polónia and Johan García Zaldúa

Theoretical Set Up: "When Worlds Collide They Also Intermingle"

The "early modern age" (1400–1800) is conventionally seen as a time of growing interconnectivity among several continents and oceans. This opened the door for the creation of a world economy as well as for environmental impacts and landscape transformations resulting from global transfers. During this period, Europeans invaded regions of the old and new worlds, aiming for a quick, effective and profitable use of their resources. The Europeans moved towards the other old-world continents of Africa and Asia and projected themselves into newly discovered continents and subcontinents, including the Americas, Australia and Oceania. Overseas settlements and long-distance trade were established, and economic emporia and political empires were created, changing the world system for good.

A. Polónia (✉) · J. García Zaldúa
CITCEM, University of Porto, Porto, Portugal

© The Author(s) 2019
R. Pieper et al. (eds.), *Mining, Money and Markets in the Early Modern Atlantic*, Palgrave Studies in Economic History,
https://doi.org/10.1007/978-3-030-23894-0_6

For more than thirty years, the notion of "ecological imperialism" has informed research into the history of European overseas expansion. According to this paradigm, Europeans tried to replicate their way of living as much as possible in new territories, implying an intense projection of their influence upon them. A colonial economy, ruled by European markets, introduced new patterns of territory management, property regimes (Prem 1992) and soil exploitation. On the Spanish American mainland, gold and silver exploitation, extensive livestock rearing and epidemic outbreaks, alongside the impact of forced labour organisation through the *Encomienda* and the *Repartimiento* system, provoked both the depletion of local ecosystems and human depopulation (Reff 1991, pp. 9–32; Cook 1998). Formerly unexploited lands were increasingly used by Spanish settlement and ranching, changing the natural and environmental equilibria (Richards 2003, pp. 334–379). Mining in Spanish America emerges as particularly responsible for the exhaustion of resources (Powell 1952) and furthermore for long-lasting pollution, exacerbated by the use of mercury (Bakewell 1990, pp. 131–153).

In line with this view, environmental colonialism seems much more important than any other. Ecological and environmental equilibria were unbalanced, not in the course of a long-term process, but in a short and invasive onslaught of transformation and depletion (Crosby 1986, 1988, pp. 114–115). The "ecological imperialism" thesis claims that the aggressive behaviour of European agents towards pre-existing environments led to a heavy appropriation of primary products for human use by land appropriation against the needs of other species and other cultures; the depletion of natural resources; the extinction of vegetable and animal species; the destruction of ecosystems; and the drastic changes in landscape—all driven by European needs.

This paradigm disregards as irrelevant questions such as the concrete models of colonisation, e.g. the Spanish, the Portuguese, the French, the British and the Dutch, when applied to America, including their dissimilarities. It disregards also the role of self-organising networks, which tend to be by nature cross-cultural and frequently trans-imperial. Similarly, local agents—the colonised—are usually excluded from the dynamics of colonial processes, with global interpretations centring almost exclusively on the actions of European powers, agents and policies. In doing so, it ignores the important processes of adaptation and evolution that resulted precisely from the entanglement of nature and nurture, which necessarily accrued to all the peoples and environments involved. Therefore, the ecological imperialism perspective needs to be reviewed.

Examples come from the output of so-called post-colonial studies, developed since the 1980s. More recent perspectives centred on a connected history of the colonial empires (Subrahmanyam 2007) and the agenda of a highly prolific world or global historiography have been contributing to a revision of Eurocentric interpretations of colonial phenomena (e.g. Boyajian 2008; Darwin 2008; Andrews 1984; Polónia 2012), reflected in recent publications (Antunes and Polónia 2016; Polónia and Antunes 2017), and the organisation of scientific panels and conferences on the subject.[1]

This historiographical revision is built upon the adoption of concepts and models of analysis stemming from cooperation theories (Nowak 2006; Santos et al. 2006), first developed in economics, biology, anthropology, psychology, physics and mathematics (Hammerstein 2003; Fischbacher 2004; Hagen and Hammerstein 2006; Ostrom 1990; Richerson and Boyd 2005) and now applied to history. In the language of cooperation, the fact that interactions are generally repeated and bidirectional in time offers those interacting the opportunity to reciprocate, sharing the costs and benefits accrued to both entities, in our case, colonisers and colonised.

The application of this theory and model of analysis by colonial and environmental historians clearly has something to offer to a reanalysis of the environmental effects, including the human impacts, of

---

[1]Such panels included: "The Power of the Commoners: Informal Agent-Based Networks as Source of Power in the First Global Age", organised by Amélia Polónia, Social Science History Conference (Chicago 2010); "Beyond Empires: Self Organizing Cross Imperial Networks vs Institutional Empires, 1500–1800", coordinated by Amélia Polónia and Cátia Antunes, European Social Science History Conference (Glasgow 2012); "Fighting Monopolies, Building Global Empires", coordinated by Amélia Polónia and Cátia Antunes in conference on "Colonial (Mis)Understandings: Portugal and Europe in Global Perspective (1450–1900)" (Lisbon 2013). As for conferences, see, e.g., cooperation under the Premise of Imperialism, coordinated by Tanja Bührer, Flavio Eichmann and Stig Förster (Bern 2013).

European colonialism in Asia, Africa, America and Oceania during the First Global Age. Indeed, spatio-temporal models of cooperation, which go well beyond strict collaborative efforts between equal parties, permit an assessment of how far the unequal roles played by the parties involved affected cooperation, adaptation and reciprocity. New avenues of research can be defined in order, first, to question how local actors and Europeans interacted in order to use and manage the available natural resources and, second, to ask which mechanisms of adaptation existed, both for Europeans to survive in totally different and frequently hostile environments and for autochthonous people and environments to react, resist or voluntarily adapt to the new ecological elements. This new approach—already employed in the economic, social and cultural analyses of empires—remains to be applied in the framework of history.

Spontaneous or imposed cooperation between colonisers and colonised, negotiation, resistance—they all are dimensions involved in the analysis of colonial dynamics. The rationale of such historiographical analyses has to include other cultures, other civilisations (in the plural use of the concept) and the plurality of pre-colonial landscapes. They are not only *required* but *essential* to the study of European empires. This will not mean an extended economic history; questions such as productivity and efficiency will not be addressed here.

A predominantly global, transnational historiography does not centre around one-sided imperial impacts, but instead on the results of reciprocal encounters and the corresponding mechanisms of interaction. This includes questions of dissent, collaboration and resistance, and of hybridisation phenomena, based on syncretic mechanisms (Bührer et al. 2017, p. 2).[2] In any case, these encounters definitely merit further investigation which cannot happen without non-European contributions.

---

[2]"Why did indigenous actors engage in negotiations with imperial interlopers at all? To what extent did their interests overlap? Was a faithful and mutually beneficial relationship possible or could empires only produce contingent accommodations? How far did these cross-cultural interactions create imperial situations on the ground, and to what extent did pre-colonial cultures, socio-political and economic realities determine co-operative structures?"—those are specific questions raised by Bührer.

The insertion of local realities into global processes is another priority in colonial and global studies. It will perceive the autochthonous and indigenous actors[3] as active participants in European colonial territories. All of this requires contributions from a historiography based on the former colonised territories and cultures, in Africa, America and Asia.

This will not only answer, but amplify the European historiographic tendencies on colonial studies. Let us take the premises of the 1970s and 1980s British historian Ronald Robinson, who challenged the hitherto predominant Eurocentric theories of imperialism, by formulating an approach according to which indigenous collaboration represented both a formative and a continuous factor of imperialism. Robinson's theory, expressed in his "Non-European Foundations of European Imperialism" particularly emphasised that by collaborating with the colonial state, indigenous actors contributed to the creation of empires, to their preservation, and, eventually, to their dissolution (Robinson 1972, pp. 117–142; 1986, pp. 267–289). This follows and goes beyond "The Imperialism of Free Trade" as presented by John Gallagher in 1953 (Gallagher and Robinson 1953, pp. 1–15). Robinson's emphasis on indigenous actors opened the way for perspectives (Peers 2002, p. 456) and theories of indigenous "agency" (Clancy-Smith 2004, p. 126). According to this, colonial subjects are not helpless victims of superior forces and institutions, but historical actors who negotiate terms and gain centrality in historical processes.

By accepting the interactions between "colonised" and "colonisers" as essential to the mechanisms of empire building, the broker, the intermediary, acting as a go-between, a translator or a mediator, emerge as essential (Burbank and Cooper 2010, pp. 13–14). This perception goes beyond recognising access to local knowledge about politics, economies and revenue systems, as well as the view of cultures as levers used by colonisers to achieve more effective exploitation of local resources (Bayly 1998, p. 7). It implies, in fact, the actual transference

---

[3]The latter understood, according to the proposition of Wayne E. Lee, as those forming part of "generations of experience with the local climate, terrain, and subsistence system" which "operated according to different cultural systems" (see Lee 2011, pp. 1–16).

of knowledge and cultural resources, and eventually the demonstration of the dependence of colonisers upon the colonised. This is research that requires intensive empirical collaborative work in order to be able to identify those that are in the mainstream of empire building.

The impact of pre-colonial, autochthonous and indigenous realities on patterns of cooperation has traditionally been neglected by imperial history. That disregard for pre-colonial structures was and still is caused by the fact that pre-colonial heritage has been obscured and transformed by colonial categorisations and representations. That is why one needs the contributions of non-European historians in order to understand the different rationalities at stake. In fact, pre-colonial realities were transformed and often simplified along colonial categorisations according to imperial interest or because imperial interlocutors simply did not understand the complex cultural patterns and societies they encountered.

Summing up, a more global, transnational and transcultural approach to colonial dynamics, as well as a more complex analysis based on mechanisms of cooperation and on the role of brokers and go-betweens (both European and non-European), requires a cross-cultural, transnational perspective. In the Iberian case, this assessment applies to Africa and America alike.

Based on these assumptions, rather than the classic understanding of European colonialism as a single all-encompassing process, "the Columbian exchange", this paper argues that one has to differentiate between distinct colonial experiences. In the domain of mining, from the sixteenth to the eighteenth centuries, the experience was quite different in West Africa, East Africa and Brazil. This is relevant in the evaluation of ecological impacts because nature and culture (the local cultures) matter in processes of empire building. A quite different model of colonisation; a quite different way of empire building; and a quite different degree of regulation were responsible for diverse ecological and human impacts. Different models of interaction with local populations and different models of urbanisation, nucleation or aggregation of autochthonous inhabitants (stimulated by induced or forced migration patterns) affect disease transmission by the spreading of viruses and bacteria. *Encomiendas, repartimientos congregaciones* in Spanish America, Jesuit missions in Brazil, nucleation of population in

Africa (Weiskel 1988; Garcia Bernal 1978; Neto 2012; Bernier et al. 2014), different degrees of co-habitation and intermingling between colonised and colonisers, are critical variables in these complex and dynamic processes. Their form and extent depend on colonial models. For instance, when Douglass North et al. (2000) argue that the differences in development between Latin America and Anglo-Saxon America derived from the inefficiency and inadequacy of Iberian institutions to promote modern growth, they inadvertently demonstrate the role that time and the environment of contact plays in the development of institutional systems and property models. Daron Acemoglu et al. went further, with their claim that economic development in former colonies is a function of the institutions imposed by the colonists (2001). However, the equation is much more complex than that.

We can easily accept that different models of colonisation involving quite different ways of state and empire building, different degrees of regulation and different property regimes, necessarily and naturally led to very different ecological dynamics and diverse human and ecological impacts. This does not mean, however, that the determining force of institutions is to be taken as granted. One must take into consideration that environmental differences between regions largely accounted for institutional differences (Engerman and Sokoloff 2012). To conclude: overstating the clarity of state intentions and the capacities of the colonisers creates a bias in the historical discourse that obscures rather than revealing the processes that occurred; no single, general model of analysis seems adequate in this domain.

Interdependencies between worlds necessarily run deeper and were more dynamic than typically stated. Instead, adaptation—of the colonised to the presence and methods of the colonisers and vice versa—prevailed. Survival and the success of economic enterprises in such different worlds as those which made up the Americas inevitably required adaptation and acculturation, for Europeans too. In other words, the lives of the first settlers, or groups of settlers, would most probably accelerate reciprocal acculturation processes, different from those expected or described by the traditional imperial historiography. These circumstances should have led, in fact, to inevitable mechanisms of exchange, namely in the processes of resource identification, location and appropriation.

More often than not, colonisers depended on autochthones to provide them with the resources they needed, sometimes counting on their own methods, sometimes transferring technologies that would unbalance the ecological standing equilibrium. Those are, however, domains in which we often lack measurable testimonies, precisely because they occurred out of the frame (or at least the focus) of the conventional "empires". Only a systematic analysis of these dynamics can provide an understanding of the long-term ecological impact of cooperation between colonisers and colonised, with the former benefitting from the environmental knowledge of the latter. A one-sided view focusing on the action of colonisers alone is far from sufficient for such understanding. In fact, transfer flows, interaction, adaptation and assimilation processes were never unidirectional.

Two main ideas will be stressed in this paper: reciprocity, syncretism and evolvability are paramount to understanding social and ecological processes alike (no species survives without assimilation by the receiving ecosystem and cultures) and, besides destruction patterns and stressful mechanisms projected onto the ecosystems by colonial activities, we should also look at the mechanisms of adaptation, both by the humans and by the environment, and analyse the resilience of ecosystems and human communities to different kinds and degrees of stress. As part of this reasoning, we introduce the case study of copper production in Michoacán, its particularities and its impacts not only on the physical landscape but also on the human geography of production, and on the sociopolitical and economic organisation of the social groups associated with the different stages of exploitation and transformation. "Natural" and human landscapes alike will be scrutinised in this paper.

A Case Study: Human, Social and Environmental Impacts of Colonial Copper Production in Mexico (1521–1630)

The arrival of Spanish conquistadors in the territory of modern Mexico in the early sixteenth century brought deep sociopolitical, demographic and economic changes that permanently affected the lives of the native societies that inhabited this territory. One of the main agents of change in New Spain was the introduction of a new economic paradigm based partly on the large-scale exploitation and transformation of metallic ores. It is well known that silver production was the main activity of this type carried out in New Spain, followed by the mining of alluvial gold, although the latter took place on considerably smaller scale.

Colonial New Spain also produced an abundance of other, less precious but very important strategic metals vital for the functioning of the internal economy, among which lead, tin and copper stand out as the most important. Archaeology tells us that all these metals were mined, collected and transformed during the pre-Hispanic period of Mexico; however, the scale and range of exploitation implemented during the colonial period far surpassed any prior operations. The permanent increase in the scale of production and the introduction of new technologies and reagents during the colonial era left an indelible mark on the populations and environments subjected to its influence. Regarding the production of silver, for instance, the human and environmental impacts of mercury use and the large-scale deforestation produced by charcoal making have been successfully analysed (Guerrero 2017; Studnicki-Gizbert and Schecter 2010).

Less noteworthy than silver and gold, copper was an essential material used and demanded extensively in New Spain. In a period, when iron was scarce and expensive, copper and bronze fulfilled, to a certain extent, the needs for simple tools, especially among the indigenous population who used them in agriculture. It was employed in the manufacturing of all types of household items, such as cauldrons, kettles, pans and knives (among many others), used by all social groups throughout New Spain. Furthermore, the availability of copper facilitated the development of other important economic activities like sugar production and silver refining (Barrett 1987, pp. 1–4). The greatest demand for copper, however, came from artillery production and coinage, both activities carried out at the foundry and at the mint of Mexico City.

Development of the Indigenous-Spanish Copper Production in Sixteenth-Century Mexico

Despite the importance of copper and its heavy use during the colonial period, the Spaniards that arrived in New Spain had a very limited knowledge of copper production. At that moment, no copper mines had been worked in Spain for several centuries. All the copper consumed in the Iberian Peninsula was bought in its metallic form from production centres in Central Europe and Scandinavia through trade intermediaries in Venice and Antwerp (Sánchez Gómez 1989). Without an existing copper-mining tradition, the Spaniards of the time were not familiar with copper ores, and with no ores to smelt, the knowledge and skills necessary for turning copper ore into metallic copper were not part of their metallurgical repertoire. This does not mean they were not skilled coppersmiths; on the contrary, Spanish specialists, given access to metallic copper, were as good as any other coppersmiths in Europe. Nonetheless, the lack of knowledge and practical skills of mining and metallurgy limited the development of Spanish copper production in New Spain and defined the characteristics of the productive relationship that Spanish authorities established with the native specialists of Michoacán.

The indigenous metallurgical tradition that Spain encountered in Mexico was based on copper, and alloys that had been developed during the course of at least eight centuries around the geological availability of vast, varied and easily accessible ore deposits. Within its own sociocultural context, it was a highly efficient metallurgy, complex and advanced enough to fulfil the cultural and utilitarian demands of native societies (Hosler 1994). Although at the time of the encounter indigenous metallurgy was present in several parts of Mexico, the core of copper production was located in Southern and Central Michoacán and the area along the middle portion of the Balsas River (Maps 6.1 and 6.2). According to early historical accounts, native metal workers from this region were simultaneously experienced prospectors, miners, smelters and metalsmiths. Although the communities in the area were specialised in metal production, this was not a full-time activity but one carried out only when indigenous elites demanded specific amounts of metal (Warren 1968).

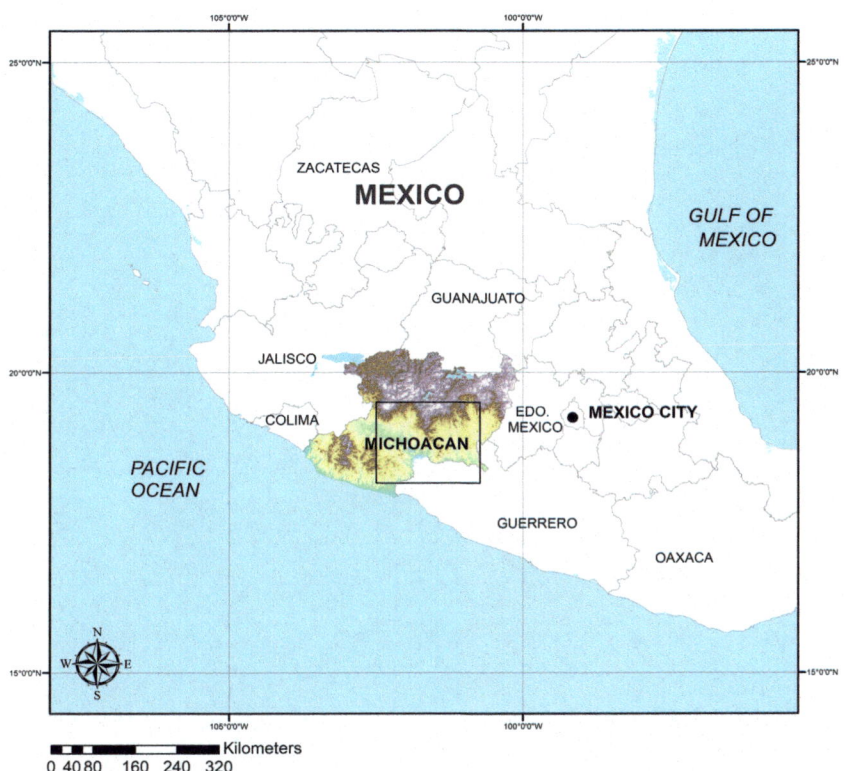

Map 6.1 Location of the area of study within Mexico (*Sources* Topographic and geologic maps, scale 1:250,000 [E14-1 Morelia and E14-4 Ciudad Altamirano] produced by INEGI)

The increasing need for copper during the first years of colonisation, the Spanish inability to produce it autonomously, and the difficulties and high costs of importing the metal from Spain created the conditions for establishing a productive relationship with the copper producers of Michoacán. The Spanish Crown had received early reports about the availability of copper resources in the newly acquired territories and the existence of a local metallurgical tradition.

The chronicles of the conquest of Mexico narrate several initial episodes of interaction between indigenous and Spanish metallurgies. Bernal Díaz del Castillo in his *The True History of the Conquest of*

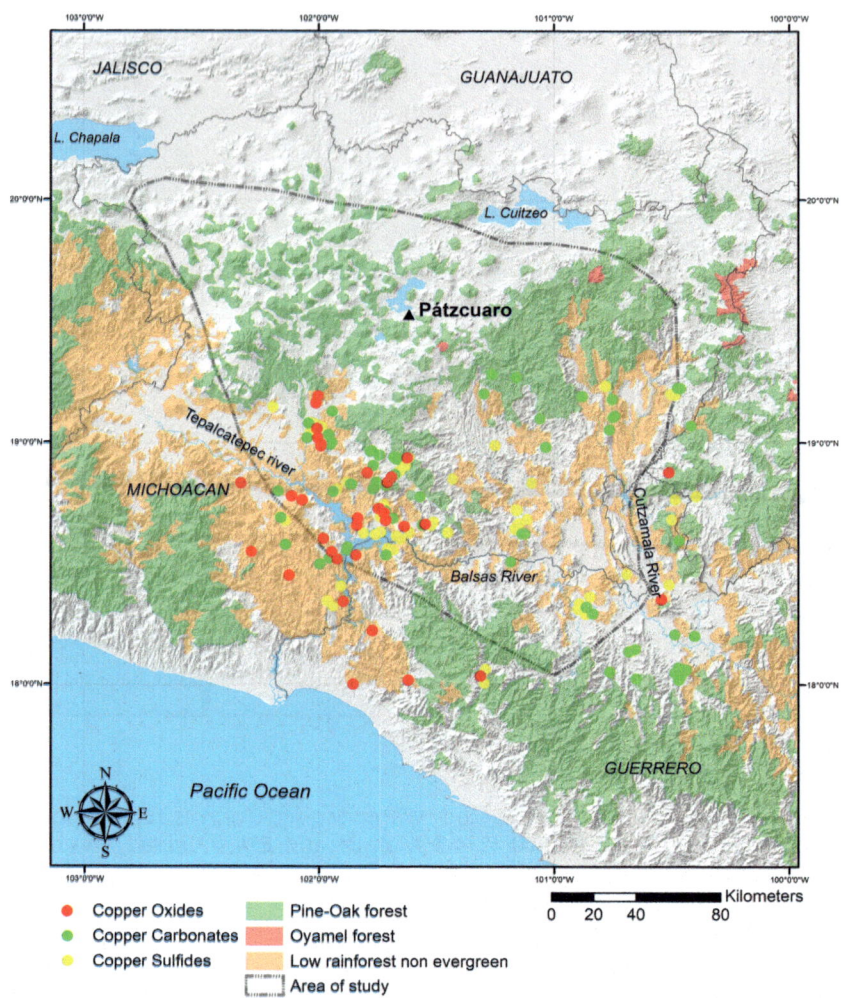

Map 6.2 Area of study in relation to elevation, ore deposits and vegetation (*Sources* Topographic and geologic maps, scale 1:250,000 [E14-1 Morelia and E14-4 Ciudad Altamirano] produced by INEGI)

New Spain mentions how Cortés, preparing his army to face Pánfilo de Narvaez in battle, asked his *Chinantec* native allies to provide several hundred of their spears but instead of knapped obsidian, Cortés instructed them to make the spearheads out of, metal. Cortés sent a Spanish spearman and examples of Spanish metal spearheads. The result, according to Bernal Díaz, was a group of spears with metal points technically superior to the original Spanish models (Díaz del Castillo 2010, pp. 185–186). Díaz del Castillo narrates a similar episode that took place when Cortés was preparing the siege of Tenochtitlan. He instructed his native allies to make 8000 metal arrowheads based on the Spanish models provided. Just as the previous example, Bernal Díaz says that the arrows were even better than the ones brought from Spain (Díaz del Castillo 2010, p. 263).

In 1521, Hernán Cortés sent word to the Crown about the availability of resources for making artillery. In one of his letters to the sovereigns, dated the 15 October, he explains the situation after the conquest of Tenochtitlan and complains about the lack of artillery to defend the recently occupied city. In the letter, he reports that he made five pieces: two medium-sized culverins, two smaller, and a serpentine cannon using copper and tin bought from the natives supplemented by some mined by the Spaniards in the province of Taxco (Cortés 1806, pp. 306–307).

In 1531, metal workers from the town of Sinagua in the copper region of Michoacán delivered a shipment of 40–50 loads of copper (roughly 28–36 quintals or 1280–1600 kg) as tribute to the Crown.[4] This large quantity of metal did not pass unnoticed by the authorities, as we can infer from the documents of the time. At that moment, the administration of the town of Sinagua was in the hands of the Crown; hence, this tribute went straight to Mexico City and not to the regional *encomenderos* as was the case for the other copper-producing towns in the region. Apparently, news of this type of large shipments reached Spain because in December 1532, the Empress Isabella of Portugal summoned Luis Fernández to the court in order to discuss the artillery to be made in New Spain. In her letter, the Empress expressed that

---

[4]This data is mentioned in a document of 1533 published by Warren (1968) with the name "Minas de Cobre de Michoacán". The original is held at the AGI, INDIFERENTE, 1204. The document offers the testimony of Alonso de Escobar, *corregidor* of the town, appointed by the king, who says that the shipment had occurred two years before the hearings of 1533 and the quantity of copper shipped was the result of two years of native's work.

"she has been told that in New Spain there is great *aparejo*[5] for making artillery" and commanded Fernández to make the arrangements necessary to travel there and start the works ("Real cédula a Luis Fernández").

The plan with Fernández did not come about, and in 1533 the court summoned another expert, Rodrigo Martínez. He was an artilleryman who had been involved in manufacturing the artillery for the *Moluccas* fleet under the command of Simón de Alcazaba, as well as some other pieces for the city of La Coruña ("Real cédula al corregidor de La Coruña"; "Real Cédula a Rodrigo Martínez"). According to contemporary documents, Martínez was also a former resident of Mexico City ("Real Cédula a la Audiencia de México").[6] Martínez was called to the court in February 1533 to discuss the artillery of New Spain ("Carta acordada"), and apparently the Crown and Martínez made an agreement and signed a contract to manufacture artillery in New Spain ("Mandamiento"). Unfortunately, Martínez served as captain of one of the vessels in the expedition of Simón de Alcazaba that departed in 1534 towards Cape Horn, and he died somewhere in Patagonia.

Within this context, in 1533 the Crown ordered a more detailed report on the availability of copper ore, location of deposits, and the state of indigenous production in Michoacán (Warren 1968). It is possible that this report was commissioned in order to be ready for the arrival of Rodrigo Martínez. The official in charge of gathering information and producing the report was the judge and member of the second *audiencia* Vasco de Quiroga and the hearings were conducted during the fall of 1533. The final document was sent to Francisco Ceynos, then judge of the Crown in Mexico City, who in turn submitted it to the *Consejo de Indias* (Council of the Indies) in Spain. The finished document included the location of the main deposits (Map 6.2), the state

---

[5] *Aparejo* is a Spanish word that defines a set of certain things necessary to make something.

[6] It is possible that this Rodrigo Martinez was the same Rodrigo Martín who was the captain of artillery in the fleet of Pánfilo de Narvaez and who was convinced by Cortés to join his forces before the conquest of Tenochtitlan (Díaz del Castillo 1862, p. 528). If he is the same person, then it is possible that he was the one in charge of making the artillery pieces that Cortés made with the copper and tin he found and bought from the natives of Tenochtitlan and Taxco in 1522 after the fall of the city.

of indigenous technology, estimated rates of production and suggestions for increasing those rates as well as facilitating the mining and the work on extractive metallurgy.

The document included the testimonies of two groups of informants, one of Spanish *encomenderos*, miners and crown agents; and another group of indigenous authorities and native copper specialists, among them miners and smelters. The data contained in Quiroga's report gives us a clear idea of the state of technology and spatial distribution of copper production at this time (Map 6.3). With regard to the native process of production, it mentions that mining was done using stone tools and that the smelting was conducted by blowing through canes, noting that this was a very labour-intensive process that could not produce important quantities of metal. The Spanish informants suggested that by introducing bellows and forges to the techniques that natives used, the whole production could be greatly improved. The indigenous informants, for their part, suggested that by introducing Spanish iron tools the miners would be able to access parts of the mines that were very difficult to work with their traditional tools. They also suggested that, in order to increase production, the communities of specialists could be exempted from other types of tribute so that they could focus producing copper.

The information in the report was not immediately put to use, nor were the suggestions for enhancing the production implemented, perhaps because of the death of Rodrigo Martinez. Nonetheless, this initial report and the data gathered from the two groups of informants set up the foundations for an indigenous-Spanish copper production that would start to operate with hybrid technology at some point between 1538 and 1542.

The first Viceroy of New Spain, Antonio de Mendoza, arrived in 1535 with orders to create a mint in Mexico City, an endeavour which, along with the creation of an artillery foundry, demanded considerable quantities and an ongoing supply of copper. With an increasing demand for metal and for certainty of its future availability, the Viceroy asked the Crown several times to send specialists from Spain who knew how to mine and smelt the reported copper ore in Michoacán ("Dos cartas a la Emperatriz"). The requests, however, were unsuccessful, and

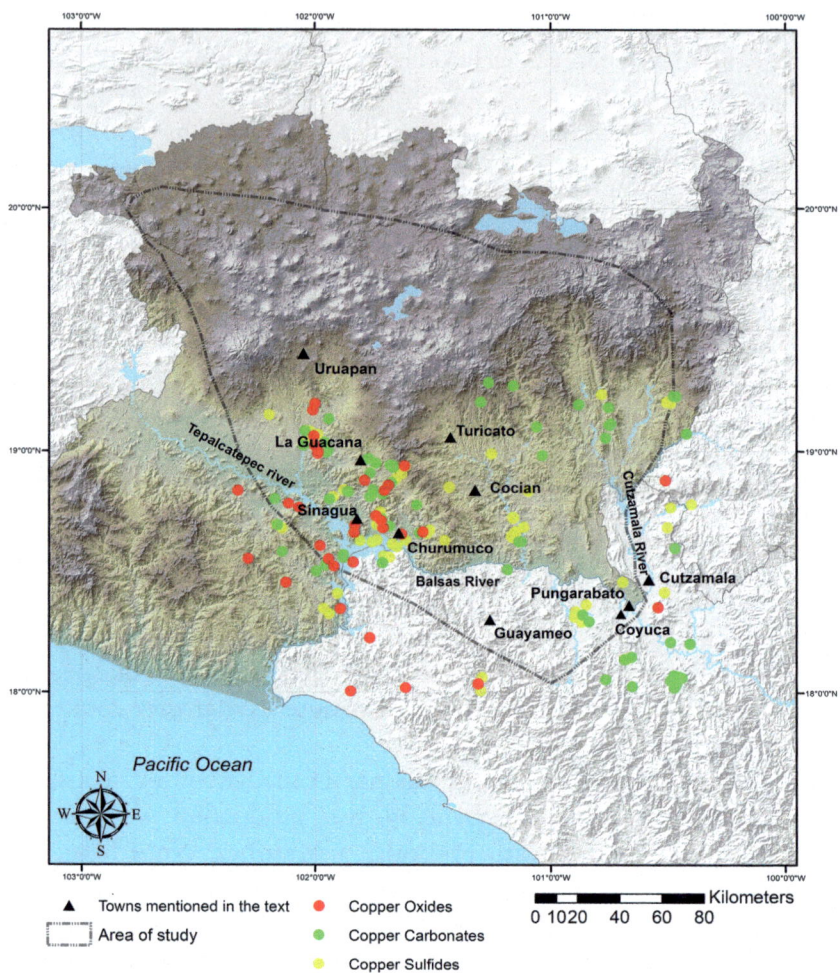

Map 6.3 Geocoding of the places mentioned in the document of 1533 in relation to elevation and presence of copper ore deposits in the area of study (*Sources* Topographic and geologic maps, scale 1:250,000 [E14-1 Morelia and E14-4 Ciudad Altamirano] produced by INEGI)

in 1537 the Empress Isabella of Portugal wrote to the viceroy about the impossibility of the Crown sending such specialists. She, however, suggested that, considering the vastness of the ore deposits, the viceroy should try to find knowledgeable masters there ("Real cédula a Antonio de Mendoza"). Upon this suggestion, the Spanish authorities turned their attention to the technology, expertise, and labour force of the indigenous metallurgists in the area as a way to autonomously produce copper for New Spain.

We believe that between 1538 and 1542 arrangements were made with the local specialists in Michoacán to introduce bellows, forges, iron tools and beaten copper techniques as a way to adapt the native technology of production to the colonial demands. In 1542, the viceroy granted a privilege (*merced*) to the copper producers from the city of Michoacán (Pátzcuaro), for a partial tribute exemption as payment for the delivery of an initial shipment of copper to the artillery foundry in Mexico, simultaneously setting the price for future shipments ("Que a los indios de Mechuacan").

According to the document, the natives of the city of Michoacán provided 53 quintals of copper to the artillery foundry in Mexico City. Given the considerable quantity of metal, the viceroy granted the *merced* and ordered the officials to consider the metal from the natives partial payment of their tribute. The shipment was valued as equivalent to 500 pieces of cotton clothing (*mantas*). The text mentions that, for future shipments, this should be the price to pay to them. Since the *merced* had to be preceded by a formal request, this implies that the native producers were already negotiating with colonial authorities based on their metallurgical knowledge and skills.

The amount of copper delivered by the natives of Pátzcuaro is considerable. 53 quintals were around 2500 kg, and this suggests a considerable increase in the productive capabilities of the native producers. The document does not mention how often this tribute was delivered, but considering the fact that at this time tributes in the region were demanded every 80 days, it is not beyond imagining that the annual quantity of copper was much higher. This is especially interesting with regard to an area that is not mentioned in Quiroga's 1533 report as a production zone. However, the archaeological research made by Blanca Maldonado (2006) and the survey and dating conducted by José Luis Punzo Díaz et al. (2015) suggest there were several copper smelting sites just a few kilometres south of Pátzcuaro.

The amount of copper sent to Mexico in 1542 is comparable with the 1531 amount (Warren 1968), with the difference that those 28–36 quintals of copper were the product of two years of native labour. This increase in productivity could be explained by a hybridisation of the technology in Pátzcuaro and its surroundings between 1538 and 1542, for example, by introducing and adapting European forges, bellows and iron tools into the indigenous metallurgical practice as had been suggested in the report of 1533. Bellows and forges were in use in Mexico City from the 1520s and, given the colonial authorities' need for copper, a logical step would have been to integrate these techniques with existing indigenous technology.

In this regard, it is worth mentioning the information provided by the friar Toribio de Benavente (Motolinía), who between 1536 and 1542 wrote his famous *History of the Indians of New Spain*. In his chapter dedicated to the mechanical arts that the natives knew themselves and the ones they learnt from the Spaniards, he mentioned the manufacture of blacksmith's bellows (*fuelles de herreros*) by the natives of Michoacán "where there is a great deal of work with deer hides" (De Benavente 1914, pp. 216–217, translation by the authors).

In the following year, the natives of Michoacán also started to supply copper to the mint of Mexico City for the manufacture of copper and silver coins, establishing a close, long-lasting relationship with this entity as they became the sole suppliers of all the copper at the mint. This moment saw the beginning of a new episode of technological exchange. When the workers at the mint could not produce copper blanks of sufficient quality to be struck for coinage, through the intercession of the viceroy, the natives of Michoacán took the job. Specialists from the area were taken to the mint and instructed in the technical characteristics of the task. From 1543 until 1551, local Michoacán producers not only supplied copper ingots to the mint and the foundry, but also made and cut the blanks used in the coinage of copper currency.

These initial processes of exchange, negotiation and cooperation between the native producers of Michoacán and the Spanish authorities left the monopoly on copper production in the hands of indigenous communities until at least 1588, when the Crown decided to introduce a royal officer (*juez administrador de cobres*) to the region, responsible for the procurement of copper on behalf of the Crown. The end of the monopoly, which was heavily influenced by the massive native population loss after the epidemics of 1576–1578, brought Spanish entrepreneurs to the region, who inserted themselves into the chain of production. This series of events, in addition to a new period of high imperial demand for copper, triggered new processes of exchange and cooperation between the natives and the Spaniards.

One such process was the extensive interaction between the native producers and the Spanish private entrepreneurs who, since 1585, had become a permanent presence in the regional dynamics of copper production. The documents mention several Spanish private enterprises in which positive cooperation with the native producers was at the core of the issue, and the success or failure of the business enterprises depended on the fluidity of this relationship. The same situation is present in the pairing between Crown and native producers; unsurprisingly, the different commissions given to the copper administrators from 1588 to 1616 highlighted the importance of maintaining a fluid, cordial and trusting relationship with native copper producers.

From Mining to Currency

As mentioned above, metallic copper was typically demanded artillery-related purposes but at the same time a large part of its production was channelled towards the production of official and unofficial currencies that circulated in New Spain during the sixteenth century. Metallic copper was essential for the mint and its proper functioning depended partly on a constant supply of quality metal. Every silver coin produced by the mint contained a small quantity of copper that varied, depending on the quality of the silver, but was typically around 60 ounces for every 100 silver marks (Castro Gutiérrez 2012, p. 33). In this regard, the mint in Mexico City received a steady supply of copper from Michoacán to be used in the coinage of official silver and copper currency from at least 1543.

Between 1542 and 1551–1552 the mint coined its own copper currency, and valuable data about its production and origin of the metal appears in a report on a 1545 visit to the mint by the Crown's inspector Francisco Tello de Sandoval (Pradeau 1953). He was a prominent figure of the court, inquisitor of Toledo, canon of the Cathedral of Seville and member of the Supreme Council of the Indies (Pradeau 1953, p. 9). He had been sent to accomplish two main tasks: the first was to inspect and report on how well and in what way Antonio de Mendoza and the royal court of Mexico (*Segunda Audiencia*) were performing their duties; and the second was to proclaim the enactment of the New Laws of 1542–1543. During his inspection, Sandoval produced a detailed report on the functioning of the Mexico City mint.

The report of his visit is dated 27 May 1545. Tello de Sandoval conducted several hearings and summoned different officers of the mint to provide their testimony. One of the questions was about the coins that were being made. At that moment two types of coins were produced: silver coins and copper coins. The silver coins were silver *reales* with denominations of four, two, one and half a real. The copper coins on the other hand were made with denominations of four and two *maravedís*. These coins were produced from 1542 until 1551–1552. Their production stopped due to widespread rejection by the natives, and the coins were taken out of circulation by a royal order in 1556 (Nesmith 1955, pp. 41–43).

According to the testimonies of the officers, the coinage of copper *cuartos* had proven to be rather difficult. In the beginning, the mint workers tried to use the copper provided by the natives of Michoacán, but the blanks produced in the mint ended up being too brittle and hence unsuitable for being struck. A practical solution for this problem was provided by the coppersmiths of Michoacán. Through the intervention of the viceroy, the blanks were going to be cut and produced in Michoacán by the same natives in charge of smelting, who afterwards would send them to the mint ready to be struck.

In order to do this, natives of Michoacán visited the mint where they were instructed with the specifications necessary for the job. This solution appears to have been successful, as Tello de Sandoval's detailed report does not include any complaint about the copper coins.

One of the witnesses was Juan Gutiérrez, an assayer at the mint, who, asked about the copper coins, declared:

He said that copper coins have been and are made in *cuartos* worth four *maravedís* and [in *cuartos* worth] two *maravedís* are being made at present; and [he said] that they have made *cuartos* in said house, but now those are brought already made from Michoacán and they have nothing to do but to verify the weight, to check if they are good, and coin them; this is currently done in said *casa de moneda*. And it has been for more or less two years that this is done in this way; and this is done by the Indians of Michoacán who were taught in this house, from the copper of extracted from the province of Michoacán. (Pradeau 1953, p. 41, translation by the authors)

Only a couple of years later, the relationship between the producers from Michoacán and the mint took an interesting turn. In volume 1 of their *Historia General de la Real Hacienda*, Fonseca and Urrutia (Fonseca and Urrutia 1845, p. 115) recovered a decree from the Viceroy Mendoza, dated 9 February 1546. The decree is an order to the chief magistrate of Michoacán and other colonial authorities to purchase all the metallic copper needed in the mint in order to keep producing copper currency. According to the decree, the money for the operation was to be taken from the regional tributes and the copper to be purchased had to be in a specific state, well refined and ready for coinage—which we suppose was referring to the delivery of the blanks mentioned in Tello de Sandoval's report. The amount to be paid was 18 pesos for each quintal of these characteristics delivered in Mexico City.

Although short, this is an interesting piece of data which stands out not only as the earliest reference to copper trade between the Crown and the native producers of Michoacán outside the system of tribute, but also as evidence of what looks like a new round of indigenous negotiations based on their metallurgical skills. This new decree comes only a few years after the initial *merced* of 1542 and it appears to be related to the new suite of technical skills acquired by the native producers through their interaction with the mint during the three previous years.

In addition to the coins produced at the mint, other types of unofficial disc-like currencies containing copper (*tejuelos*) circulated in New Spain. The *tepuzque* gold was the main parallel currency of this type. In his Numismatic History of Mexico, Pradeau mentions that its use began around 1522 when tradesmen and merchants melted down their gold dust and flakes into disc-like items for an easier handling. According to Pradeau, as their circulation became popular and due to the lack of control, the gold discs almost immediately began to be debased by adding different amounts of copper; conscious of this trickery, the natives called it *tepuzque*, meaning *copper* (Pradeau 1938, p. 21).

Only a handful of examples of *tepuzque* have been recovered, all of them from sixteenth-century shipwreck sites (Rivero Franyutti 2016, pp. 302–303). However, none have been thoroughly studied. Nonetheless, through examination of some early historical accounts, such as the *Historia Verdadera de la Conquista* (Díaz del Castillo 1862, vol. 2, pp. 418–419), the *Suma de Visitas* (García Castro 2013) or the *Libro de las Tasaciones* (González de Cossio 1952) we can infer that it was a widely used copper–gold alloy with a variable percentage of copper as its base metal. According to these sources, it had a concentration of copper between 75 and 30%; in other words, gold fineness between 6 and 19k. References to this alloy can be found across the tax records of native towns: for instance, the 1541 tax record of the town of Xochimilco, not far from Mexico City, mentioned that every 80 days the natives had to give 50 *tejuelos* of gold with a fineness of 10k (41.7% of gold), and that each *tejuelo* had to weight 10 *pesos* (González de Cossio 1952, p. 303).

The *tepuzque* was rapidly incorporated as a currency years before the establishment of the mint of Mexico City and remained in use until at least the end of the seventeenth century despite various attempts by the Crown to stop its circulation. *Tepuzque*-gold pesos became popular in New Spain, and it was one of the currencies demanded from the native communities as part of their tributes. Using the data contained in García Castro (2013) and González de Cossio (1952), we managed to identify 28 *tepuzque* tributary towns, more than half of them in the region of Michoacán, not far from the copper production area. What is interesting here is that geocoding the data from the tax records of the sixteenth century reveals what appears to be a correlation between the regions where copper was available and produced, and the use of *tepuzque* as tribute. This is particularly relevant for the area of Michoacán and the modern states of Colima, Guerrero, Mexico and Oaxaca; all of which are areas with copper deposits (Map 6.4).

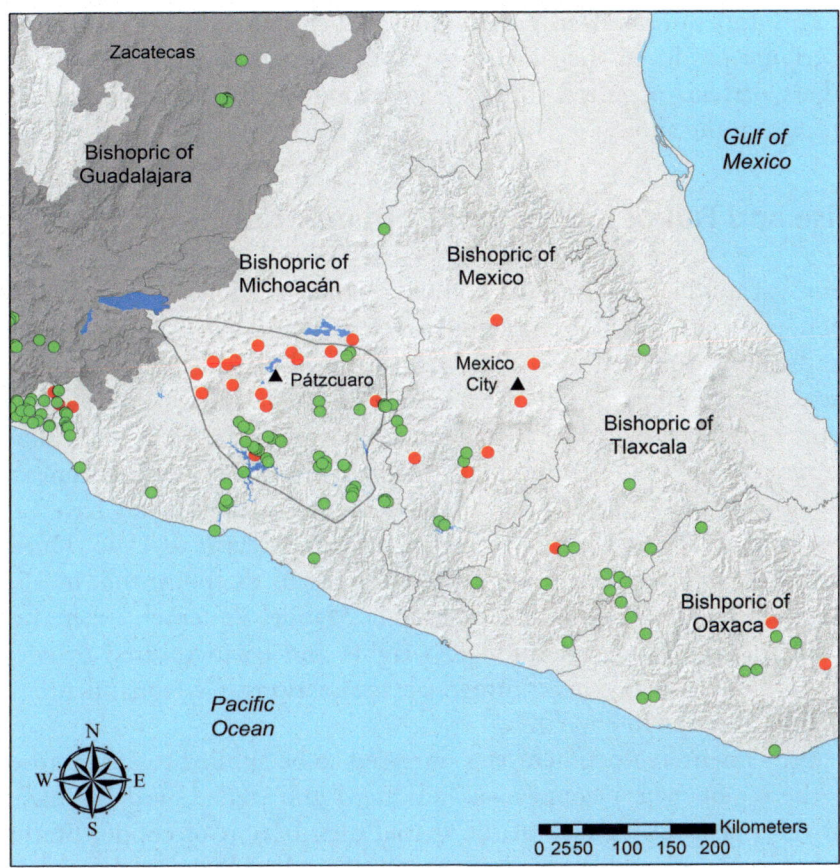

- ● Tepuzque Tributary Town According to 16th Century Tax Records
- ● Copper Deposit
- ▨ Audiencia de Guadalajara
- ☐ Audiencia de Mexico
- ⬡ Area of Study

Map 6.4 Geocoding of the identified Indian towns that gave tepuzque as tribute and their relation to the general distribution of copper ore deposits (*Sources*: Tributary towns: [Suma de visitas; Libro de las Tasaciones]; Copper deposits: [Web portal of the Servicio Geologico Mexicano, https://www.sgm.gob.mx/GeoInfoMexGobMx/]; administrative borders: Stangl [2018])

This does not necessarily mean that all the towns which gave *tepuzque* as tribute produced their own pesos, but it suggests that some did and others perhaps acquired them through trade with the ones that were producing the alloy.

Rise and Fall of Indigenous Copper Production

From around 1545 on, the native producers of Michoacán held a monopoly on copper production in New Spain. From mining to the production of refined ingots and items, the whole chain of production, including transport, was in the hands of native producers. Around 1552 copper disappeared from tax records in the region, an event that Elinore Barrett considers the beginning of copper production as a commercial activity in New Spain (Barrett 1987, p. 14), although we have seen that the initial stage of this process can be dated back to 1546. During the following years, the trade in copper grew to the point that in 1570 the main authority of Pátzcuaro, García Manuel Pimentel, ordered the standardisation and regulation of weights and balances used to weigh the ore and metal in commercial transactions ("Mandamiento de Manuel García Pimentel").

The Pimentel's document was intended to be applied to eleven towns in the region where copper was produced and traded, and the data it provides gives us an idea of the spatial distribution of copper production centres, data that can be contrasted with earlier and later documents. If we compare the 1570 data with that from 1533, we can see a deep change in the spatial configuration of production. While the 1533 document clearly states that the smelting of ore was conducted in places around the mines located in the lowlands and the foothills of the Balsas-Tepalcatepec depression, the data from 1570 places most of the smelting sites on the high plateau, where there were abundant pine-oak forests for making quality charcoal (Map 6.5).

This shift has two important implications: first, the data from 1570 shows a tendency towards specialisation. While the people around the ore deposits did the mining, some communities on the plateau made the charcoal and others smelted the ore, often working the metal as well. This implies an important change in relationships from the data from

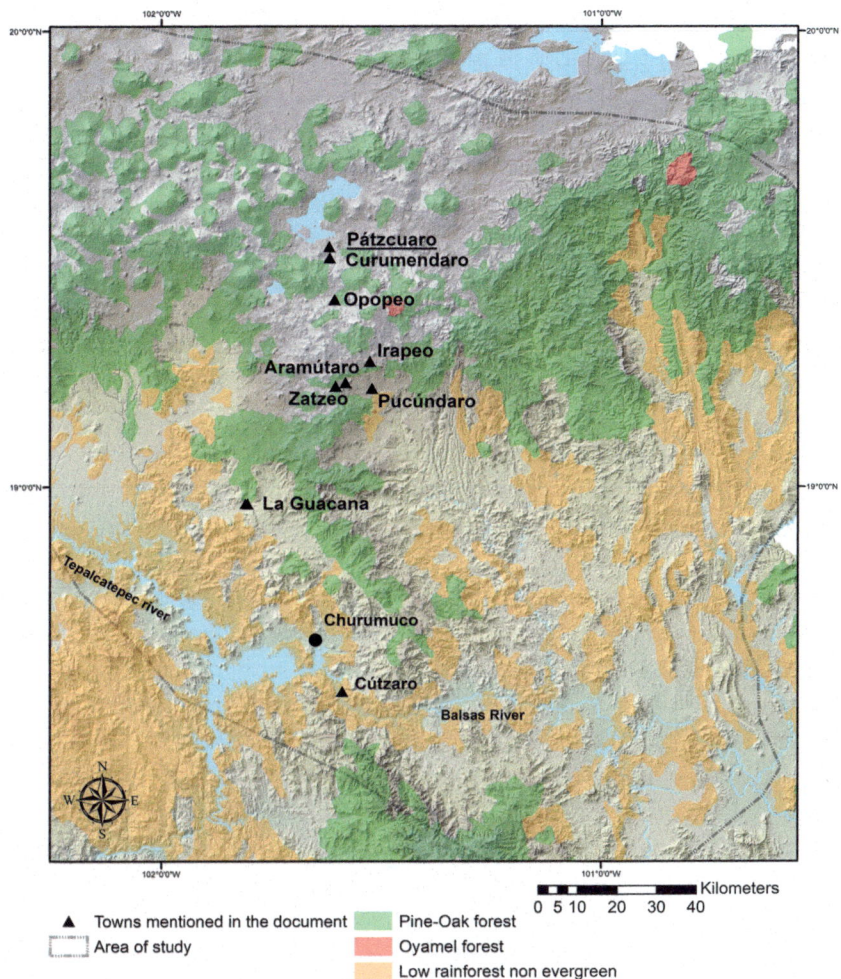

Map 6.5 Geocoding of the places mentioned in Pimentel's 1570 document and their relation with a new spatial pattern of distribution associated with the predominant pine-oak vegetation (*Sources* Topographic and geologic maps, scale 1:250,000 [E14-1 Morelia and E14-4 Ciudad Altamirano] produced by INEGI serie 2015–2017)

1533, in which the native metallurgists were simultaneously prospectors, miners, smelters and metalworkers. Later documents not only confirm that tendency but emphasise labour specialisation at a greater scale.

The second issue to consider is that the change in the spatial pattern and the increasing number of production centres near fuel sources seem to relate to an important rise in copper production. The ratio of charcoal consumption in the smelting and refining operations oscillated from 3:1 to 15:1, meaning that for each quintal of ore three to fifteen quintals of charcoal were required, depending on the type and quality of the ore. In economic terms, it was more efficient to place the smelting sites in or around the fuel sources in order to reduce the high cost of transport over considerable distances. We do not have an estimate of copper production for the sixteenth century, but the estimate of copper demand in New Spain made by Barrett for the end of the eighteenth century was of 20,300 quintals per year (Barrett 1987, pp. 46–49).

The epidemics of 1576–1578 devastated New Spain in general and the native population of the copper region of Michoacán in particular. Demographers of the event estimate that at least one-third or even half of the native population died during the outbreak (Acuña-Soto et al. 2004). This catastrophic event had a profound impact on the dynamics of production. With a permanent increase in demand for copper, exacerbated by the maritime conflict against England and the urgent need to supply bronze artillery for ports and ships, the Crown in 1587 decided to take control of copper production, which had become more than ever a strategic resource. This decision brought a number of changes that modified not only the dynamics of production, but also the settlement patterns and the sociopolitical and economic organisation of copper-producing indigenous communities.

The Crown introduced its own agents to supervise charcoal making and the mining and smelting of ore ("Comisión a Alonso Delgado Guzmán"). Around the same time, it allowed and encouraged the presence of Spanish and mestizo entrepreneurs who organised their own mining and smelting operations, most of them as joint enterprises that included individuals and groups of indigenous producers ("Solicitud para que se pague a Francisco de Ayala"). Around the 1590s, the Crown began to establish its own mining and smelting operations in order to secure a constant flow of metal for the artillery foundries. It placed a ban on the private trade of copper, meaning that all the metal produced in the region during this period was to be handed to the Crown in order to face the extraordinary demands of the moment.

These measures in themselves were profound enough to have a significant impact on local colonial-indigenous relations. However, the most important upheaval of indigenous communities came with the implementation of a policy of resettling dispersed indigenous communities, under the so-called *congregaciones de indios*. Although this policy was widely applied across New Spain (Cline 1949), in the copper-producing regions, the decisions were guided by the Crown's need for ensuring a permanent flow of metal ("Parecer que don García Rodríguez de Valdez"). The policy of *congregaciones* was implemented between 1598 and 1607 and its outcome was a systemic change in the social and spatial organisation of the native people involved in the production process. Three congregations were established in the region, two of which, located in the lowlands, specialised in mining and ore procurement, and one (Santa Clara) on the high plateau specialised in charcoal making, smelting and refining. The number of people resettled during this process was not massive, but it included most of the remaining specialists who held the knowledge and skills for copper mining and smelting (Lemoine 1962, pp. 695–702) (Map 6.6).

The consequence of the resettlement process for the local communities of producers was the reduction of the specialised labour force in three towns, the loss of some of their ancestral lands and the gradual loss of the commercial freedom they had previously held. With the implementation of tight state control, these communities were no longer entrepreneurs and business owners but became in most cases wage-earning employees at the service of the Crown and, from 1620, at the service of the successive private concessionaires. In the mining district, the *cerro de Inguarán*, the main copper ore deposit of the region—until that moment property of the local communities—was expropriated by the Crown and claimed by the king as part of his royal heritage (Barrett 1987, pp. 25–42).

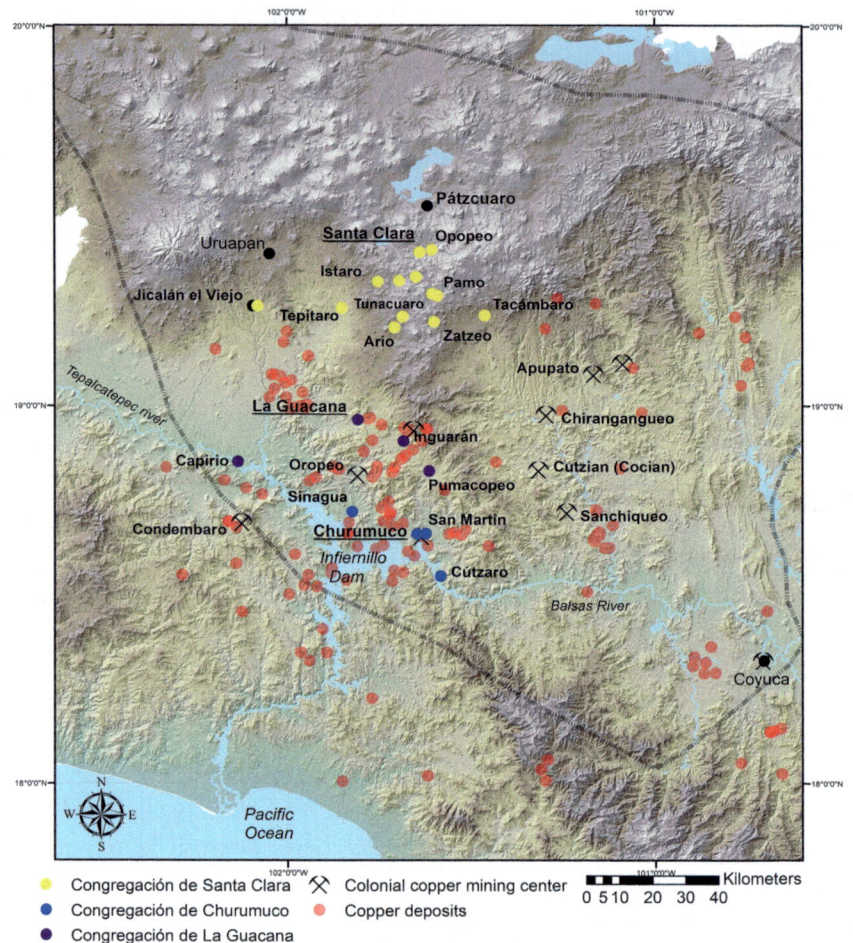

Map 6.6 Outcome of the policy of congregations associated with copper production (*Sources* See Map 6.5)

Final Remarks

The dynamics of copper production in Michoacán is one example of how the encounter between different groups of colonial society and the introduction of a new economic model of production deeply modified a deep-rooted technological tradition and its natural landscape. The main changes, however, were in the sociocultural aspects of production, and in the human geography of mining and extractive metallurgy.

This case study highlights that technology—in this case metallurgical technology—served as a common ground for cultural interaction, dialogue, exchange, negotiation and cooperation between indigenous and Spanish communities. These processes, however, were immersed in the context of asymmetric power relations that was inherent to the colonial structure. This means that the Spanish attitudes towards the material and the scale of operation they needed to fulfil colonial demands defined the course of development followed by the copper production technology.

In technical terms, the best example of interaction and exchange is the development of a hybrid technology for copper production based on the indigenous pre-Hispanic metallurgical tradition but with the adaptation of Spanish techniques. This new hybrid technology was exclusive to indigenous communities of producers from its beginning around 1538 until 1588. However, during the last decade of the sixteenth century, in a context of dynamic interaction between the native communities and the Crown, this way of producing was eventually transmitted, accepted, learnt and used by the emerging *mestizo* and Spanish communities of the late sixteenth and early seventeenth centuries, who slowly but steadily introduced themselves into the different stages of copper production.

The indigenous communities of specialised metalworkers used their expertise to negotiate economic and political privileges from the colonial authorities, and to benefit from business activities established in cooperation with Spanish entrepreneurs. Nonetheless, the power asymmetry between the Spanish colonial system and the native communities defined the parameters of production. For instance, the colonial authorities determined what would be produced, by whom, in what quantity and how often. In the same way, it restricted indigenous metallurgical practice to copper production in a region that was historically characterised by a metallurgical tradition that involved the extensive working of other metals (silver, gold, tin and arsenic) besides copper.

Another example of these power dynamics is visible in the gradual transformation of native labour: from part-time metallurgical specialists in 1533 to the full-time wage labourers of 1607; and from the native specialist who was at the same time miner, smelter and coppersmith in 1533 to the specialist miner (*barretero*), smelter (*fundidor*) and copper-smith (*calderero*) of the late sixteenth century. The human geography associated with copper production was also massively modified across the sixteenth century, reaching its peak in the first decade of the seventeenth century with the policy of congregations.

The scale of production and the permanent increase in demand prompted specialisation of the labour force based on the availability of resources from different ecological niches. The concentration of smelting activities around the fuel sources, for example, triggered processes of intensive deforestation, remodelling of the physical landscape and the rise of pastoralism within the pine-oak area.

We do not know for certain the production rates during the sixteenth, seventeenth or even eighteenth centuries but we do know that production was enough to make New Spain self-sufficient in copper and maintain a surplus that was regularly exported to other places in the empire such as Cuba, the Philippines, Peru and more frequently Spain.

Bibliography

Primary Sources

Archivo General de Indias

"Real Cédula a Luis Fernández, para que si está preparado, venga al Consejo de Indias a tratar de la artillería que se ha de hacer en Nueva España." AGI, INDIFERENTE, 422, L.15, F.211V.

"Real Cédula a Rodrigo Martínez, maestro artillero, vecino de Cuéllar, para que, en plazo de tres días, vaya al Consejo de Indias a dar cuenta de cierto cobre que recibió para hacer artillería, con destino a la armada de la Especiería, al mando de Simón de la Alcazaba." AGI, INDIFERENTE, 422, L.15, F.110V.

"Real cédula al corregidor de La Coruña, para que averigüe si Rodrigo Martínez, artillero, entregó como dice a dicha ciudad, un tiro de metal de 28 quintales de peso." AGI, INDIFERENTE, 422, L.15, F.151V–152R.

"Real cédula a la Audiencia de México para que haga justicia en la reclamación presentada por Rodrigo Martínez, fundidor de artillería, para se le devuelvan los bienes que dejó en aquellas tierras al cargo de Sancho de Frías, vecino de esa ciudad." AGI, MEXICO, 1088, L.2, F.101V–102R.

"Carta acordada del Consejo de Indias a Rodrigo Martínez para que venga de nuevo a la Corte a tratar sobre la fundición de artillería en Nueva España." AGI, INDIFERENTE, 422, L.15, F.217R.

"Mandamiento del Consejo de Indias a Diego de la Haya, cambio de la Corte, para que pague 20 ducados a Rodrigo Martínez, artillero, por haber estado en la Corte por mandato de S.M. para hacer cierto asiento sobre artillería en Nueva España." AGI, INDIFERENTE, 422, L.15, F.227R.

"Dos cartas a la Emperatriz de Antonio de Mendoza: 1ª, notificando que viene a España Cabeza de Vaca y Francisco Dorantes, que se habían escapado de la armada de Pánfilo de Narváez, a hacer relación de lo que en ella sucedió; 2ª, tratando de asuntos de buen gobierno, justicia y Real Hacienda de Nueva España." AGI, PATRONATO, 184, R.27.

"Real cédula a Antonio de Mendoza, virrey de Nueva España, en respuesta a su carta duplicada de 30 de abril de 1537 sobre distintos temas." AGI, MEXICO, 1088, L.3, F.75–77.

"Parecer que don García Rodríguez de Valdez dio acerca de las diligencias que hizo en la provincia de Mechuacan sobre las minas de metales de cobre que hay en aquella provincia." AGI, MEXICO, 258, N.12.

Archivo General de la Nación

"Que a los indios de Mechuacan se les tenga en cuenta cincuenta y tres quintales de cobre que dieron para la fundición." AGN, MERCEDES, Vol. 1, Exp. 112, F.56V.

"Comisión a Alonso Delgado Guzmán para lo tocante al cobre de la provincia de Mechuacan." AGN, General de Parte, Vol. 5, Exp. 1427.

"Solicitud para que se pague a Francisco de Ayala el pago de los metales de cobre que se le deben." AGN, Indiferente Virreinal, Caja 4371, Exp. 027.

Archivo Municipal de Pátzcuaro

"Mandamiento de Manuel García Pimentel, alcalde mayor, para que los tratos de cobre se hagan por peso y medida y en pesas de hierro marcadas." AMP, Siglo XVI, Caja 131, Legajo 4, Año 1570.

Source Editions and Old Prints

Cortés, Hernán, 1806. *Cartas y Relaciones de Hernán Cortés Al Emperador Carlos V*, edited by Don Pascual De Gayangos. Paris: Imprenta Central de los Ferro-Carriles.

De Benavente, Toribio, 1914. *Historia de Los Indios de La Nueva España*. Barcelona: Herederos de Juan Gili.

Díaz del Castillo, Bernal, 1862. *Verdadera Historia de Los Sucesos de La Conquista de La Nueva España Por El Capitan Bernal Díaz Del Castillo Uno de Sus Conquistadores*. Madrid: Imprenta de Tejado.

Díaz del Castillo, Bernal, 2008. *The History of the Conquest of New Spain by Bernal Díaz Del Castillo*, edited and with an introduction by David Carrasco. Albuquerque: University of New Mexico Press.

Díaz del Castillo, Bernal, 2010. *The True History of the Conquest of New Spain*, edited by Genaro Garcia, vols. 2–3. Cambridge: Cambridge University Press.

González de Cossio, Francisco, ed., 1952. *El Libro de Las Tasaciones de Pueblos de La Nueva España, Siglo XVI*. Mexico City: Archivo General de la Nación.

Lemoine, Ernesto, 1962. "La Relación de La Guacana, Michoacán, de Baltasar Dorantes Carranza. Año de 1605." In: *Boletín Del Archivo General de La Nación* 3, 4: 669–702.

Secondary Sources

Acemoglu, Daron; Johnson, Simon; Robinson, James A., 2001. "The Colonial Origins of Comparative Development: An Empirical Investigation." In: *The American Economic Review* 91, 5: 1369–1401.

Acuña-Soto, Rodolfo, et al., 2004. "When Half of the Population Died: The Epidemic of Hemorrhagic Fevers of 1576 in Mexico." In: *FEMS Microbiology Letters* 240, 1: 1–5. https://doi.org/10.1016/j.femsle.2004.09.011.

Andrews, Kenneth R., 1984. *Trade, Plunder, and Settlement: Maritime Enterprise and the Genesis of the British Empire, 1480–1630*. Cambridge: Cambridge University Press.

Antunes, Catia; Polónia, Amélia, eds., 2016. *Beyond Empires: Global, Self-Organizing Cross-Imperial Networks, 1500–1800*. Leiden: Brill.

Bakewell, Peter, 1990. "La minería en la Hispanoamérica Colonial." In: *América Latina en la Epoca Colonial*, edited by M. Léon Portilla, vol. 2. Barcelona: Editorial Crítica: 131–153.

Barrett, Elinore M., 1987. *The Mexican Colonial Copper Industry*. Albuquerque: University of New Mexico Press.

Bayly, Christopher A., 1998. *Empire and Information: Intelligence Gathering and Social Communication in India, 1780–1870*. Cambridge: Cambridge University Press.

Beltrán Martínez, Román, 1952. "Primeras Casas de Fundicion." In: *Historia Mexicana* 1, 3: 372–394.

Bernier, Marc A.; Donato, Clorinda; Lüsebrink, Hans-Jürgen, 2014. *Jesuit Accounts of the Colonial Americas: Intercultural Transfers, Intellectual Disputes, and Textualities*. Toronto: UCLA and University of Toronto Press.

Boyajian, James, 2008. *Portuguese Trade in Asia Under the Habsburgs, 1580–1640*. Baltimore: Johns Hopkins University Press.

Bührer, Tanja; Eichmann, Flavio; Förster, Stig; Stuchtey, Benedikt, 2017. *Cooperation and Empire: Local Realities of Global Processes*. New York and Oxford: Berghahn.

Burbank, Jane; Cooper, Frederick, 2010. *Empires in World History*. Princeton: Princeton University Press: 13–14.

Castro Gutiérrez, Felipe, 2012. *Historia Social de La Real Casa de Moneda de México*. Mexico: UNAM.

Clancy-Smith, Julia A., 2004. "Collaboration and Empire in the Middle East and North Africa: Introduction and Response." In: *Comparative Studies of South Asia, Africa and the Middle East* 24, 1: 126.

Cline, Howard, 1949. "Civil Congregations of the Indians in New Spain, 1598–1606." In: *The Hispanic American Historical Review* 29, 3: 349–369.

Cook, Noble D., 1998. *Born to Die: Disease and New World Conquest, 1492–1650*. Cambridge and New York: Cambridge University Press.

Crosby, Alfred, 1986. *Ecological Imperialism: The Biological Expansion of Europe, 900–1900*. New York: Cambridge University Press.

Crosby, Alfred, 1988. "Ecological Imperialism: The Overseas Migration of Western Europeans as a Biological Phenomenon." In: *The Ends of the Earth Perspectives on Modern Environmental History*, edited by D. Worster. Cambridge: Cambridge University Press.

Darwin, John, 2008. *After Tamerlane: The Rise and Fall of Global Empires, 1400–2000*. London: Bloomsbury.

Engerman, Stanley L.; Sokoloff, Kenneth L., 2012. *Economic Development in the Americas Since 1500: Endowments and Institutions*. New York: Cambridge University Press.

Fischbacher, Urs, 2004. "Social Norms and Human Cooperation." In: *Trends in Cognitive Sciences* 8: 185–190.

Fonseca, Fabian de; Urrutia, Carlos de, 1845. *Historia General de La Real Hacienda*, tomo 1. Mexico: Vicente G. Torres.

Gallagher, John; Robinson, Ronald, 1953. "The Imperialism of Free Trade." In: *The Economic History Review* 6, 1: 1–15.

Garcia Bernal, Manuela C., 1978. *Población y encomienda en Yucatán bajo los Austrias*. Sevilla: Escuela de Estudios Hispano-Americanos.

García Castro, Leopoldo René, 2013. *Suma de visitas de pueblos de la Nueva España, 1548–1550*. México, DF: El Colegio Mexiquense.

Guerrero, Saul, 2017. *Silver by Fire, Silver by Mercury: A Chemical History of Silver Refining in New Spain and Mexico, 16th to 19th Centuries*. Leiden: Brill.

Hagen, Edward H.; Hammerstein, Peter, 2006. "Game Theory and Human Evolution: A Critique of Some Recent Interpretations of Experimental Games." In: *Theoretical Population Biology* 69: 339–348.

Hammerstein, Peter, ed., 2003. *Genetic and Cultural Evolution of Cooperation*. Cambridge, MA: MIT Press in cooperation with Dahlem University Press.

Hosler, Dorothy, 1994. *The Sounds and Colors of Power: The Sacred Metallurgical Technology of Ancient West Mexico*. Boston, MA: MIT Press.

Lee, Wayne E., 2011. "Projecting Power in the Early Modern World: The Spanish Model?" In: *Empires and Indigenes: Intercultural Alliance, Imperial Expansion, and Warfare in the Early Modern World*, edited by W. E. Lee. New York: New York University Press: 1–16.

Maldonado, Blanca, 2006. *Preindustrial Copper Production at the Archaeological Zone of Itziparatzico*. PhD dissertation, University of Pennsylvania.

Nesmith, Robert I., 1955. *The Coinage of the First Mint of the Americas at Mexico City 1536–1572*. New York: The American Numismatic Society.

Neto, Manoel J. de Miranda, 2012. *A utopia possível: missões Jesuíticas em Guairá, Itatim e Tape, 1609–1767, e seu suporte econômico-ecológico*. Brasília: Fundação Alexandre de Gusmão.

North, Douglass C.; Summerhill, William R.; Weingast, Barry R., 2000. "Order, Disorder, and Economic Change: Latin America vs. North America." In: *Governing for Prosperity*, edited by B. Bueno de Mesquita and H. Root. New Haven: Yale University Press.

Nowak, Martin A., 2006. "Five Rules for the Evolution of Cooperation." In: *Science* 314, 5805: 1560–1563.

Ostrom, Elinor, 1990. *Governing the Commons: The Evolution of Institutions for Collective Action*. Cambridge: Cambridge University Press.

Peers, Douglas M. 2002. "Is Humpty Dumpty Back Together Again? The Revival of Imperial History and the Oxford History of the British Empire." In: *Journal of World History* 13, 2: 451–467.

Polónia, Amélia, 2012. "Indivíduos e redes auto-organizadas na construção do império ultramarino português." In: *Economia, Instituições e Império. Estudos em Homenagem a Joaquim Romero de Magalhães*, edited by A. Garrido, L. F. Costa, and L. M. Duarte. Coimbra: Almedina: 349–372.

Polónia, Amélia; Antunes, Catia, eds., 2017. *Mechanisms of Global Empire Building, 15th–18th Centuries*. Porto: CITCEM/Afrontamento.

Powell, Philip W., 1952. *Soldiers, Indians, and Silver: The Northward Advance of New Spain, 1550–1600*. Berkeley and Los Angeles: University of California Press.

Pradeau, Alberto Francisco, 1938. *The Numismatic History of Mexico: From the Pre-Columbian Epoch to 1823*. Los Angeles: Western Printing Company.

Pradeau, Alberto Francisco, 1953. *Don Antonio de Mendoza y La Casa de Moneda de México en 1543*. Mexico: Antigua Librería Robredo, de José Porrúa y Hijos.

Prem, Hans-Jürgen, 1992. "Spanish Colonization and Indian Property in Central Mexico, 1521–1620." In: *Annals of the Association of American Geographers* 82, 3: 444–459.

Punzo Díaz, José Luis; Morales, Juan; Goguitchaichvili, Avto, 2015. "Evidencia de Escorias de Cobre Prehispánicas En El Área de Santa Clara Del Cobre, Michoacán, Occidente de México." In: *Arqueología Iberoamericana* 28: 46–51.

Reff, Daniel T., 1991. *Disease, Depopulation and Culture Change in Northwestern New Spain, 1518–1764*. Salt Lake City: University of Utah Press.

Richards, John F., 2003. *The Unending Frontier: An Environmental History of the Early Modern World*. Berkeley: University of California Press.

Richerson, Peter J.; Robert Boyd, 2005. *Not by Genes Alone: How Culture Transformed Human Evolution*. Chicago: University of Chicago Press.

Rivero Franyutti, Agustín, 2016. "Las acepciones del nahuatlismo tepuzque en el español de México." In: *Anuario de Letras, Lingüística y Filología* 4, 2: 297–334.

Robinson, Ronald, 1972. "Non-European Foundations of European Imperialism: Sketch for a Theory of Collaboration." In: *Studies in the Theory of Imperialism*, edited by Roger Owen and Robert Sutcliffe. London: Longman: 117–142.

Robinson, Ronald, 1986. "The Excentric Idea of Imperialism, with or without Empire." In: *Imperialism and After: Continuities and Discontinuities*, edited by W. J. Mommsen and Jürgen Osterhammel. London: Allen & Unwin: 267–289.

Sánchez Gómez, Julio, 1989. *De Minería, Metalúrgica y Comercio de Metales: La Minería No Férrica En El Reino de Castilla, 1450–1610.* Salamanca: Universidad de Salamanca.

Santos, Francisco C.; Pacheco, Jorge M.; Lenaerts, Tom, 2006. "Cooperation Prevails When Individuals Adjust Their Social Ties." In: *PLoS Computational Biology* 2, 10: e140. https://doi.org/10.1371/journal.pcbi.0020140.

Subrahmanyam, Sanjay, 2007. "Holding the World in Balance: The Connected Histories of the Iberian Overseas Empires, 1500–1640." In: *American Historical Review* 112, 5: 1359–1385.

Stangl, Werner, 2018. "Obispado_Dissolve_1701 [v. 2018-09-15]". In: *HGIS de las Indias* (Austrian Science Fund, P 26379-G18: 2015–2019). https://www.hgis-indias.net/index.php/download-data. Accessed October 29, 2018.

Studnicki-Gizbert, Daviken; Schecter, David. 2010. "The Environmental Dynamics of a Colonial Fuel-Rush: Silver Mining and Deforestation in New Spain, 1522 to 1810." In: *Environmental History* 15, January: 94–119.

Warren, J. Benedict, 1968. "Minas de Cobre de Michoacán 1533." In: *Anales Del Museo Michoacano* 6: 35–52.

Weiskel, Timothy C., 1988. "Toward an Archaeology of Colonialism: Elements in the Ecological Transformation of the Ivory Coast." In: *The Ends of the Earth: Perspectives on Modern Environmental History*, edited by Donald Worster. Cambridge: Cambridge University Press: 141–172.

7

American Silver and Its Repercussions on the Old World: The Curious Case of the Loss-Making Spanish Precious Metal Sector, 1590s–1640s

Domenic Hofmann

In the spring of 1603, Juan Córdoba Canales was supposed to travel from Spain to America to tackle some of the problems facing the silver mines at Potosí, in Peru. The rich silver mines were on the verge of depletion and Don Córdoba Canales was one of the most able mining experts of his time. He had considerable experience in the American mining sector and was a particular expert in the amalgamation process, which used mercury in the refinement of silver (Sánchez Gómez 1989). Canales's mining expertise even brought him to Vienna, where he sought to introduce his amalgamation process into the silver mines of Central Europe (Born 1786, pp. 13–17). Fearing the consequences for the Spanish mining industry if such a distinguished expert was lost, the royal authorities ordered Córdoba Canales to remain in Spain (Crown 1603). The decision to restrain Córdoba Canales from engaging with

D. Hofmann (✉)
Institute of History, University of Graz, Graz, Austria
e-mail: domenichofmann@gmx.at

© The Author(s) 2019
R. Pieper et al. (eds.), *Mining, Money and Markets in the Early Modern Atlantic*, Palgrave Studies in Economic History,
https://doi.org/10.1007/978-3-030-23894-0_7

163

American mines is a remarkable incident, as the American precious metal sector is generally considered to have been of utmost importance to the Spanish Crown. On the other hand, the role of the precious metal mining sector in the Spanish mainland for the Crown has been largely neglected in the literature.

American silver rose to prominence after the invention of the amalgamation process by Bartholomé de Medina in 1554. Using mercury for refinement enabled large quantities of silver to be produced in Spanish American mines. Initially, the so-called patio process enabled mines in New Spain not only to work lower-grade silver ores but to do so much more economically. After several attempts, the amalgamation process was successfully adopted in mines in Peru during the 1570s, leading to a spike in American silver production for the next 60 years (Bakewell 1997). Consequently, more than 70% of global silver production in early modern times took place in Spanish America. The Spanish Crown relied heavily on remittances of American silver across the Atlantic, but despite the predominance of American silver, the Spanish Crown still promoted and at times even favoured the mainland precious metal sector.

Similar to Renate Pieper's approach, this article argues that American silver had an impact beyond the realm of politics. While Pieper examines the cultural shifts precipitated by American silver, this investigation analyses the Spanish precious metal sector at the time of the "first" American silver boom in an effort to uncover the industrial and technological impact of the American silver on Europe. The literature to date has overwhelmingly focused on the American mining sector in the early modern period. Mining ambitions in the Spanish mainland have only seldomly been studied. Most importantly, Julio Sánchez Gómez conducted a comprehensive evaluation of the development of the early modern Spanish mining sector (Sánchez Gómez 1989; Sánchez Gómez et al. 1997). His investigation focused particularly on the Guadalcanal mining endeavour in the mid-sixteenth century. The time after the depletion of the Guadalcanal mine in the 1570s (and the beginning of the American silver boom) has received even less attention.

As a capital-intensive and highly specialised industry, early modern mining is generally studied alongside technological matters. It is hardly surprising that the literature has paid special attention to the development of the technological outlines of the early modern Spanish mining sector. Central to this study is the work by García Tapia, who focused on the creation and development of a patent system in Spain (Garcia Tapia 1994), and Mateos Royo's recent study of technological development in the Spanish mining sector (Mateos Royo 2009). Both authors argue that technological progress in the Spanish mining sector mainly occurred in the second half of the sixteenth century. The first half of the seventeenth century, in contrast, is not identified with substantial technological change (cf. Sánchez Gómez et al. 1997).

While past research has provided general insights into the institutional and technological outline of the early modern Spanish mining industry, a more fundamental question not been adequately addressed. For more than a decade, the Guadalcanal silver mine in Andalusia was one of the most important sources of silver in the Spanish Empire. It was discovered during the 1550s and soon became a cash cow for the Spanish Crown. However, during the next few years, the silver production in the mine gradually declined and Guadalcanal was finally depleted in the early 1570s (Sánchez Gómez 1989). When the Crown realised that the mine could no longer be operated profitably, it was quick to cease its mining efforts. Consequently, for almost twenty years, there is no evidence of any ambition by royal authorities towards a precious metal sector in Spain.

It was not until 1594 that the Crown reinstated a mining administration in Spain. The most able mining experts in Spain were hired to join the royal administration to promote and support a privately run mining sector. However, unlike earlier times, no profitable precious metal mines could be found. Instead of contributing to Spain's finances, the Crown's mining ambitions drained its coffers. Contrary to the case of Guadalcanal, though, the royal authorities continued with their efforts searching for and promoting precious metal mines in Spain—and continued to do so for almost half a century. Why would the Spanish Crown support an unprofitable precious metal sector, while—at the same time—having access to large quantities of silver from America? While during the 1560s and 1570s, the royal authorities were very concerned about the profitability of the Spanish precious metal sector, after the 1590s they seem no longer to have cared, or at least they were not inclined to act decisively upon the matter.

Up until now, historiography has only superficially dealt with this matter. Sánchez Gómez presented the argument that actors often engaged in the early modern Spanish mining industry because they believed rich metal deposits existed in Spain (Sánchez Gómez et al. 1997, p. 387ff.). While this line of reasoning explains why private actors might have been tempted to invest in mines in Spain, it does not explain why the Crown continued to promote institutions engaged in a loss-making Spanish precious metal sector for several decades. Another periodically advanced argument points to the lack of copper mines in Spain (Goodman 1988, p. 115). It suggests that the problems in the coinage of *vellón* copper coins at the beginning of the seventeenth century led to the promotion of precious metal mines in Spain. However, this argument, similarly, does not hold up: it was silver rather than copper mines being promoted at the time. One has to look beyond financial reasons to understand the nature of the royal mining efforts at the time. While financial concerns doubtless played a role in the Crown's inclination to promote precious metal mines in Spain (cf. Naharro Quirós 1991), there were other developments driving the Crown's efforts to promote Spanish mining at the end of the sixteenth century.

This investigation aims to find the reasons for the Crown's promotion of the loss-making Spanish mining sector. It gains new insights into this matter by evaluating the Spanish mining sector at the end of the sixteenth century and beginning of the seventeenth century in the light of the American silver boom at the time. American silver created a range of new opportunities and challenges for the Spanish motherland. The Crown in Spain had to react to those new developments, and the Spanish mining sector played a major role in that royal strategy. As a consequence, profound changes in Spain at the time are also noticeable.

Spanish Precious Metal Mining as a Reaction to the Dependence on American Silver

After the adoption of the amalgamation process, American silver production rose precipitously. The sharp increase in silver output is especially evidence after the 1580s, when the amalgamation process arrived in Peru, and lasted until the 1640s when silver production in America began to decline. The renewed interest in Spanish precious metal mines emerged in almost exactly the same period. From 1594 onward, the Spanish Crown set up institutions to attract investment to precious metal mines in mainland Spain. In particular, the office of General Administrator of all mines in Spain played a major role in the promotion of mainland mines (cf. Tejada Hernández 2017). At the time, this position was almost exclusively held by senior mining experts. The General Administrator served as the link between private mining actors and the Spanish Crown and was pivotal in implementing the Crown's strategy for the mining sector. Their main task was to convince actors to invest in precious metal mines, to provide basic technological advice, and make sure that mining laws were respected (Sánchez Gómez 1989, p. 716). The General Administrator also had a team of assistants supporting him in his work (Ortíz 1599). Evidently, the Crown invested quite a large sum of money to promote precious metal mining in peninsula Spain. This is paradoxical: one would expect that in a time when large quantities of American silver were available, no attempt to broaden the silver supply would be necessary. However, on closer examination of the political and financial during the second half of the sixteenth century, it becomes apparent that the Crown was heavily reliant on a constant influx of American silver. The dependence on American silver was a source of much concern for the royal authorities at the time (Table 7.1).

At the beginning of the American silver boom in the mid-sixteenth century, around 12% of the Crown's finances in Spain came from American silver remittances. By the 1580s, this figure had risen to 20% (Thompson 1976, p. 288; Drelichman and Voth 2010, p. 5). As a result, the reliable arrival of "treasure ships" from America in the Spanish port of Seville was of paramount importance for the Spanish

Table 7.1 American silver production 1500–1670 (in pesos)

Decenial silver production (in pesos)	
1561–1570	29,857,103
1571–1580	39,630,395
1581–1590	90,605,672
1591–1600	106,291,974
1601–1610	104,942,794
1611–1620	97,658,111
1621–1630	115,200,563
1631–1640	95,277,153
1641–1650	91,128,127
1651–1660	75,821,958
1661–1670	70,758,447

Source Garner, SpAmSilverOutputex.xls. Colonial Table 1. www.insidemydesk. com. Last accessed December 19, 2018

Crown, and any disruption of the transportation of silver was highly disturbing. Especially during the time of the naval war with England at the end of the sixteenth century, the secure route across the Atlantic was thought to be compromised. It was not, however, hostile corsairs but rather the unpredictability of the weather in the Atlantic shipping lanes that was the determining factor in the decision to renew mining institutions in the Spanish peninsula (cf. Chaunu 1959).

In the summer of 1594, as in any other year, the fleet—filled to the brim with silver from mines all over Spanish America—was supposed to meet in the port of Havana to make its final passage to Seville. That year, however, the ships from New Spain were delayed and did not arrive until mid-August. Unable to make the round trip in time before the harsh winter weather started, it was decided that most of the ships would overwinter in Cuba. While some ships made the dangerous journey across the Atlantic, the precious silver fleet stayed in America and only returned to Spain at the beginning of the following year (De La Fuente 2008, p. 64). The silver was not lost and arrived in Spain in 1595; but in the meantime, the Crown had to compensate for the shortfall of the desperately awaited silver from America. Interruptions in transportation from America were not new. For instance, the first two years of the 1590s seem to have seen small delays in the passage of the fleet from America, but the events of 1594 were the most drastic interruption in the Carrera de Indias at the time. When the Crown noticed the shortfall of the silver fleet, they acted immediately. Within a few weeks, they employed Karl Schedler, a German mining expert closely tied to the Fugger Company, as General Administrator of mines in Spain (Schedler 1594). As one of the most senior and experienced mining experts in Spain at the time, he was instructed to set up a mining sector and convince as many people as possible to invest in Spanish precious metal mines. The Spanish Crown hoped to minimise their reliance on American silver and limit their vulnerability to problems in the transportation of silver across the Atlantic. The concern about reliance on transported silver is best demonstrated by the royal officials who, on several occasions, praised the Spanish mining sector for avoiding the risks that attended precious metals from the Indies (for instance: cf. Junta de Minas 1606; Ayanz 1603).

The looming dangers of corsairs and storms, several losses of ships in the Atlantic and the eventual delay of the silver fleet in 1594 led the Spanish Crown to re-establish a mining sector in Spain. Their hope was that Spanish mines would be serviceable within a short period of time.

The Continued Push to Promote a Fruitless Mining Sector

When the Crown set to promote Spanish mines at the end of the seventeenth century, everybody expected a quick and phenomenal success. The boom of American silver and its influx to Europe also fuelled high hopes for the discovery of extraordinary riches in the Spanish mainland. Thanks to the activities of royal officials, many people were soon engaged in the search for new mines (Fernández De Pinedo Fernández 2007, p. 56). Stories soon began to circulate that people had found pieces of gold or silver and had become wealthy almost overnight (Sánchez Gómez 1989, p. 697). The recent success of the Guadalcanal mine, traces of ancient Roman mining sites in Spain and at times even the Bible were cited as proof that rich metal deposits existed in Spain (cf. Alonso Carranza 1629, p. 14). Some officials were duped into claiming that Spain had richer silver mines than those in America

(Granvela 1584). However, while ships from America made the annual passage across the Atlantic to provide the Crown with much needed silver, the fabled riches in Spanish precious metal mines failed to materialise[1] (cf. Sánchez Gómez 1989).

Although no successful mining endeavour took root, the attempts to promote a Spanish precious metal sector continued into the first years of the seventeenth century. The Crown eagerly sought to engage the most capable mining experts, often themselves originating from America, and to bind their knowledge into the Spanish mining sector. The most telling example is that of Don Córdoba Canales, one of the most senior experts, who was prevented from returning to Peru. Instead, the royal officials granted him generous funding and asked Canales to work the once rich silver mines in Guadalcanal. The exclusive licence to work the mines in Guadalcanal was a special permission by the Crown and only a very few people were granted access there. The contract for his mining endeavour lasted six years and Córdoba Canales received an additional credit of 10,000 ducats from the *Consejo de Hacienda*, the financial council in Spain. The contract specifically mentioned that, due to the high costs and low output in the first years of his contract, Juan Córdoba Canales only had to pay the *diezmo*—a tenth of the silver output as taxes. He was also granted access to the mine and given materials which he was supposed to return at the end of the contract. Most remarkable however was the allowance from the Crown that he would be provided with as much mercury as he needed, which he could buy at market price (González 1831, pp. 642–648). This guarantee was issued at a time of serious mercury shortage in America. The Crown was particularly worried about access to mercury in America for the refinement of silver at the time and usually sent as much Spanish mercury as possible to America. Nevertheless, the precious metal mining sector in Spain was considered to be of such importance that mercury was provided for Spanish mining even in times of mercury crisis in America.

---

[1]At the end of the sixteenth century and in the first half of the seventeenth century no long-lasting precious metal mine endeavour existed in Spain. Royal officials on several occasion also complained about many disappointments in precious metal mines and the high costs of the mining sector. For instance, cf. Crown (1623).

Don Córdoba Canales was the ideal person to receive such an allowance. His family was directly linked to Antonio Mosén Boteller, a collaborator of Bartholomé de Medina and one of the earliest experts in the amalgamation process in New Spain. Without a doubt, knowledge of the amalgamation technique was passed on in his family and Don Córdoba Canales himself was also the holder of a patent for a "secret technique" to refine silver ore with mercury which he planned to employ in the mines of Guadalcanal (Sánchez Gómez et al. 1997, p. 99). The Canales endeavour was provided with mercury via the Fugger Company in Almadén as well as other ingredients, such as salt, which were necessary for his refinement process. However, like almost any other precious metal mining business in Spain at the time, Don Córdoba Canales's endeavours failed to turn a profit. Within a few years, Canales ran out of money and had to cease his mining efforts. Unable to repay the money he had been granted, Córdoba Canales had no choice but to flee and go underground (Molina 1607). The once-celebrated mining expert never recovered, and soon afterward died a poor man.

While Spain's precious metal mines were unable to make a profit, the Spanish mining institutions soon turned out to serve another important purpose. At the beginning of the seventeenth century, the Spanish Crown was informed about problems with the refinement of a special sulphurous ore composition called *negrillos* in the mines of Potosí in Peru. The officials in Peru reported that the mines of Potosí were on the verge of depletion because the Peruvian miners could not come up with a technical solution. The Crown assigned the job to their mining institutions and the task was given to the General Administrator of mines, Jerónimo Ayanz. The inability to refine *negrillo* ores in Potosí was perceived as a serious risk: the most valuable mining sector in the Spanish Empire was on the verge of depletion (García Tapia 2010, pp. 147–149).

Metal samples from the mines in Potosí were sent to Spain and Ayanz was ordered to conduct experiments to devise a solution for the Peruvian mines. In addition, a number of experts in America, Spain and Central Europe were contacted to assist the General Administrator in this matter. To engage other miners in the quest, the Crown promised a reward to experts who could come up with a solution. For more than two years, Ayanz and his officials worked on the *negrillo* problem in Spain (García Tapia 2010, pp. 147–148). Nevertheless, the fear remained that no solution could be found to the problems in Peru and the Crown worried that it was about to lose the most precious silver mine in the Spanish Empire. In an effort to counteract the technological problems, the Crown specifically asked their mining administration and in particular General Administrator Ayanz to continue seeking new mines in Spain. Local officials were asked to send rock samples from potential precious metal mines to the Court to be investigated by the General Administrator and his team (Ayanz 1603). Thus, in response to critical technological problems, the Spanish Crown chose to continue promoting the precious metal sector in Spain.

The technological and logistical aspects of American silver production and the perceived risks of reliance on American silver played a considerable part in the formation of the Crown's strategy regarding Spanish precious metal mines. However, there was also a third factor driving the promotion of mining institutions at the beginning of the seventeenth century: the reliable distribution of mercury to America. Ever since the invention of the amalgamation process, mercury was a critical ingredient for the refinement process. With the rise of American silver production, the required quantity of mercury also rose drastically, and a reliable supply of mercury came to be of the highest importance.

In early modern times, the Spanish monarchy relied on two main mercury mines to supply the silver mines. The Huancavelica mine in Peru was, at the time, the most important producer of mercury. Due to its location in Peru, it had the special responsibility of supplying mercury to Potosi, the richest silver mine of the time. In addition, mercury from Huancavelica could be delivered to New Spain as needed. The second important mercury mine was located in Almadén in Spain. However, during the sixteenth and seventeenth centuries, Almadén played a relatively minor role in the supply and distribution of mercury; it only delivered mercury to New Spain (cf. Lacueva Muñoz 2010, p. 72). The mining sector in New Spain had a subordinate role in the production of silver at the time and mines there required less mercury, as the local silver ores contained more lead. As a result, simple smelting refinement prevailed throughout the sixteenth and seventeenth centuries in New Spain.

Things changed when grave problems surfaced in the mines of Huancavelica. At the turn of the seventeenth century, the death rate among the (indigenous) workforce was unusually high. The death rate stemmed from poor labour conditions and close contact with toxic mercury and caused a drastic decline in mercury output at Huancavelica (Lohmann Villena 1949, p. 195). The Crown in Spain ordered the viceroy of Peru to abolish the *mita*, the indigenous draft labour system, in favour of a combination of wage and convict labour. In addition, the Crown prohibited shaft mining in order to reduce exposure to mercury; only open pit mining was allowed. Nevertheless, the mercury production in the mines of Huancavelica was seriously affected. In 1604, less than 900 quintals was produced in Huancavelica. In the following five years, the annual quota of 5000 quintals also fell seriously short. While in 1605 the mine produced around 3000 quintals of mercury, a mere 1700 was registered in 1607 and the warehouses, once filled with mercury, were almost completely depleted (Lohmann Villena 1949, pp. 195–204).

Once again, royal officials on both sides of the Atlantic were quick to act. By 1605 a massive exchange of letters regarding the situation at Huancavelica had taken place between the Crown and the viceroy in Peru and New Spain. The Crown tried to support the officials in Peru in overcoming the problems: in 1608, a team of experts from the mercury mine of Almadén in Spain went to Huancavelica to support the miners in Peru (Lohmann Villena 1949, p. 194).

At the same time, immediate measures were taken in peninsular Spain to overcome the problems of mercury production. Mercury from Almadén was identified as one way to solve the difficulties. Even though the mine was leased to the Fugger company, the Crown sent royal mining experts to Almadén to raise the mercury output (Crown 1606). The Crown asked the Fuggers to collaborate in this matter and offered them additional remuneration for their assistance.

In addition, experts were encouraged to contribute to the technological advancement of the mercury mining sector. The first two decades of the seventeenth century—the time of the mercury crises in America—saw a number of inventions for mercury mines and a variety of improvements in the refinement of silver without the use of mercury (Sánchez Gómez et al. 1997, p. 237).

After the re-establishment of efforts to promote precious metal mining in Spain from 1594, the Crown supported the Spanish mining sector for more than half a century. The only break occurred during the second decade of the seventeenth century, when the effort focused on the precious metal industry in Spain declined. This decline coincided with the time of the Twelve Years' Truce when hostilities with the Netherlands were put on hold. In that case, the connection between the Spanish mining sector and the risks of American silver similarly point to the entwined nature of war efforts and the silver industry in early modern times (cf. Stein and Stein 2000).

Attempts to promote a Spanish precious metal sector finally came to an end during the 1640s, accompanied by two major developments: first, Spain came under heavy political pressure during the 1630s. In 1635, France entered the Thirty Years' War on the side of the Netherlands against Spain. In 1640, an uprising in Portugal opened the way for Portugal to break away from Spanish rule, and in the same year, a revolt broke out in Catalunya. Those pressing political problems drew attention away from internal circumstances at a time of many changes for the Spanish mining sector.[2] Unsurprisingly, it was at the same time that the long relationship with the German Fugger Company, who had held several important mining concessions in Spain, came to an end.

The political pressure at the time coincided with another important development: American silver production began to decline during the mid-1630s. While in the first decade of the seventeenth century almost 30 million pesos were shipped to Spain, in the decade after 1641 this number almost halved to around 18 million (Garner 2007). The decline of remittances had begun at the beginning of the seventeenth century, but it was not until the end of the 1630s that silver production in America was invariably falling and would only recover in the first half of the eighteenth century. The rich silver

---

[2]In 1643 the King's *valido*, the royal favorite of Philip IV was overthrown, and consequently, the Junta system was altered. The *Junta de Minas* was revised and the main strategy in the mining sector shifted to the promotion of more useful metals such as lead or sulphur.

deposits in America had an exceptional allure for the Spanish monarchy and the situation in American silver mines seemed to have a direct connection with the peninsular Spanish precious metal mining sector. The Crown followed a pro-cyclical approach: in times of access to large quantities of American silver, a Spanish mining sector appeared and received significant support; when silver remittances from America dried up, the efforts to promote silver mines in Spain declined accordingly.

The Mining Sector as a Forerunner to Progress in Spain

The Spanish monarchy invested in the precious metal sector in Spain in an effort to counteract the risks inherent in its dependence on American silver. Despite strenuous efforts, the Spanish mines never revealed rich ore deposits. Nevertheless, the promotion of Iberian mines had profound implications for Spain at the time.

Most importantly, Spanish mining efforts led to major technological advancements throughout the late sixteenth and early seventeenth centuries in Spain and possibly beyond. Historiography regarding the Spanish mining sector has already described a number of technological advancements during the early modern period as many new technologies were introduced to Spanish mines, especially during the sixteenth century (cf. Matilla Tascón 1958; cf. Sánchez Gómez 1989). A particularly strong indicator for technological progress is the emergence of a patent system (cf. Davids 2011) and the sheer number of registrations for new patents. Recent research regarding early modern Spain has therefore tended to focus on the second half of the sixteenth century as a time of technological progress in the Spanish mining sector. On the other hand, very little advancement has been reported in the seventeenth century (Garcia Tapia 1994; Mateos Royo 2009). This view of Spanish development falls in line with the concept of *decadencia*, the overall decline of Spain during the seventeenth century.

However, archival accounts from the *Junta de Mina*, the highest mining body in Spain, suggest a somewhat different picture.[3] Initially, the *Consejo de Hacienda* was in charge of awarding patents, but from 1624 on the *Junta de Minas* became a central institution for the handling of patent applications. According to those accounts the *Junta de Minas* awarded 26 patents between 1624 and 1650.[4] The majority of those patents regarded the refinement of (silver) ores, mills, and the drainage of mines. In order to make the Spanish precious metal sector profitable, the Crown was interested in strengthening the technology available in Spain. Patents were a means of diffusing inventions as well as improving the technological standards in mines. Before an invention was approved, the new procedure or technique was scrutinised by the officials of the *Junta de Minas*. The invention was presented to the commission and examined for its operational capability, uniqueness and benefit. Inventors often had to convince the royal officials of the utility and necessity of their discovery. Upon receiving a patent, inventors were able to monetise and spread their invention. While the actual impact of new technologies is often difficult to determine due to the private character of precious metal mines, there are indications that patents improved some mining sites. In 1644, a patent which improved the refinement of lead was awarded to Diego Felipe de Cuadros (Patent 1644). This improvement was directly applied to the royal lead mines in Linares and had a major impact on the operation of the mine. Thanks to this invention, lead could be produced in much greater quantities than before (Pretel 1646).

The technological improvements in early modern Spanish mining were a direct consequence of the reliance upon and risks of American silver. However, America often also played a pivotal role in the advancement of the Spanish mining sector. A steady flow of mining experts from America came to Spain to engage in the Spanish mining sector

---

[3]Archivo General de Simancas, Contadurías Generales, 852.

[4]This would seem to be a substantial number of patents; Mateos Royo and José Antonio have previously identified only 14 patents aimed at the mining sector in Spain for the period between 1522 and 1622 (Mateos Royo 2009, p. 153).

(Sánchez Gómez et al. 1997, p. 202). Due to their knowledge and the experience, they had gained in America, they often received important positions in the mining administration or specific mining concessions from the Crown.

In addition to the influx of American experts, the spatial hurdles of the Atlantic had a similarly significant impact on the technological and institutional landscape in Spain. As with the case of General Administrator Ayanz and the *negrillos* in Potosi, inventions often had to come across the Atlantic. Because such shipments were costly and took a great deal of time, procedures for examining new technologies before their actual application had to be developed and optimised (Ayanz 1603).

The most notable aspect of the technological developments in the early modern Spanish mining industry was the fact that mining institutions also drove technological developments in other branches of industry. Especially after 1624 the *Junta de Minas*, the highest body in the mining sector, also approved inventions including grain mills, ovens for the refinement of sugar and new techniques in the refinement of sugar canes.[5] The exact circumstances of this development still require more research, but these findings suggest that the *Junta de Minas* might have taken a lead role in the overall evaluation of new techniques, at least in the case of patents for industries related to mining.[6] What is clear, however, is the fact that technological progress in the mining sector spilled over into other branches of industry in early modern Spain.

---

[5]About 20% of all patents found in the accounts of the *Junta de Minas* were intended for inventions beyond the mining sector.

[6]One approach to explain these spillover effects might be the versatility of the early modern mining sector. The mining sector required large machinery such as (stamp) mills, wheels and ovens as well as smaller tools and machines such as pumps, carts and winches. Machines in the mining sector also featured a variety of materials such as wood, stone and even masonry. As a consequence, it is quite likely that the *Junta de Minas* was comprised of experts with an ample knowledge of different skills and techniques. As the mining sector was one of the most central sectors of industry, at the time it might have become the main place of contact for a many inventors of different sectors at the time.

Technological advancements in the Spanish motherland were driven by developments in the Americas. Mining experts from America brought their knowledge back to Spain. Most importantly, the risks inherent in Spanish dependence on American silver drove the initial promotion of mining institutions. Those institutions, in return, powered advancements in industries well beyond the mining sector.

How Spanish Mining Entangled the World

This investigation argues that Spanish mining efforts led to technological as well as institutional developments, but the risks of American silver also had another profound consequence. The desire to minimise the risks of reliance on American silver led to the entanglement of actors across the Atlantic as well as in Central Europe. Mining experts from America frequently travelled across the Atlantic to engage in Spanish mining activities. As the example of Córdoba Canales shows, American experts sometimes even came to Central Europe to introduce new technologies or to work in Central European mines. In addition, many experts from Germany travelled to Spain and even tried to travel on to America to engage in the rich overseas mining sites (cf. Ghilarducci 2011).

Besides the entanglement of actors such as mining experts, there was also an economic entanglement between America and Europe. An important example is visible in the crises that struck American silver mines at the beginning of the seventeenth century. As well as sending teams of experts to American mercury mines and providing additional mercury supplies from Almadén in Spain, the Spanish Crown sought to import mercury from outside of the Spanish Empire in an effort to cover its mercury shortages. One attempt was to import mercury from the mine in Idrija in Carniola (today's Slovenia). Via the Habsburg connection, Spanish officials contacted officials in Idrija and entered negotiations to import mercury from Central Europe to America (Valentinitsch 1981). While the initial negotiations proved unsuccessful, an agreement was reached in 1620, allowing for the import of 16,000 quintals from Idrija to Seville over the following four years, to be shipped on to America.

The curious entanglement between America, Spain and Central Europe at the beginning of the seventeenth century is also directly reflected in the institutional setting of the *Junta de Minas*, the most important mining body in early modern Spain. Historically, *Consejo de Hacienda*, the Spanish financial council, had always administered Spanish mining. At the beginning of the seventeenth century, however, the affiliation of the *Junta de Minas* changed. To the annoyance of the President of the *Consejo de Hacienda*, Juan de Acuña, I. marqués de Vallecerrato, oversight of the *Junta de Minas* was granted to the *Consejo de Indias* (Acuña 1602). The Spanish mining sector was now directly administered by the highest administrative organ in charge of the Americas and the Philippines. The reasons for this shift are probably twofold. On the one hand, it connected the American and Spanish mining sectors and allowed them to react more efficiently to developments in both regions, a strategy which, again, points to Spanish mining ambitions as a way to ameliorate the risks posed by dependence on the American silver sector. Secondly, the official name of the *Junta* was *Junta de Minas y de Carintia*. The addition "de Carintia" is a clear reference to the mercury mine in Idrija and (indirectly) to the shortfalls in the supply of mercury for the American mining sector. While the mine in Idrija was in fact located in the Duchy of Carniola, the royal officials referred to the area as "Carintia", which was in fact a more northerly region of Carniola.

The entanglement of mining sectors not only incorporated America, Spain and Central Europe but reached as far as China. Reports from the Junta de Minas show that royal officials did not only look at mercury mines in Central Europe, but also contemplated importing mercury from mines in China to solve the crises in America (Bakewell 1997). Pedro de Baeça de Silveira, a senior merchant in the Portuguese *carreira* trade, was commissioned to import mercury from Canton, China to America via the Philippines. The contract with Pedro de Baeça, signed in 1606, provided for the haulage of 16,000 quintals of mercury within four years (De Baeza 1607). While there are, unfortunately, no indications that any Chinese mercury actually arrived in America, the example of the Junta de Minas shows that the American mining sector entangled a wide range of actors, commodities and institutions between Spain, America, Central Europe and China.

Summary and Discussion: America as a Trigger for the Advancement of Europe

At the height of the (first) American silver boom in the late sixteenth and early seventeenth centuries, the Spanish Crown promoted a precious metal mining sector in Spain. The most able mining experts, often originating from America, were employed by the Spanish Crown to search for and promote Spanish precious metal mines. These efforts met with little success and Spanish mining endeavours turned out to be largely fruitless and loss-making, but the royal authorities continued to subsidise the precious metal sector and promote mining institutions for almost half a century. While historiography has given valuable insights into the technological and administrative landscape of the early modern Spanish mining sector, it has not sufficiently addressed the reasons for this paradoxical strategy.

This article demonstrates that the Spanish Crown maintained and promoted a precious metal sector in Spain as a response to the influx of American silver, having recognised its reliance on the American mining sector and the risks stemming from its dependence. In an effort to ameliorate the risk of losing this important but unreliable financial lifeline, the Crown pursued a strategy of diversification, investing in a loss-making mining sector in the Spanish mainland.

The initial trigger for the promotion of mining institutions in Iberia was the insecurity inherent in transporting silver across the Atlantic, demonstrated in 1594 when the silver fleet was unable to make its return to Seville due to bad weather. The Crown—already under financial strain due to its expensive military endeavours—did not receive any remittances from America, resulting in a budget shortfall of a quarter. Painfully aware of its weakness, the Crown immediately chose to re-establish the peninsular Spanish mining administration. The newly appointed General Administrator, together with a team of officials, aimed to discover precious metal mines and prompt people to work them. At the same time, technological deficiencies and problems in American mines created and exacerbated existing risks. One example was the case of a special ore composition in Peru called *negrillos*. The local miners were unable to refine those ores and rumours began to spread that the most valuable mine in the Spanish Empire was about to be depleted. In response, the royal mining experts in Spain searched for a technological solution and the effort to promote Spanish mainland mines continued in the hopes of swiftly offsetting the shortfall. Besides the technological and logistical risks of American silver, the supply of mercury was also understood to be problematic. After the discovery of the amalgamation process, the distribution of mercury to silver mines in America gained great importance. However, when problems hampered production in the primary American mercury mine at Huancavelica, the Spanish Crown once again made use of their mining institutions in Spain to improve the mercury output in the Spanish mine of Almadén as well as contemplating the import of mercury from Central Europe and even China.

The promotion of Spanish precious metals lasted until the 1640s when Spain was hampered by political chaos and the depression that had settled into silver production in America.

While the Spanish Crown was unable to turn a profit from the Spanish mining sector, the strategy to promote a mining sector nevertheless had far-reaching impacts on the technological and institutional landscape of the Spanish Empire. Firstly, Spanish mining ambitions drove technological advancement. Many mining experts from America and Central Europe were attracted by the activities in the Spanish mining sector. Further, to overcome the unprofitability of the Spanish mining sector, the Crown sought mining experts to raise the technological profile of peninsular Spain. Evidence suggests that these attempts directly translated to mining sites.

The technological advancement of the mining sector led to spillover effects, bringing technological improvements to other branches of industry in Spain at the time. The spatial hurdle of the Atlantic, also, demanded a professionalisation of mining institutions and actors. In addition, the Spanish mining institutions and the demands of the nascent mainland Spanish mining industry brought about an entanglement of actors from both sides of the Atlantic and linked together far-flung mining areas.

America was at the heart of these developments. The role of America in early modern times went far beyond that of a mere supplier of commodities. Thanks to its rich silver deposits, America drove the

technological and institutional developments in the Spanish motherland. This research gives a first glimpse into the tangible impact of
America on the development of Spain. Further research will have to
address whether these effects had lasting implications for Spain in early
modern times and perhaps even beyond. As the influx of American silver was not only limited to Spain but reached into Central Europe and
Asia, it will have to be established whether America also had similar
impacts on regions beyond Spain.

Bibliography

Primary Sources

Archival and Manuscript Sources
Acuña, Juan de, 1602. Letter to the Crown, November 18. Archivo General de
    Simancas (AGS), Consejos y Juntas de Hacienda (CJH), 441-8-12.
Ayanz, Gerónimo de, 1603. Response regarding *negrillos* in Potosí, April 28.
    AGS, Contadurías Generales (CCG), 854.
Crown, 1603. Order to Córdoba Canales, May 5. Archivo General de
    Simancas (AGS), Patronato, LEG. 87, DOC. 97.
Crown, 1606. Letter to the Marques de Montesclaro, June 28. Archivo
    General de Indias, INDIFERENTE, 541, L.1M.
Crown, 1623. Letter to Markus Fugger, November 3. AGS, CJH, 612-1-8-2.
Granvela, Antonio de, 1584. Letter to Christoual de Salazar, September 18.
    AGS, Estado, 1531-28.
Junta de Minas, 1606. Letter to the Crown, December 5. AGS, CJH, 477.
Molina, Melchior de, 1607. Letter to the Crown, May 17. AGS, CCG, 854.
Ortíz de Rodríguez, Pedro, 1599. Letter to the Crown, August 20. AGS, CJH,
    388-3.
Patent, 1644. Patent Awarded to Diego Felipe de Cuadros, December 21.
    AGS, CCG, 852, 48.
Pretel, Andrés, 1646. Record from About the Lead Mine in Linares on June
    12. AGS, CCG, 852, 13.
Schedler, Karl, 1594 Appointment as Administrator General, December 31.
    AGS, CCG, 850.

Source Editions and Old Prints

De Baeza, Pedro, 1607. *Traslado del memorial que se hizo con el licenciado don Francisco de Tejada, oydor del Consejo Real de las Indias: Para tratar el assiento del açoge con Pedro de Baeça.* Madrid: [s.d.].

González, Tomás, 1831. *Noticia histórica documentada de las célebres minas de Guadalcanal. Desde su descubrimiento en el año de 1555, hasta que dejaron de labrarse por cuenta de la real hacienda,* vol. 2. Madrid: Miguel de Burgos.

Von Born, Ignaz, 1786. *Über das anquicken der gold- und silberhältigen Erze, Rohsteine, Schwarzkupfer und Hüttenspeise.* Wien: Wappler.

Secondary Sources

Bakewell, Peter J., 1997. *Mines of Silver and Gold in the Americas.* Brookfield: Aldershot.

Carranza, Alonso, 1629. *El Ajustamiento y Proporcion de las Monedas de Oro, Plata y Cobre, y la Reduccion destos metalles a su debida estimación.* Madrid: Francisco Martínez.

Chaunu, Pierre, 1959. *Seville et l'Atlantique (1504–1650).* Paris: S.E.V.P.E.N.

Davids, Karel, 2011. "Dutch and Spanish Global Networks of Knowledge in the Early Modern Period: Structures, Connections, Changes." In: *Centres and Cycles of Accumulation in and Around the Netherlands During the Early Modern Period,* edited by Lissa Roberts. Berlin: Lit: 29–51.

De La Fuente, Alejandro, 2008. *Havana and the Atlantic in the Sixteenth Century.* Chapel Hill: University of North Carolina Press.

Drelichman, Mauricio; Voth, Hans-Joachim, 2010. "The Sustainable Debts of Philip II: A Reconstruction of Castile's Fiscal Position, 1566–1596." In: *The Journal of Economic History,* 70: 813–842.

Fernández de Pinedo Fernández, Emiliano, 2007. "Antecedentes de la Mineria Espanola contemporánea: La minería en la Corona de Castilla (1515–1715)." In: *Minería y desarrollo económico en España,* edited by Miguel Ángel López Morell, Miguel A. Pérez de Perceval Verde, and Alejandro Sánchez Rodríguez. Madrid: Síntesis: 47–68.

García Tapia, Nicolás, 1994. *Patentes de invención españolas en el siglo de oro.* Madrid: Ministro de Industria y Energia.

García Tapia, Nicolás, 2010. *Un inventor navarro, Jerónimo de Ayanz y Beaumont (1553–1613).* Pamplona: Universidad Pública de Navarra.

Garner, Richard, 2007. http://www.insidemydesk.com/hdd.html. Accessed September 19, 2018.

Ghilarducci, Eleonora, 2011. "Las composiciones de extranjeros en la Nueva España, 1595–1700." In: *Cuadernos de Historia Moderna. Anejos 10. Los Extranjeros y la Nación en España y la América española*: 177–193.

Goodman, David C., 1988. *Power and Penury: Government, Technology and Science in Philip's II Spain*. Cambridge: Cambridge University Press.

Lacueva Muñoz, Jaime J., 2010. *La plata del rey y sus vasallos. Minería y metalurgia en México (Siglos XVI y XVII)*. Sevilla: Escuela de Estudios Hispano-Americanos.

Lohmann Villena, Guillermo, 1949. *Las minas de huancavelica en los siglos XVI y XVII*. Sevilla: Escuela de Estudios Hispano-Americanos de Sevilla.

Mateos Royo, José Antonio, 2009. "State Policy, Institutional Framework and Technical Monopoly in Early Modern Spain: Invention Patents in the Crown of Aragon During the Seventeenth Century." In: *History and Technology*, 25, 2: 147–162.

Matilla Tascón, Antonio, 1958. *Historia de las Minas de Almadén*. Madrid: Consejo de Administración de las Minas de Almadén.

Naharro Quirós, Elena, 1991. "La búsqueda de metales preciosos y la ordenación legal de la minería peninsular en el reinado de Felipe II." In: *Anuario de la historia del derecho español*, 61: 165–203.

Sánchez Gómez, Julio, 1989. *De Mineria, metalurgia y comercio de metales. La minería no férrica en el reino de Castilla. 1450–1610*. Salamanca: Ediciones Universidad de Salamanca.

Sánchez Gómez, Julio, et al., 1997. *La savia del imperio. Tres estudios de economía colonial*. Salamanca: Ediciones Universidad de Salamanca.

Stein, Stanley J.; Stein, Barbara H., 2000. *Silver, Trade, and War: Spain and America in the Making of Early Modern Europe*. Baltimore: Johns Hopkins University Press.

Tejada Hernández, Francisco José, 2017. *El derecho minero romano ante la ilustración hispanoamericana*. Madrid: Dykinson.

Thompson, I. A. A., 1976. *War and Government in Habsburg Spain, 1560–1620*. London: The Athlone.

Valentinitsch, Helfried, 1981. *Das landesfürstliche Quecksilberbergwerk Idria 1575–1659. Produktion, Technik, rechtliche und soziale Verhältnisse, Betriebsbedarf, Quecksilberhandel*. Graz: Historische Landeskommission für Steiermark.

8

Information and Decision Making: The Logic of Spanish Mining Administration, 1675–1700

Peter Paul Marckhgott-Sanabria

Introduction

In the summer of 1697, Miguel de Unda y Garibay, "knight of the order of Calatrava, and superintendent and General Administrator of the royal factory and mine […] of Almadén" (Unda y Garibay 1697), encountered a problem. This comes as no surprise: dealing with problems related to running Almadén, an enterprise of substantial proportion and, according to King Charles II., the "most precious jewel of [his] monarchy" (Consejo de Hacienda 1682), was essentially his job description. This time, the cause of Unda's predicament was that the number of labourers at his disposal did—once again—not meet the number he deemed necessary to keep operations running in an orderly fashion. Up to several hundred men were employed at the mine at any

P. P. Marckhgott-Sanabria (✉)
Austrian Academy of Sciences, Vienna, Austria

© The Author(s) 2019
R. Pieper et al. (eds.), *Mining, Money and Markets*
in the Early Modern Atlantic, Palgrave Studies in Economic History,
https://doi.org/10.1007/978-3-030-23894-0_8

given moment, most of them contracted or forced unskilled labourers. Among their most important tasks was manning the pumps and windlasses, and keeping the shafts and galleries of the mine drained. It was only when the water that entered the mine through the surrounding rock—and through the mine's accesses—could be kept at a minimum that skilled miners could enter and mine the precious cinnabar, the base for smelting pure mercury. Draining had to be undertaken continuously, sometimes literally day and night, and when for whatever reason it stopped, entering water could destroy in a few weeks the progress of many months. As serious as this sounds, such a lack of labourers was not unusual in late seventeenth-century Spanish mining, when labour was scarce, and it seemed that administrators could never get enough workers for their liking. Reports from throughout the last decades of the seventeenth century frequently complained about a lack of labourers (Fernandez de Portalegre 1694).

In the same summer of 1697, the Consejo de Hacienda, or "Council of the Treasury", received not only Miguel de Unda y Garibay's complaints, but also letters from all over Castile, reporting the departure of forced workers for Almadén. The numbers that appear in those letters are small, generally only a handful of forced workers were sent from one location at any one time. To name just a few examples, Fernando Caniego (1697) reported the acquisition of seven slaves and two convicts for Almadén from Cordoba, while the Marquis de Valbermoso (1697) wrote from Seville that three "gypsies" and one Turk had arrived at the mine and the Marquis de San Vicente (1697) confirmed the delivery of "seven black slaves".

It seems astonishing that a relatively trivial matter such as the movement of forced workers from various towns to a mine, an action that took place frequently, probably several times a month, prompted such a large number of letters, written by various actors. In some of the cases, not one but several letters were delivered, referring to the acquisition, departure and arrival of the same group of forced workers. All of this correspondence was not dealt with by Almadén's administrator, a competent public servant, but passed through the hands of the Consejo de Hacienda, the highest authority of all of Spain's fiscal administration, which supposedly had better things to do than be concerned with such subordinate matters.

If one believes the half-joking remarks of historians conducting research in Spanish archives, the reason is clear: according to "common knowledge" among researchers, early modern Spain was a monarchy made of paper, where reports about every conceivable topic were written by members of the administrative apparatus, but those reports served more than anything as a proof that their authors were dedicated civil servants doing a good job, and thus deserving a promotion or at least a pay raise. The aim of this chapter is to show that, on the contrary, in the late seventeenth century, the Spanish Consejo de Hacienda gathered information on mining and mining-related topics because it needed to understand the mining sector in order to better influence its development.

The administrative aspects of Spanish mining history, and especially the role higher echelons of Spain's administrative apparatus played within the mining sector, have not been very extensively covered by historiography. This is because they lie in a "blind spot" between two fields of research. Matilla Tascón's (1987) and Sánchez Gómez' (1989) books are still among the most important monographs on Spanish mining, each covering a wide array of topics. More recently, others like Gil Bautista (2015) have added to this type of historiography. Of special interest are topics like working conditions, especially the fate of the forced workers present at several mines (Hernandez Sobrino 2010), or mining technology (Sánchez Gómez and Mira Delli-Zotti 2000). And although mines' administrators, whose central position within a mine's personnel made them very visible in the sources, have received some scrutiny since the time of Matilla Tascón, the role of higher-level administrative bodies like the Consejo de Hacienda is hardly within the scope of mining historians.

On the other hand, the Consejo de Hacienda and its activities are at the centre of attention for scholars studying early modern Spanish fiscal administration. In the last decades, the transitional period from Habsburg to Bourbon rule in Spain has been studied by researchers like Sánchez Belén (1996), readjusting the traditional image of the final phase of Habsburg rule in Spain as a period of decay. However, research in this area is focused on the Consejo de Hacienda's core responsibilities, which included the collection of taxes and other contributions, but not the management of public enterprises like mines.

This chapter will show that the establishment of a knowledge base on the state of Spanish mines run by the Consejo de Hacienda was a crucial element in enabling closer control over publicly owned mines, as demanded by Charles II.[1] In order to achieve this, it used an array of complementary instruments for acquiring mining-related knowledge. To demonstrate the importance of the highest echelons of Spanish fiscal administration for Spanish mining, correspondence between the Consejo de Hacienda and the other actors within Spain's mining sector, particularly administrators of mines administered by the royal treasury, as well as correspondence between fiscal administration and the monarch, has been analysed. The results are presented in three sections. In the first section, the relationship between the Consejo de Hacienda and the Spanish mining sector is scrutinised. The close involvement of the Consejo in mining matters, prevalent at the end of the seventeenth century, resulted from a series of contingencies, which need to be addressed to in order to properly assess its role in late seventeenth mining administration. The second section focuses on the reasons why the Consejo de Hacienda increased its control over select Spanish mines during the last decades of the seventeenth century. Finally, the third section addresses the instruments that were available in order to gather information on the mines controlled by the treasury.

The Consejo de Hacienda and Mining

As part of early modern Spain's system of councils each of which was responsible for administering a certain aspect of the Spanish monarchy, the Consejo de Hacienda's task was overseeing the Castilian economy. According to its *ordenanzas*, which since the sixteenth century provided the legal basis of the Consejo de Hacienda's work, the organisation was tasked with deciding all matters related to the royal treasury (Garzón Pareja 1984).

---

[1]In the light of Charles II supposed inability to rule on his own, and dependence on advisors, whenever the terms "Charles II" or "monarch" are used, they are to be understood more as a reference to the office and not the person.

In practice, the Consejo de Hacienda's work mostly involved controlling the collection of taxes and other duties. To this end, it employed a force of expert tax collectors. However, in theory, every economic activity in Castile was ultimately part of the Consejo de Hacienda's responsibilities (Bermejo Cabrero 2016).

Among the many sectors of economy in which the Consejo de Hacienda was involved, the mining sector was one of the most important. Although according to the mining regal, as part of the medieval concept of *jura regalia*, mining as well as the commerce of mining products was a privilege reserved for the monarch alone, this did not necessarily mean that the crown or its institutions managed all mines directly. In principle, there were three ways of exercising control over mines: mines could be run directly by the royal treasury; be rented out to privileged businesspeople, with the Spanish term *asiento* referring to both the legal institute and the contract; or they could be left for private entrepreneurs to run without much public involvement, except for the granting of mining licences and the collection of duties.

The division between direct control, renting out and private economy was not limited to the mining sector, but reflects the general constitution of European early modern economy, which relied heavily on the practice of renting out royal privileges. It is therefore interesting that in different mining regions, different methods of managing mines were preferred. While in Central Europe, princely administrations regularly acted as mining entrepreneurs, as for example, A. Westermann and E. Westermann (2009) have shown, Hispano-American mines tended to be owned by private investors in the overwhelming number of cases (Brown 2012). In contrast, the Spanish mining sector followed the middle path, and mines were usually rented out. At the end of the seventeenth century, however, the mercury mines of Almadén and the silver mines of Guadalcanal were under the direct administration of the Consejo de Hacienda. During the heyday of Spanish mining in the late sixteenth and early seventeenth centuries, both Almadén and Guadalcanal, like the other Spanish mines, were rented out. The process that led to the exceptional situation of their direct control was not a planned one. Rather, it was the result of circumstance—the demise of the Fugger trading company—and of the importance of those mines for the American mining sector. As an understanding of how and why the mines came under direct control of the Consejo de Hacienda is crucial for the analysis of its consequences, both mines shall briefly be treated individually.

Almadén

The mercury mines of Almadén, located west of Ciudad Real in what is today Castilla-La Mancha must be considered the core of the early modern Spanish mining sector. By providing the mercury desperately sought after by Spanish American silver miners, they formed the basis for the production of American silver, which not only constituted the foundation of the Spanish Empire's economy, but also, via the taxes the miners had to pay, a major source of income for the Spanish crown (Pieper 2009).

Even though, in the jurisdiction of the crown of Castile, the production of mercury had been a royal privilege since medieval times, this fact does not necessarily suggest the extent to which the Almadén mercury mines were controlled by the institutions of Castile's royal treasury. In fact, the history of the Almadén mines provides a prime example of how, over the course of centuries, control over Spanish mines could shift several times between different forms of administration.

The tradition of renting out the working of Almadén was established before the mines were formally controlled by the Castilian crown, when, after the conquest of New Castile by the Christian states, the mines were put under the control of the Order of Calatrava. The shift in formal control from the Order to the monarch took place in the last decades of the fifteenth century, when during the reign of Fernando and Isabella, the Order's properties were taken over by the crown. In the early sixteenth century, the mines of Almadén were for the first time managed directly by the royal treasury, albeit only for a short period. According to Matilla Tascón (2005), the great chronicler of the history of Almadén, this first period of "regal administration" lasted from about 1500 to 1511. During this time, a number of royal officials managed the mines, rendering accounts to the institutions of the royal treasury.

After this first foray into direct management of Almadén by the royal treasury, the more traditional practice of renting out the mines was quickly resumed. Among the merchants who rented the mines in the following decades, the best-known are members of the Fugger as well as the Welser families. In the early sixteenth century both of these Upper German merchant families became heavily involved in financing the Spanish crown and sought to benefit from the Spanish expansion into the Americas and the newfound deposits of precious metals (Denzer 2005).

While Bartholomew Welser's leasehold of Almadén from 1533 to 1537 seems, according to Matilla Tascón (2005), to have been the only attempt by the Welser family to run the mines, the Fugger Family would be closely linked to Almadén for more than a century. From 1524 to 1645, the Fugger family were given several leaseholds over the mines. During this period, the mines benefitted substantially from the first Spanish American mining boom which took place in the second half of the sixteenth century, and, because of the newly introduced amalgamation process, dramatically increased American demand for mercury (Arduz Eguía 2000).

The first "golden period" of Almadén, linked both to the Fuggers' leasehold and the American mining boom, saw massive expansions of the mines, only briefly interrupted by disasters like the fire of 1550 (Gil Bautista 2015). It lasted until the mid-seventeenth century when the Fugger family started to experience severe economic problems which led to their bankruptcy, and, at the same time, fell out of the Spanish crown's favour. Attempts to secure a new leasehold failed, and the Fuggers had to leave the mine. Their trusted administrator, Mateo Nagelin, managed to hold out for another year or so, but was eventually replaced as well (Matilla Tascón 1987).

At this moment, the royal treasury took over the management of Almadén. While it was the second time the mine was subject to "regal administration" since it had officially been transferred to the crown's assets around 1500, it was the first time that the royal treasury, or to be more precise the Consejo de Hacienda, was to run the mines for a considerable period of time. While the first attempt to manage the mines directly through the institutions of the royal treasury—in a time of transition, when the mines had only recently become royal assets—this time control by the royal treasury was set to last. In fact, this form of managing the Almadén mines would outlast Habsburg rule in Spain and continue into the Bourbon period.

Guadalcanal

The second major mining centre, which during the last decades of the seventeenth century was at least temporarily managed by the Consejo de Hacienda, was the mines of Guadalcanal. Located about 100 kilometres north of Seville in the Sierra Norte mountain range, it contained considerable silver deposits. At the end of the seventeenth century, Guadalcanal and Almadén were remarkably unequal, to an extent that—at first sight—it might seem absurd even to compare the two mines. While Almadén was, despite the problems the mine undoubtedly faced during the seventeenth century, always an important mining centre whose economic relevance transcended Spain's borders, Guadalcanal was the opposite. During the seventeenth century, Guadalcanal was a struggling, barely productive money sink well past its prime.

Guadalcanal was not always in the deplorable state it was in the 1680s, when Spanish authorities decided to make an effort to revive the mine and make it profitable again. In fact, it had experienced a short boom in the late sixteenth and early seventeenth centuries, when it produced—by European standards, although not in comparison with the mines in Peru and New Spain—considerable amounts of silver. Sánchez Gómez (1989) portrays this first period of working the Guadalcanal mines. According to him, the way in which the mine was administered in its beginnings was the same way that Almadén was administered at the end of the seventeenth century: at the expense of the royal treasury, by the Consejo de Hacienda and officials appointed by it. So, while Sánchez Gómez claims that one of the major challenges in opening Guadalcanal was that mines run by the royal treasury were unheard of, in the case of Almadén the project of a mine managed by the Consejo de Hacienda, as put into practice from the middle of the seventeenth century onward, had a precedent in Guadalcanal.

Like its weighty counterpart, the mine at Guadalcanal experienced shifts in the way it was managed. Unlike Almadén, where the shift was from being rented out to management by the Consejo de Hacienda, in Guadalcanal it was at first the opposite. Here, too, the Fugger family was able to obtain an agreement with the crown granting them a lease-hold over the mines. Again, this leasehold would last well into the seventeenth century, when economic problems and the loss of confidence by the Spanish crown forced the Fugger family to cease operations and withdraw from Spain.

In the 1680s, half a century after the Fuggers abandoned Guadalcanal, the royal treasury started a project to revive the mine. When the attempts to revive the mine began, much of the effort that was put into re-establishing, it was geared towards draining and cleaning the mine. These attempts provoked concerns whether the mine should be run via an asiento or directly by the Consejo de Hacienda.

Almadén and Guadalcanal—An "Unequal Couple"

Despite the significant differences in size and importance between Almadén and Guadalcanal, there are similarities between the two mines that justify and demand that they be studied as an "unequal couple". Besides the obvious fact that both mines were controlled by the Consejo de Hacienda at the end of the seventeenth century, the reasons why they were handed over to it are quite similar, which means late seventeenth-century Spanish public mining must be seen as a legacy of the late sixteenth and early seventeenth centuries Fugger-controlled mining.

The history of both Almadén and Guadalcanal shows that until the late seventeenth century, having mines controlled directly by institutions of the royal treasury was very rare. Although Guadalcanal was projected as a public enterprise, and Almadén was run by the Consejo de Hacienda for a few years at the beginning of the sixteenth century, the more "natural" form of managing even such important mines was via leaseholds. Thus, running the mines via the Consejo de Hacienda, as put into practice during the last decades of the seventeenth century, must be regarded as a new method of mining management.

Lastly, both mines had connections to Spain's American possessions and American mining. The importance of Almadén for the American silver production is obvious and has often been stated (Brown 2012). In the case of Guadalcanal, the mere fact that it was a silver mine links it to America. But while in the sixteenth and early seventeenth centuries the promotion of Guadalcanal can be seen as an attempt to build a counter-weight to the American dominance (Hofmann 2018), in the late seventeenth century, Guadalcanal was purely auxiliary to Almadén and thus, American mining.

The Consejo de Hacienda Takes Closer Control Over the Mines

The Consejo de Hacienda assuming direct control over Spanish mines in the second half of the seventeenth century had two direct or immediate consequences. Firstly, from the moment the Consejo took over, a mine was run at the expense of the royal treasury. While during periods of renting out mines, it was the leaseholders' obligation to assure financing of the operations and maintenance of the mines, these tasks now fell directly within the scope of the Consejo de Hacienda and its subordinate organisations. According to the Consejo de Hacienda's documents, this more immediate approach to financing of mines was essential to the process of strengthening public control over the Spanish mining sector in the late seventeenth century. In its reports to the monarch and his ministers, the Council members even go as far as using the term "mines run at the expense of the royal treasury" (Consejo de Hacienda 1693) as a synonym for mines directly controlled by it, stressing the financial aspect of this form of management.

Although the Consejo de Hacienda stressed the financial aspect of its new responsibilities, direct financing through the royal treasury was not the only immediate consequence of taking over Spanish mines. The second of these consequences was related to the mines' leading personnel. Similar to the case of financing, the Consejo de Hacienda's influence over the mines' administrators became more direct. Until the mid-seventeenth century, when both Almadén and Guadalcanal were rented out to the Fugger family, the leaseholders had a say in choosing the officials in charge, with men like Mateo Nagelin and Guillermo Sayler, both of German descent and the best-known administrators of Almadén (Matilla Tascón 1987), overseeing the day-to-day operations of the mines. From the moment the Consejo de Hacienda took over, administrators were no longer provided by the leaseholders, but by the royal treasury, who recruited suitable men from its existing personnel.

Apart from these consequences, in the late seventeenth century, another change in mining administration took place. In contrast to the first two, this change was not a direct result of transferring control over mines to the Consejo de Hacienda, but it was a necessary condition for the establishment of closer control over the mines. The Consejo de Hacienda did not limit itself to indirectly controlling the mines via choosing administrators, but directly interfered in the mines' management, frequently telling the administrators what to do and occasionally even how to do it. This went as far as involvement in the daily business of the mines. A telling example of the extent of the Consejo de Hacienda's concern with minute details can be found in the case of the provision of forced workers for Almadén presented at the beginning of this chapter. Although the orders issued by the Consejo associated with the plethora of reports sent in this matter could not be found, they undoubtedly existed, and references to them appear in the reports. In other cases, too, the Consejo devised detailed instructions for mine administrators (Consejo de Hacienda 1684).

The elevated level of intervention in the working of publicly controlled mines was not the Consejo de Hacienda's own decision. In fact, closer control especially over Almadén was demanded by the government. Charles II was known to have been very interested in everything related to mining, a fact that becomes obvious when reading his responses to reports regarding mining-related topics. In 1682, for instance, Charles raised his concern over the state of the mine of Almadén, brought to his knowledge by a Consejo report:

> The work at this mine is very delayed, and as it belongs to the best gems of my monarchy, it causes me great concern and sentiment that the scrutiny owed to such an important business is not applied, as the resources that are dedicated to this have to be so effective that they are not exposed to the contingencies of the common miserable state of things, because if no mercury goes [to America], no silver can come back[…]. (Consejo de Hacienda 1682, translation by the author)

The image of the mines, particularly Almadén, as "gems of the monarchy" is ubiquitous in correspondence between Charles II and the Consejo de Hacienda whenever the topic of mining comes up. Its use is not limited to the monarch: even the Council members employ the same imagery in their reports. It appears that Charles II liked this image, and thus the Consejo de Hacienda kept using it. The following example, although similar in its content, is not related to the one above, preceding it by more than two years. It is taken from one of the Consejo de Hacienda's reports in which the Council members strive to convince the monarch of the importance of continued scrutiny regarding the administration of state-run mines:

> [The Consejo] sees it as its obligation to put [information on the state of Almadén] in the royal hands of your majesty, as it deems it to be a matter so important that the highest reason does not equal it, as this jewel is amongst the most precious and valuable your majesty has in his royal possessions, as through the shipments of mercury that are made to the Indies, all the silver that comes from there, and especially from New Spain, is hauled to these [Iberian] kingdoms, and to ensure today not only the persistence of this factory, but also the certain hopes of new discoveries of rich veins, from which very considerable quantities of mercury are to proceed […]. (Consejo de Hacienda 1680, translation by the author)

Interestingly, in both cases, Almadén is compared to jewels not in order to praise its glory, but rather to stress the need to care for it. In fact, for most of the second half of the seventeenth century the state of the mine, as well as that of other mines controlled by the Consejo de Hacienda, did not at all resemble that of jewels (Gil Bautista 2015). However, viewing Almadén as a "most precious jewel of the monarchy" cannot be attributed to pure sentiment. The economic reasoning behind this claim is clearly evident in both extracts. According to both the Consejo de Hacienda and the monarch, it was Almadén's role as provider of mercury for the American silver mines that made it worthwhile for the royal treasury to engage in Spanish mining to the extent that it did in the late seventeenth century.

Especially when spring came to an end, all of Spain's mining sector was oriented towards providing the highest amount of mercury possible for America: The day of San Juan, June 24, was the day when the fleet to the Indies was supposed to depart. Until sometime before that date, all the mercury designated to be shipped to New Spain had to arrive at the stores in Seville. At the end of the seventeenth century, still, the fleet was the preferred method for shipping mercury. If mercury from Almadén did not arrive in time, it had to sit in the stores until the following year, or, in times of great need for mercury, was sent with a lone ship, which was costlier and more dangerous. The timely arrival of Almadén mercury at Seville was therefore of the utmost importance, as evidenced by repeated statements from the Consejo de Hacienda.

All this points towards one of the most important questions when assessing early modern Spanish mining: its purpose. For much of the early modern period, Spanish mining for minerals other than iron was not profitable. Guadalcanal, the crown's silver mine reactivated in the 1680s, was dwarfed by the Spanish American silver mines. What is more, in the period in question, the mine hardly ever reached a state where one could even think of mining silver ore in it; most of the time was spent draining and repairing the mine that had been abandoned for decades. Almadén did not directly turn a profit either. The mine's operations were financed partly by the town and villages in its surroundings and partly by aid from the Consejo de Hacienda whenever the towns' contributions did not suffice (Consejo de Hacienda 1699). The mercury it produced was sold to American miners at a price that was kept artificially low in order to support the production of American silver. Almadén did not benefit from these mercury sales.

Spanish mining thus makes only sense when seen as an auxiliary sector to the American silver mining sector. In this sense, the "jewels of the monarchy", by serving American mining, served the good of the royal treasury. And the good of the royal treasury is what the Consejo de Hacienda existed for. "Good of the royal treasury" is one of the arguments the Consejo de Hacienda used when trying to convince the monarch of the merit of its actions. It appears time and time again in the Consejo de Hacienda's correspondence, such that it was almost formulaic. Every decision the Consejo de Hacienda approved, be it a decision taken by itself, by its subordinates at the mine, or—predictably—by the monarch, are presented as being beneficial to the royal treasury. However, more often than not, it is not specified why exactly a certain measure is supposed to benefit the treasury. As we have seen, it was not always clear whether a decision would benefit the treasury's income directly: in the Spanish mining sector, much depended on indirect profits from the Spanish American silver mines. The phrase is thus to be seen more as an affirmation of the Consejo de Hacienda's dedication to serving its purpose than a proper explanation of its reasoning.

Spanish mining only makes sense when understood as an auxiliary to America. For the Consejo de Hacienda, this made it more difficult to properly judge the benefit of its mines as it could not rely on a simple calculation of the mines' earnings against their costs, but also had to consider the American mining sector. On the other hand, having the mines run by the Consejo instead of privates also proved to be an advantage: keeping the big picture in mind was easier for Madrid bureaucrats than for businessmen who had been awarded a lease. The focus on economy, due to Spain's subsidiary role to American mining, left the administration and government little space to use the mines to show off the splendour and innovation of their realm.

Early modern mining is often associated with the idea of progress, above all technological progress. The specific needs of the mining sector, especially the need for draining the shafts and galleries of the mines, prompted the introduction of quite complex machinery. Also, the further refinement of the minerals once they had been mined demanded the use of ovens, which were subject to technological innovation too (Mansilla Plaza 2000). This technological and innovative aspect of mining inspired not only historiography, but also the princes that owned and financed the mines. In Central Europe, princes had pride in the technology used in their mines and used it to represent their power, progressiveness and scholarship (Sánchez Gómez and Pieper 2000).

In contrast to rest of Europe, Spain's mines were not a playing field for princes but an auxiliary to actually profitable American silver mines. It is therefore logical that the documentation of the Consejo de Hacienda, including documents directed to the monarch, seldom contains references to the innovativeness of the mining sector as a whole, or of a certain measure, for example, the construction of an innovative machine (Unda y Garibay 1697). Even though Spanish mines were regarded—just like those in Central Europe—as "jewels of the monarchy", in Spain this seems to have been based on the mines' economic importance rather than their technological leadership. It can be argued that, due to the lesser importance of Central European mines in comparison with Spanish and especially Spanish American mines, Central European princes had to resort to emphasising their mines' role as technological showcase rather than economic powerhouse.

Instruments of Communication Between the Consejo de Hacienda and the Mines

In order to effectively exercise their control, the Consejo de Hacienda needed to know what was happening in the mines under its control. This is particularly true as the members of the Consejo generally had no previous knowledge of mining-related matters. Even the administrators in charge of managing the mines were chosen from the royal treasury's personnel and were generally lawyers by training and tax collectors by profession (Sanabria 1681). In order to stay informed about the mines, the Consejo de Hacienda employed of several instruments, which can be divided into two broad categories: those which were used routinely and those which were used extraordinarily.

Routine correspondence between the mines and the Consejo de Hacienda came in the form of regular—mostly weekly—reports written by the administrators of the mines and sent to the Consejo de Hacienda. In the documentation of the section, "Consejo y Juntas de Hacienda" analysed for this chapter, those regular reports originated exclusively from the mines that were at the time controlled directly by the Consejo de Hacienda—that is, Almadén and Guadalcanal. It must therefore be assumed that regular reports were an instrument of comparatively close control over mines.

Apart from regular reports, the Consejo de Hacienda relied on other methods of gaining knowledge about the state and dealings of the mines under its control. There are two dimensions along which to distinguish those additional instruments of gathering information from regular reports. Firstly, while ordinary reports were sent and received on a regular basis—as has been mentioned above, reports were sent as much as once per week—other instruments of information gathering were sent at irregular intervals. We can therefore distinguish between regular and irregular instruments of information. Secondly, we can distinguish between reports that were sent without being specifically requested by the Consejo de Hacienda and those that were requested. Typically, regular reports were not requested specifically, whereas such irregular instruments of information were.

In addition to the different forms of reports, the Consejo de Hacienda employed more thorough, but also more exceptional, means of obtaining information. Visual inspections of the mines were conducted by mine administrators, often in the company of other expert officials, and served to get a first-hand account of the state of the subterranean parts of the mine. Visits by members of the Spanish fiscal administration were an even closer means of control, which saw experts sent out to the mines and put all of its dealings, as well as the conduct of its administrators, under scrutiny.

Reports by the Administrators

Among the sources of information the Consejo de Hacienda could rely upon when judging the technical and financial state of the mines under its control, as well as the activities that were undertaken, tasks that were tackled or problems that might have arisen, the reports by the mines' administrators were the most numerous and arguably the most useful. Among the Consejo and Juntas de Hacienda's documentation analysed for this study, several dozen such reports have been found. There is a striking disparity in the origins of those reports. From all reports found as attachments to *consultas* crafted by the Consejo de Hacienda for the monarch and his government between 1675 and 1700, about two-thirds concern the mercury mine at Almadén, whereas about a quarter concern the Guadalcanal silver mine and the remaining ten per cent other mines such as Linares (Jaén, modern-day Andalusia) or Hellín (Albacete, modern-day Castilla-La Mancha).

One must of course take this information with a grain of salt. As these percentages do not stem from reports sent to the Consejo de Hacienda by the administrators of the mines, but from those the Consejo relayed to the monarch, they demonstrate the importance either the Consejo or the monarch attributed to the different mines rather than the actual number of reports the Consejo de Hacienda received from each mine. What can be said with certainty, however, is that the mine of Almadén, with its strategic importance for American silver production, was the main concern of the Consejo de Hacienda, whereas the other mines were considered less crucial. From the communication that took place between Almadén and the Consejo de Hacienda in 1697, which has been analysed in detail, it can be said that reports were sent as often as every week. Again, this does not necessarily translate to other mines, nor to other times. Nevertheless, this fact gives a hint at the intensity with which the Consejo de Hacienda was informed, and demanded to be informed, about the mines it controlled.

The reports typically consist of several parts. The main part is the description of events since the last letter, the state of the mine, problems they encountered and how they solved them, or if they could not solve them, petitions for help. With these reports, the Consejo de Hacienda gained a quite detailed and up to date summary of what happens at the mines. At the same time, the level of detail in each report enables us to reconstruct the main problems faced by mines in the late seventeenth century Spain. The most pressing among them was lack of funds, especially to pay the workers, lack of workers, both free and forced (Consejo de Hacienda 1699), draining the mines (Montoya 1693) and other weather-induced problems such as illness (Consejo de Hacienda 1684).

Another regular element of the reports was certification of the amount of metal produced, signed by the mine's chief accountant. Reports by the *veedores*—foremen—were a means of gathering information which complemented the administrator's reports. References to additional reports by *veedores* included in *consultas* by the Consejo show that the Consejo did not limit itself by trusting only the word of the highest-ranking official. The same is true when the mine's accountant or the town's scribe sent certifications that complemented the administrator's reports.

The Consejo de Hacienda did not limit itself to simply receiving administrators' reports but actively requested additional information, as references made in reports by the mines' administrators show. This was primarily the case when the Consejo de Hacienda was not satisfied with the contents of the routine reports it received, either because it found that the reports were lacking information, or because the reports alluded to important matters that needed to be resolved based on as much information as possible.

Weekly reports, apart from being quite formulaic in the sense of following a fixed form, as we have shown, were often very brief. This was partly due to fact that sometimes there was not much to report. Among the material analysed for this chapter, there are periods where entire weeks were spent pumping water and cleaning galleries. The corresponding reports, therefore, do not contain much information apart from the change in water height in the mine. However, such a lack of information could also mean the administrator was not keen, for instance, on reporting problems to the Consejo de Hacienda when he thought he could handle them without bothering his superiors.

From a formal point of view, there is not much difference between regular reports and those written because they were requested: both follow the same basic structure of a letter and both contain the same type of information, given by the same person; the administrator of the mine. There are usually some cues that tell whether a report was requested, for example, references at the beginning of the report referring to the Consejo as having asked about some topic. The reason why we nevertheless stress the difference between regular and requested reports is because it shows that the Consejo de Hacienda genuinely cared about what was written in the reports and sent to it. Early modern Spanish administration is known for producing huge quantities of correspondence on a variety of topics. Some scholars argue that there is a significant disparity between the amount of material produced by this "monarchy of paper", and how little it was put to use. According to them, many reports were archived without having led to any activity, completely defeating their purpose. I claim that in mining administration reports *did* have an impact, and the fact that the Consejo de Hacienda obviously read them and demanded additional information supports this claim.

Visits and *Vistas de Ojo*

Reports by the administrators of the mines controlled by the Consejo constitute the vast majority of documents sent from the mines to the Consejo de Hacienda, informing the Consejo about the state and the dealings of the mines. Although they were undoubtedly an adequate instrument for describing the day-to-day operations of mines, they came with a couple of disadvantages for the Consejo de Hacienda's effort to build a complete picture of the state of the mines (Consejo de Hacienda 1684). In order to achieve this, the Consejo de Hacienda relied on additional instruments complementing the reports. Those instruments were meant to add a more hands-on perspective to the standard reports, which could be somewhat detached from the mine, in the sense that they required hacienda officials to actually enter the subterranean portions of the mine. It seems that management entering the mines did not happen regularly; otherwise, it would not be mentioned explicitly the way it was in the documentation of the Consejo de Hacienda. Especially in cases where the officials involved had passed a certain age or had health problems, entering the mines proved quite cumbersome and dangerous.

There were two distinct instruments that involved entering the mines, although the terms used in the sources in order to refer to them sound quite similar in Spanish: the *visita*, which translates to visit, and the *vista de ojos*, which could be translated as visual inspection.

Of the two instruments, the *vista de ojos* was the less extraordinary one. It involved the administrator of the mine that was to be inspected, a hacienda official, and his subordinates entering the shafts and galleries of the mine under his control in order to investigate the state of the mine with his own eyes. In contrast, during normal operation, the administrator would rely on his subordinates to inform him about what was going on underground. Thus, a *vista de ojos* provided him with a more immediate picture of the mine, which he could then relay to the Consejo de Hacienda through an extended report. In contrast to normal reports, written by the administrator, in the case of *vistas de ojos* other participating individuals such as foremen and supervisors could be required to hand in reports, which made for a more complete coverage of the inspection. That requirement could also serve as a means of increased control over the administrator, as the Consejo de Hacienda, and by extension the monarch, would now hear not only his words. The fact that the town scribe was also present at such occasions reinforced this aspect of a *vista de ojos*.

In contrast to a *vista de ojos*, executed by the mine's own administrator, albeit aided by other individuals, a visit to the mine was an instrument that involved external personnel. In a mining context, a *visita* was an inspection of a mine by higher-ranking officials of the fiscal administration. Compared to *vistas de ojos*, *visitas* were an extraordinary, major event at the mines. More than other instruments discussed in this chapter, visits had a pronounced controlling or inspecting character. They were ordered when the Consejo de Hacienda was displeased with the reports that the administrators sent or suspected malversations at the mine.

The documentation produced during a visit could be quite considerable. Visits could involve one or more visitors, experienced officials who often had experience in managing mines themselves. These visitors could be aided by additional individuals, such as expert miners from America or other specialists. Additionally, the individuals who would also participate in ordinary *vistas de ojos*, that is the mine's administrator, his subordinates, and the town's scribe, would join the inspecting party. All of them would later file a report about the visit. This fact proves the usefulness of the information gathered during those visits, which provided the Consejo de Hacienda—and the monarch—with the most meticulous information about the mines.

When speaking about the phenomenon of the *visita* in the context of early modern Spanish (fiscal) administration, we must distinguish between visits that aimed to control the management of the finances of the entity subject to the visit and those which sought to ensure good practice in the activities of the entity. As the usual duties of the Spanish fiscal administration included activities such as collecting taxes and administering monopolies—fiscal administration in its narrower sense—management of finances is the core activity of that entity. Public enterprises like state-managed mines were an exception in this regard. The nature of the visit, for example, decides who conducts the visit.

In the mining sector, there were two types of visits. One type was visits by accountants in order to control the accounting of the mine. Such a visit took place at the end of the seventeenth century, when Almaden's famous superintendent Miguel de Unda y Garibay was suspected of mismanaging the mine. The investigation ended in autumn of 1700, possibly because of the death of King Charles II.

The second type of visit was more peculiar for the mining sector. As Guadalcanal and Almaden were managed directly by the Consejo de Hacienda, it was not only interested in proper accounting, but also in the technicalities of running a mine. This type of visit was undertaken when the Consejo de Hacienda noticed some kind of problem in the working of the mine. An example of this type of visit took place in Guadalcanal in 1691, when the Consejo realised that Rafael Gomez had not drained the mines as expected.

Conclusion

In this chapter, it was shown that the transfer of control over Spanish mines to the Consejo de Hacienda, which took place in the second half of the seventeenth century, was not the result of a planned process but in fact due to a series of contingencies, above all related to the demise of the Fugger family, which had up until that point been heavily involved in Spanish mining. The Consejo de Hacienda used the opportunity to establish a closer control over the mines than would have been the immediate result of this transfer and got involved even in low-level decision making in mining matters. This dedication to the mining sector was not rooted in the Consejo de Hacienda's own intentions, but was demanded by the monarch, Charles II, who had an avid interest in the mines and tried to revive the Spanish mining sector, primarily because of the vital importance of the Almadén mercury mine for the American silver mining sector. Smaller mines like Guadalcanal were dragged along in the process.

The new method of having the mines managed by the Consejo de Hacienda caused a number of problems, not only due to the distance between the mines and Madrid, but also—mostly—to the lack of prior knowledge about mining matters that the officials involved possessed. A system of complementary information instruments was applied to keep the Consejo informed about the state and dealings of the mines under its control. Thanks to this information, the Consejo de Hacienda could fulfil the expectations placed upon it and take sensible decisions promoting the Spanish and thus the American mining sectors. In this context, the Spanish mining sector could be more effectively geared towards serving American mining, as it proved easier for the Consejo de Hacienda to keep the bigger picture in mind and not to focus on the immediate profitability of the mines under its control, as had been the case when the mines were run by leaseholders.

Bibliography

Primary Sources

Caniego, Fernando, 1697. Letter to Gil Pardo de Najera, Cordoba, August 10: Archivo General de Simancas, CJH,LEG,1994.

Consejo de Hacienda, 1680. Consulta, Madrid, February 21: Archivo General de Simancas, CJH,LEG,1410.

Consejo de Hacienda, 1682. Consulta, Madrid, November 25: Archivo General de Simancas, CJH,LEG,1476.

Consejo de Hacienda, 1684. Consulta, Madrid, January 18: Archivo General de Simancas, CJH,LEG,1476.

Consejo de Hacienda, 1693. Consulta, Madrid, April 28: Archivo General de Simancas, CJH,LEG,1620.

Consejo de Hacienda, 1699. Consulta, Madrid, July 18: Archivo General de Simancas, CJH,LEG,1693.

Fernandez de Portalegre, Francisco, 1694. Letter to Baptista Rivas, Guadalcanal, July 8: Archivo General de Simancas, CJH,LEG,1979.

Montoya, Antonio de, 1693. Letter to Baptista Rivas, Guadalcanal, November 26: Archivo General de Simancas, CJH,LEG,1979.

Sanabria, Diego de, 1681. Relación de Servicios, Madrid, July: Archivo General de Simancas, CJH,LEG,1979.

San Vicente, Marques de, 1697. Letter to Gil Pardo de Najera, Badajoz, September 6: Archivo General de Simancas, CJH,LEG,1994.

Unda y Garibay, Miguel de, 1694. Letter to Gil Pardo de Najera, Almadén, August 21: Archivo General de Simancas, CJH,LEG,1994.

Unda y Garibay, Miguel de, 1697. Letter to Gil Pardo de Najera, Almadén, July 17: Archivo General de Simancas, CJH,LEG,1994.

Valbermoso, Marques de, 1697. Letter to Gil Pardo de Najera, Seville, November 12: Archivo General de Simancas, CJH,LEG,1994.

Secondary Sources

Arduz Eguía, Gastón, 2000. "Sobre la Metalurgia Colonial de la Plata en Pototsí." In: *Hombres, Técnica, Plata. Minería y Sociedad en Europa y América. Siglos XVI–XIX*, edited by Julio Sánchez Gómez and Guillermo Mira Delli-Zotti. Sevilla: Aconcagua Libros: 105–128.

Brown, Kendall, 2012. *A History of Mining in Latin America: From the Colonial Era to the Present*. Albuquerque: University of New Mexico Press.

Bermejo Cabrero, José Luis, 2016. *Organización Hacendística de los Austrias a los Borbones: Consejos, Juntas y Superintendencias*. Madrid: Agencia Estatal Boletín Oficial del Estado.

Denzer, Jörg, 2005. *Die Konquista der Augsburger Welser-Gesellschaft in Südamerika (1528–1556)*. München: C. H. Beck.

Garzón Pareja, Manuel, 1984. *Historia de la Hacienda de España*, vol I. Madrid: Instituto de Estudios Fiscales.

Gil Bautista, Rafael, 2015. *Las minas de Almadén en la Edad Moderna*. Alicante: Publicaciones Universidad de Alicante.

Hernandez Sobrino, Ángel, 2010. *Los esclavos del Rey: los forzados de su majestad en las minas de Almadén, años 1550–1800*. Ciudad Real: Fundación Almadén.

Hofmann, Domenic, 2018. *'Plata o plomo?': How American Silver Led the Spanish Crown Astray: Promoting the Spanish Mining Sector (1550–1650).* PhD Dissertation, University of Graz, Graz.

Mansilla Plaza, Luis, 2000. "La Metalurgia del Mercurio en Almadén Durante la Época Colonial." In: *Hombres, Técnica, Plata. Minería y Sociedad en Europa y América. Siglos XVI–XIX,* edited by Julio Sánchez Gómez and Guillermo Mira Delli-Zotti. Seville: Aconcagua Libros: 63–76.

Matilla Tascón, Antonio, 1987. *Historia de las Minas de Almadén, Vol. II: Desde 1646 a 1799.* Madrid: Instituto de Estudios Fiscales.

Matilla Tascón, Antonio, 2005. *Historia de las Minas de Almadén, Vol. I: Desde la Epoca Romana Hasta el Año 1645.* Ciudad Real: Fundación Almadén.

Pieper, Renate, 2009. "Die Spanische Krone und der Silberbergbau in Amerika." In: *Wirtschaftslenkende Montanverwaltung – Fürstlicher Unternehmer – Merkantilismus. Zusammenhänge zwischen der Ausbildung einer fachkompetenten Beamtenschaft und der staatlichen Geld- und Wirtschaftspolitik in der Frühen Neuzeit,* edited by Angelika and Ekkehard Westermann. Husum: Matthiesen: 281–300.

Sánchez Belén, Juan A., 1996. *La política fiscal en Castilla durante el reinado de Carlos II.* Madrid: Siglo XXI de España.

Sánchez Gómez, Julio, 1989. *De minería, metalúrgica y comercio de metales. La minería no férrica en el Reino de Castilla (1450–1610).* Salamanca: Ediciones Universidad Salamanca.

Sánchez Gómez, Julio; Mira Delli-Zotti, Guillermo, eds., 2000. *Hombres, Técnica, Plata. Minería y Sociedad en Europa y América. Siglos XVI–XIX.* Seville: Aconcagua Libros.

Sánchez Gómez, Julio; Pieper, Renate, 2000. "¿Tras las huellas de un espejismo? La minería en Nueva España y Europa Central en la segunda mitad del siglo XVIII." In: *Jahrbuch für Geschichte Lateinamerikas* 37: 49–72.

Westermann, Angelika; Westermann, Ekkehard, eds., 2009. *Wirtschaftslenkende Montanverwaltung – Fürstlicher Unternehmer – Merkantilismus. Zusammenhänge zwischen der Ausbildung einer fachkompetenten Beamtenschaft und der staatlichen Geld- und Wirtschaftspolitik in der Frühen Neuzeit.* Husum: Matthiesen.

Part IV

Money

9

Some Determinants of Local Exchange Rates and in Early Modern Mexican Mining Sites, Sixteenth and Seventeenth Centuries

Claudia de Lozanne Jefferies

Introduction

Official communications between the Crown and mine owners in sixteenth-century and early seventeenth-century Mexico depict the latter as overwhelmed by a series of uncontrollable events, both global and local. The falling international price of silver, a consequence of an increase in its supply, was having negative effects on mine owners' profits. Further, they faced increases in the extraction costs of silver ores, located deeper and deeper underground as the mining industry moved through its life cycle. Petitions for reduction in the price of mercury, an indispensable production input, were numerous. According to local representatives, mine owner bankruptcies and imprisonments due to debt were frequent ("Carta de Alonso de Peralta" 1603, f. 559r–564v; "Petición" 1587, f. 331r). Silver merchants carried a considerable portion of the blame

C. de Lozanne Jefferies (✉)
School of Arts and Social Sciences,
City, University of London, London, UK
e-mail: claudia.jefferies.1@city.ac.uk

© The Author(s) 2019
R. Pieper et al. (eds.), *Mining, Money and Markets*
in the Early Modern Atlantic, Palgrave Studies in Economic History,
https://doi.org/10.1007/978-3-030-23894-0_9

for the financial problems endured by mine owners. More than once, the depiction of the silver merchant as driven exclusively by greed prompted royal action: In 1598, the Crown proposed to take over the provision of goods to towns from merchants. The plan did not materialise due to the unwillingness of viceregal authorities to finance it (Hoberman 1991, p. 77). A second attempt by the Crown to render the role of the silver merchant redundant was carried out in 1604, when the Crown tried to take over the supply of specie to mining towns. This action was equally unsuccessful, but it is of special interest as it revealed important features of the silver market in early modern Mexico.

Louisa Hoberman discusses the "image of the merchant as predator" (Hoberman 1998, p. 80) and points out that the cultivation of such an image underestimates the risk entailed in lending to mine owners, whose frequent bankruptcies had causes well beyond the activities of silver merchants. There was a dependency on silver merchants in sixteenth-century and seventeenth-century Mexican frontier towns; the scarcity of specie was a serious problem and credit was a partial solution. Mine owners needed the credit offered by merchants in order to start and safeguard the operations of their mining enterprises. Such is the outcry against silver merchants in contemporary sources; it is easy for the historian to support the stance of mine owners and depict the greed of silver merchants as one of the causes of the financial peril suffered by mining entrepreneurs. This is the case for example with Pedro Pérez Herrero, who refrains from criticising the statements issued by mine owners and seems to agree with the idea that silver merchants set their charges arbitrarily (Pérez Herrero 1988, p. 71).

High-ranking merchants had monopoly power over the means of exchange as they were in control of the Mexico City mint, the only mint in all of New Spain, which determined the monetary supply of the whole viceroyalty. Mint officials came from the silver merchants' circle and the two groups maintained strong links. Mint offices were sold at very high prices, ranging between 8000 and 130,000 *pesos de a ocho*, and delivered yields of between 1600 and 14,000 *pesos* per annum (Hoberman 1991, p. 85). For perspective, the largest single producers of silver had an annual output of between 8120 and 16,240 *pesos*. The price of mining enterprises (*haciendas de minas*) in seventeenth-century

Zacatecas ranged between 10,000 and 70,000 *pesos* (Hoberman 1991, pp. 74–76).

The levels of yield from high mint offices were high, but the volatile nature of silver production influenced the profits that mint officials could expect to obtain. Those profits were not as volatile as the individual profits of mining entrepreneurs, as the mint was a hub for much of the silver production of the whole viceroyalty. Tax evasion and smuggling added to the incomes of the mint officers. According to some estimates, one-third of the silver reaching Spain was smuggled (Hoberman 1991, p. 91). Profits of mint officials are likely to have been much higher than those registered. Lower down the merchant hierarchy were the silver wholesalers, who supplied mint merchants (who took bullion to the mint to be turned into coins) with the bullion that silver merchants had transported to Mexico City.

Specie in Mexico was scarce throughout the sixteenth and seventeenth centuries. To meet the excess demand for currency, the use of pre-Hispanic means of exchange continued in most urban centres. Cotton sheets, gold dust held in feather quills, tin and wood tokens, jade beads and cocoa beans were some of the means of exchange circulating in colonial Mexico. Barter, circulation of bullion and the payment of tributes in kind also helped counteract monetary scarcity (Giráldez 2012, pp. 147–161). Frontier mining regions were particularly affected, as only a portion of the silver produced in frontier mines returned to frontier towns as specie. Mine owners had to pay mine workers' salaries, among other mining-related expenses. During the first years of the seventeenth century, mine owners complained frequently to the viceroys, asking for the Crown's help to address the problem of scarcity of specie.

Silver merchants delivered specie to mine owners. For their services, they charged a discount rate, which was 1 *real per peso* or 8.1 *reales* per mark of 65 *reales*. The total rate of discount was about 12.5% and fluctuated according to the level of specie abundance. The rate of discount per mark fluctuated between 1 and 3 *reales*, in addition to the 8 *reales* charged as rate of discount, as shown in Table 9.1.

Silver merchants also linked the credit market to the goods and money markets. During the early years of the colonial period, *aviadores*

Table 9.1 Revenue obtained by silver merchants through the exchange of specie for bullion in 1604

Source of revenue	Value of revenue in *reales*	Revenue in percent
Median merchant profit due to higher Ag purity	2 *reales* per mark	3.07
Rate of discount	8.1 *reales* per mark	12.46
Median fluctuation of rate of discount	2 *reales* per mark	3.07
Total before costs	12.1 *reales* per mark	18.6
(minting costs)		(4.6)
Total before transport costs		14.0

Source "Carta del Factor" (1604, f. 1r–v)

were merchants who transported goods to be sold in remote regions. Their merchandise often included European goods, sold at a premium in Mexican towns. Given the high demand for specie in frontier mining towns, the role of *aviadores* diversified into one of the *rescatadores*: silver merchants, whose role was to supply the mine owners with specie, which they were to repay adding the rate of discount and 5% interest, within a time frame of 40–50 days (Bakewell 1971, p. 211). The legal price of the silver mark was set at 65 *reales*, regardless of its purity. As silver produced by amalgamation was purer than silver produced by smelting, a mark of amalgamated silver was worth 67–68 *reales* and not 65; the merchant or *rescatador* received the equivalent of 75 or 76 *reales* (taking into consideration transport costs and minting charges) for every 65 *reales* of minted silver.[1] The commission charged by silver merchants was, besides interest on mine owner debts, the main cause of mine owner complaints to the Crown against merchants.

Specie tended to flow out of circulation, as payment for imported goods and Crown taxes. The shortage of specie was partially covered by smelted silver bullion (*plata de rescate*), which circulated regionally. There were, however, legal restrictions on the use of bullion as a means of exchange. Mine owners' liquidity depended on the services offered by silver merchants.

---

[1] The risk factor has not been taken into consideration.

The rate of discount is important in the history of Mexican mining in the sixteenth and seventeenth centuries and, although briefly, it has been treated in the literature from the point of view of the mine owners (Bakewell 1971, pp. 211–212) as well as from that of the merchants (Hoberman 1998, pp. 75–77; Perez Herrero 1988, pp. 127–128; García Ruiz 1954, pp. 33–39; Orozco y Berra 1855, pp. 915–917). An in-depth analysis of the determinants of the rate of discount has not yet been carried out. For this purpose, the rate of discount will be placed within the context of the quantitative data available for the period in question and within national and international credit and exchange chains. A better picture is needed of the factors that determined not only the level of the rate of discount, but also its fluctuation. The characteristics of frontier markets and their stakeholders were an important factor underpinning the rate of discount.

Silver Merchants, Mercury and Mine Owner Debt

Frontier towns in colonial Mexico were typically newly founded urban centres whose *raison d'être* was the presence of rich silver ores in their vicinity. Provisioning farms were founded on the few fertile spots that could be found in mostly arid regions, and produced food for the mining towns. The limited number of suppliers of agricultural goods prevented the formation of markets such as those in areas where lands were more fertile and crops more abundant. Goods prices were higher than in other regions (Borah 1994, p. 68) because, in addition to being in short supply, they had to be transported across long distances into the towns. Salaries were high, as labour was scarce, and mine owners faced difficulties retaining nomadic Indians as part of their labour force. The Chichimec were not familiar with the use of means of exchange, let alone money, so high salaries were not always an incentive strong enough to recruit and retain them as a workforce. As a whole, the frontier economy was subject to "ups and downs" (García Ruiz 1954, p. 28), periods of scarcity and abundance of goods, and fluctuations in the production of silver.

Demand for specie was higher in mining regions than in predominantly agrarian regions, as money was needed to pay for miners' salaries, as well as the other costs related to mining. Mine owners faced a choice between either taking the silver they produced to Mexico City to be minted themselves, or buying coins from a silver merchant. The dependency of the mine owners on the merchants increased as contracting merchant services proved to be more convenient than transporting silver to the mint themselves. However, mine owners found it difficult to come to terms with the idea of paying the rate of discount demanded by merchants.

The provision of mercury was a Crown monopoly. There were three sources of the metal available to Spain and its colonies: the mines of Almadén near Cordoba, Huancavelica in Peru and Idrija (in modern Slovenia) (on the issue of mercury supply, cf. the contribution of Domenic Hofmann in this volume). Most of the mercury consumed in sixteenth-century and seventeenth-century Mexico came from Almadén, although it was sometimes insufficient and mercury from Idrija had to cover shortages, at least partially. Peru was suffering shortage problems of its own, as mercury production in Huancavelica was declining.

The Crown delegated the distribution of mercury to *alcaldes mayores*, who were a type of provincial administrators representing the Crown. Mercury was assigned to mine owners based on the quantities of silver that they had produced in the previous year. The Crown expected 1 *quintal* (101 pounds) of mercury to produce 100 marks of silver. The system was known as *depósito y consumido*: the mine owner received a certain volume of mercury as a loan from the Treasury, which had local subsidiaries known as *cajas reales* in every important town, and settlement was expected when the miner declared his silver production to the Treasury for taxation purposes. The way in which mercury was assigned to mine owners favoured mines with high productivity at the expense of the less productive ones. As a consequence, the bankruptcy rate of the latter increased, as well as the levels of industry concentration. In the eyes of mine owners, not only the merchants (*rescatadores*), but also the *alcaldes mayores* were perpetrators of trading abuses (Hoberman 1991, p. 77; Orozco y Berra 1855, pp. 916–917).

Through this system, as well as the uncertain nature of mining, miners' debt increased from year to year. This situation was exacerbated towards the 1630s, when the supply of mercury from Almadén was diverted to Peru in order to counteract the declining output of mercury in Huancavelica. Mine owners' dependency on credit, the existence of a rate of discount when selling their bullion to merchants and the bargaining power exerted by *alcaldes mayores* as the suppliers of mercury exacerbated the tensions between mine owners and the two other parties. As mentioned previously, official complaints from mine owners multiplied towards the end of the sixteenth and beginning of the seventeenth centuries.

Effects in Zacatecas of an Attempt to Eliminate the Rate of Discount and Mercury Intermediation

The Royal Assay, *Real Ensaye*, was in charge of determining the purity of the silver produced, as well as the rate at which it would be taxed. Mexican silver in the Habsburg period was taxed according to a two-rate system due to a concession granted to mine owners in 1572 through which the normal 20% rate levied in the rest of the Indies was reduced to 10% for silver produced through amalgamation in a mine owner's enterprise.

Silver produced through smelting by anyone who did not own a mine was taxed at 20%. The two-rate system encouraged tax evasion through informal agreements between mine owners and merchants. The latter acquired silver produced through smelting and had mine owners declare it as produced on their premises. In this way, the silver merchants would benefit from a 10% profit in tax savings. Mine owner debt held by merchants precipitated such arrangements.

Taxed silver was allowed to circulate, albeit with certain restrictions and never outside the fiscal district of its origin. Untaxed silver was also allowed to circulate and was used as a means of payment to merchants. However, merchants themselves were not allowed to carry out further transactions using untaxed silver. Taxed silver would normally bear the stamp of the assay (for which 1.5% would be charged), asserting its purity, as well as the stamp issued by the local Treasury subsidiary. Untaxed silver bore the stamp of the mine where it had been produced, in order for it to be legally used as a local means of exchange. Untaxed silver could be confiscated by royal authorities at any time, which made it a risky means of exchange.

The highest silver producer in early New Spain, Zacatecas, was by the end of the sixteenth century a town of 2000 inhabitants. Its annual production averaged 500,000 silver *pesos*. Silver production had positive spillover effects on the local economy, and there was an increasing need of specie to pay miners' salaries and cover other inputs for the mining sector.

Due to the problems caused by the scarcity of specie, mine owners made numerous petitions throughout the 1580s and 1590s calling for the establishment of a local mint in Zacatecas. The Royal Treasury in Mexico City rejected such applications, arguing that the real cause of silver scarcity was the outflow of specie as payments for imported goods (García Ruiz 1954, pp. 31–32). To address the problem, the Treasury proposed the issue of a special regional coin to prevent outflow. This proposal was rejected by mine owners, as only 30,000 marks of silver would have been minted, which mine owners regarded as far too little to solve the problem.

During his last year in office as Viceroy (1603–1604) Gaspar de Zúñiga Acevedo y Fonseca, Count of Monterrey (1560–1606), tried to address the complaints issued by mine owners in frontier regions about merchants' practices. In the Viceroy's view, such practices contributed towards the problem of scarcity of specie in New Spain (García Ruiz 1954, pp. 20–46; Orozco y Berra 1855, pp. 916–917).

The Royal Assay was the vehicle through which the 1604 plan was going to be put into practice. The plan aimed at eliminating the role of silver merchants as suppliers of specie, as well as the role of *alcaldes mayors* as suppliers of mercury, as both were the focus of complaints among mine owners. According to the plan, the Royal Assay would distribute mercury among mine owners going forward, as well as carrying out exchange transactions of bullion against silver.

Mine owners reacted positively towards the plan, although they fore-saw an increase in commodity prices as a consequence of an increase in the price of bullion, which was to be exchanged for specie at 65 *reales* per mark (García Ruiz 1954, pp. 35–37). The Viceroy's measure also implied the elimination of credit between merchants and mine own-ers, which exacerbated the problem of scarcity of specie. Contrary to viceregal plans, untaxed silver had to continue circulating after 1604. It was intended that the discrepancy between the prices of bullion and specie would disappear, because the Treasury would exchange 65 *reales* of bullion for 65 *reales* of specie. This exchange was to be carried out by a single transaction, eliminating credit. As predicted, this measure pro-voked a change in the relative prices of bullion in respect of all other commodities and in respect of specie. The price of bullion increased, while the price of silver *reales*, as well as their nominal value, remained unchanged. Hence, the relative price of specie in respect of bullion decreased, and so did the purchasing power of *reales*. As commodity prices increased, so did the demand for real cash balances. New stipula-tions were then issued to regulate the circulation of bullion (García Ruiz 1954, pp. 35–37):

- Bullion produced by smelting, which had neither been assayed nor taxed, should be stamped with an "R" and circulate within the king-dom (Nueva Galicia, in the case of Zacatecas).
- Bullion produced by amalgamation had to be assayed before circulating.
- Bullion with a high lead content produced by smelting at a mining enterprise was allowed to circulate after being taxed at 10%, only with the mine owner's name stamped on it.

Other types of lower-quality silver, which had not been assayed or taxed, were also allowed to circulate with a mine owner's name stamp as an assurance of its future assaying and taxing. The taxation and assaying requirements for bullion to legally circulate varied according to the dis-tance between the mine and the Treasury and assay offices.

In addition to the functional problems preventing the vicere-gal plan from delivering the desired outcomes, the royal authorities faced practical problems in trying to transport high volumes of silver. *Conductas* were royal carts used for transporting bullion and specie. Bearing the coat of arms of the Royal Treasury, they became the target of frequent attacks by Chichimec Indians, who resented the presence of Spaniards in their territory (García Ruiz 1954, pp. 37–38).

The viceregal attempt to eliminate the services offered by merchants as intermediaries between the goods and money markets was short-lived. The need for specie and credit, especially in mining areas, weakened the regulations introduced in 1604, and the merchant's role as a supplier of specie and credit was eventually reinstated. The number of merchants offering such services seems to have increased over time, which made the credit market more competitive, and as a consequence, the rate of discount diminished throughout the seventeenth century (Hoberman 1991, p. 90).

Analysis of the Factors that Determined the Rate of Discount and Its Varying Level

In the eighteenth century, the accomplished chemist Fausto de Elhuyar (1755–1833) classified the territory of New Spain according to its geographical and social differences (Elhuyar 1825, pp. 5–14). He highlighted the differences between what he called the "midland strip" (*faja media*) and the more arid frontier regions of the north. Focusing on the history of New Spain in the sixteenth and seventeenth centuries, Elhuyar observes that the midland strip's high plateau benefitted from a temperate climate, which suited the requirements of agrarian activities. Sedentary trading communities were established in the area prior to the Spanish conquest. As a consequence, Spaniards found it relatively easy to introduce new agrarian activities and to transform the existing markets to suit their commercial needs. In contrast, the northern frontier areas were inhabited by "savages", and agriculture was difficult, given the high temperatures and shortage of water.

The geographical and social differences highlighted by Elhuyar went hand in hand with the varied means of exchange used in the territory of New Spain, and as a consequence, with the factors that determined the demand for *reales*. Sedentary Indians, who were linked to population centres through tribute system called *encomienda* had little need for specie, as they were generally self-sufficient (García Ruiz 1954, pp. 26–27), and in any case, the value of *reales* was so high as to be out of their reach. They tended to use other means of exchange which had persisted since pre-Hispanic times. Demand for specie was higher in mining regions, which could also be divided into two types. The first one is characterised by dense populations of sedentary Indians with links to nearby agricultural areas. These mining areas were located in Elhuyar's midland strip and are known today as Guerrero, Morelos, Mexico (City and State), Hidalgo, Michoacán and Jalisco. A second type of area was the northern frontier, today's Nayarit, Sinaloa, Durango, Chihuahua, Zacatecas and San Luis Potosi, which was Chichimec territory in the sixteenth and seventeenth centuries. Miner salaries were paid in specie, which drove a higher demand for specie in mining areas. As agrarian production and the markets in frontier towns were developed to cater for the needs of the mining industry, there was little competition between agrarian producers, which lead to higher prices than in more central areas. The relatively higher goods prices determined the price of specie in respect of bullion.

Strictly speaking, the exchange rate is the price of a currency in respect of another currency. In the case of Zacatecas and other frontier mining regions, bullion was a type of local currency. It was even made to look as such, cut in a similar way as *reales* and stamped with a cross and an "R", which stood for *rescate* (Orozco y Berra 1855, p. 915). Bullion has, in this case, a dual character of commodity and local currency.

The following equation (Sargent and Velde 2003, p. 19) defines the relationship between commodity prices and exchange rate, taking the minting and melting points of a coin into consideration.

$$e_i(1 - \sigma_i)\gamma \leq p \leq e_i\gamma_i$$

where σ_i = seigniorage; e_i = exchange rate = 1^2; γ = melting point; p = the price of a basket of goods.

---

[2]$e_i = 1$, as an initial situation of equilibrium is assumed.

If prices (including the price of bullion) rise above the melting point, then the metal in coins will be more valuable than the coins' nominal value, which means that coins will be melted down and sold as silver. Hence, the price of a basket of goods must remain equal to or below the melting point of a currency to safeguard its function as unit of account and means of exchange.

Minting point is defined as the point at which metal is more valuably minted than in the form of bullion. Minting point in the case of Zacatecas and its surrounding region was low, as bullion was abundant and coins scarce. However, silver production fluctuated. In addition to this, prices were higher than in other regions and experienced significant fluctuations. There was a danger of the price level surpassing the melting point of the *reales* that implied a threat to monetary system, which at the same time would have had negative repercussions on the Crown's income from seigniorage. From the point of view of the silver merchants, it was important to safeguard the integrity and purchasing power of *reales*, as they were used to buy foreign goods.

A melting point kept high through scarcity of coins may have been a way to protect the monetary system from dramatic price-level fluctuations. Figure 9.1 depicts a comparison of corn price indices in Nueva Galicia (the colonial name of the region where Zacatecas is located) and central Mexico. The data is fragmentary, but sufficient to calculate the means of both series and test the hypothesis of equality of means.

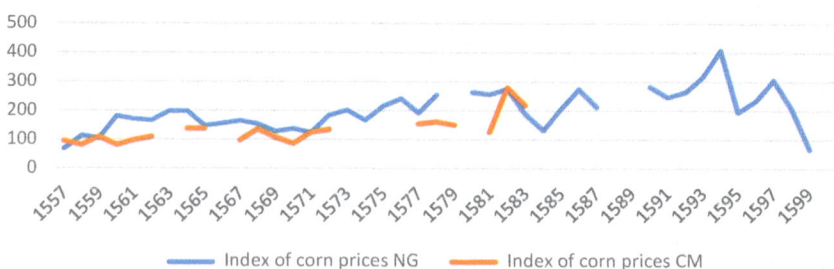

Fig. 9.1 Index of corn prices Nueva Galicia and central Mexico, 1557–1600 (base: 1557–1559) (*Source* Borah 1992)

Table 9.2 Results of difference of means test of corn prices indices Nueva Galicia and central Mexico 1557–1600

Test for equality of means between series
Date: 04/19/17; Time: 13:39
Sample: 1557–1600
Included observations: 44

Method	df	Value	Probability
t-test	59	−4.016192	0.0002
Satterhwaite-Welch t-test*	55.16861	−4.524978	0.0000
Anova F-test	(1, 59)	16.12980	0.0002
Welch F-test*	(1, 55.1686)	20.47543	0.0000

*Test allows for unequal cell variances
Source Borah (1992)

The null hypothesis that prices in Nueva Galicia were higher than in central Mexico was tested using the four different methods presented in Table 9.2: t-test, Satterthwaite-Welch t-test, Anova F-test and Welch F-test.

The four methods gave probability values of zero or close to zero. As a result, it is possible to reject the hypothesis of equality of means at a 0.05 level of significance.

Corn prices in Nueva Galicia were higher than in central Mexico, according to the available data. Since corn was the basis of mine and field labourers' diets, it would be reasonable to assume that corn price increases influenced the rest of the prices of goods consumed by labourers. High prices represented a risk to silver merchants, who would incur losses in the event of the *reales* reaching melting point. Louisa Hoberman suggested that the rate of discounts found a justification in risk, *id est damnum emergens* (Hoberman 1991, p. 69). The possibility of *reales* reaching melting point can be defined as risk.

One further source of market risk mentioned by Hoberman is a fluctuating value of silver. Silver was losing value in international markets due to an increase in global supply. According to Braudel and Spooner, silver depreciated against gold in Europe by 25% between 1500 and 1675 (Braudel and Spooner 1967, pp. 374–486). It was not only the dropping value of silver that may have motivated merchants to adapt the rate of discount by charging between 1 and 3 *reales* in addition to the discount rate of 8.1 *reales*. Currency markets are subject to varying degrees of volatility, and merchants could experience wins as well as losses.

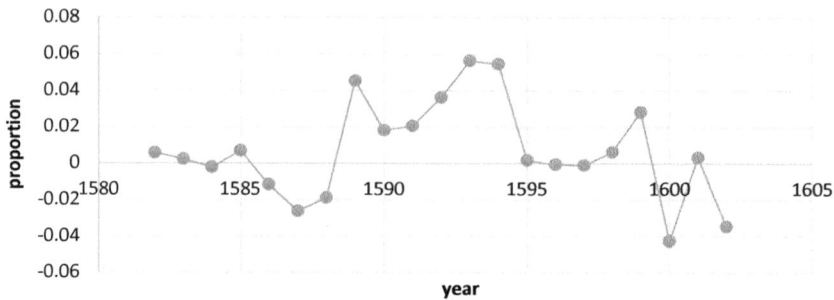

Fig. 9.2 Groot per maravedi deviation from 5-year moving mean (*Source* Denzel 1995)

Figure 9.2 depicts the deviations from the 5-year moving mean presented by the Flemish *Groot* between 1580 and 1603. During the years when the deviation was positive, it was between 1 and 6%. Such values are similar to the 1 to 3 *reales* per mark charged by silver merchants in mining towns in addition to the discount rate of 8.1 *reales*. As shown in Table 9.1, the median fluctuation of the rate of discount was 3.07%.

One further source of risk was transporting silver over the distance of more than 600 km from Zacatecas to Mexico City and vice versa. The opportunity cost of time and the transport costs (about which we know very little) should also be taken into consideration to assess the rationale behind the rate of discount. Once the merchant reached Mexico City with their silver cargo, silver wholesalers took over the task of delivering the bullion to mint merchants, who would then mint the *reales*. All these transactions are likely to have carried commissions, although this point is as yet unclear. One further source of delays was the mining process, which was very time-consuming, taking an average of six months (Pérez Herrero 1988, p. 115).

At an interest rate of 5% p.a., legitimised by *lucrum cessans*, 2.5% could have been charged for minting, in addition to the 3 *reales* (1 for seigniorage and 2 for *braceage*, i.e. minting costs) (Bakewell 1971, p. 212). Once the *reales* were minted, a portion of them returned to mining centres as specie and another portion was used to pay for imported goods. In theory, imported goods had to be paid for in specie. However, in practice, bullion is likely to have been used as a means of payment despite illegality. Sophisticated and at times illegal credit instruments were used in transatlantic trade. Dry exchange was common practice among merchants, who tended not to hold any cash at fairs (on the issue of dry exchange, among other types of transactions, cf. the contribution of Markus A. Denzel in this volume). Their capital tended to be located in Seville, or somewhere in the Atlantic, in transit between Seville and the Spanish Indies (Pérez Herrero 1988, p. 79).

Merchants would issue a bill of exchange to be payable at a fair, and the bill was subsequently exchanged, i.e. leveraged. At the end of the usance established by the bill, it would be settled in Seville with all interest accrued, often through a new bill. If the shipments from the Indies were delayed, merchants would lend their capital on interest and would make the credits appear like exchange transactions. Merchants tended to find their way around bans established by the Crown, and endorsed by the Catholic Church, which threatened to limit any commercial profits. A good example is the ban on cashless sale of merchandises payable in the Indies. Instruments of payment between Spain and the Indies were made to appear like legal exchange, hiding a credit component, often under the guise of insurance.

Credit was present throughout the silver supply chain, starting with the transactions between silver merchants and mine owners, before the silver was even extracted from the ores. At the end of the chain, as coined silver headed towards Spain, credit transactions were carried out using the travelling silver as collateral to pay for goods imported into Spain from other European countries, such as Flanders, to cater for both Spanish and Spanish American demands. Bullion was also used as credit collateral, and on its arrival, gold and silver merchants bought it in exchange for ready-minted coins. Of course, gold and silver merchants charged for their services. They committed to take the metals to the mint. As we have seen, minting was a time-consuming endeavour and hence merchants and even the Treasury preferred to buy ready-minted specie at a fee and have liquidity rather than wait for their metals to be minted.

Unlike in the Indies, bullion was not allowed to circulate as means of exchange in Spain. Because the mint did not hold any stocks of specie, buying specie from gold and silver merchants saved both merchants and the Treasury unnecessary delays as well as risk. Precious metal buyers charged 1 *maravedi* per gold peso and 4 *maravedis* per silver mark, which they exchanged at the (same) rate of 65 *reales* per mark. Gold and silver buyers exerted monopoly power over cash in Seville. There were only a few precious metal buyers in Seville, and during some periods just one.

Despite the high monopolistic or oligopolistic profits that they were bound to earn, precious metal buyers' business was uncertain; it depended on precious metal arrivals, which were at times erratic. Another issue was the fact that the Crown tended to confiscate the merchants' stock of specie in times of economic hardship (Pérez Herrero 1988, pp. 61–63). Precious metals tended to flow out of Spain as payment for government debts and imported goods from the rest of Europe. Exchange bills were issued in order to carry out these payments. The charge for transferring money from Seville to Flanders was between 6 and 7% (Pérez Herrero 1988, p. 64). On the other side of the Atlantic in Veracruz, merchants accumulated specie in order to buy merchandise as soon as shipments of goods arrived. Certain goods that could be sold with high profits were monopolised by merchants. These merchants belonged to the same network as those who issued credit to mine owners. Transatlantic debts could be paid with merchandise instead of money. And, as colonial markets consolidated, merchandise began to travel across the Atlantic in both directions. There was an exchange premium when instruments of payment were issued in Seville to pay for merchandise bought by wholesalers in the Indies. According to Tomás de Mercado, such premia were of 10% for payments in Santo Domingo, 15% in Mexico and Panama, 25% in Peru and 35% in Chile (Mercado 1571, 4, f. 69v).

It is evident that there were a number of trade-related expenses and charges inherent in the silver trade which eroded merchants' profit. After the silver was bought by gold and silver merchants, it would be used by merchants as payment for imported goods either in cash or, more frequently, an exchange bill would be issued to carry out a transfer to pay for merchandise.

Conclusions

A closer look at the rate of discount charged by silver merchants in Zacatecas for exchanging specie against bullion confirms Bakewell's statement that: "The profit of the *rescatador* or *aviador* was probably not great, although the interest seemed extortionate to his customers" (Bakewell 1971, p. 212). According to Louisa Hoberman, a commission of 12.5% was similar to the one charged by tax collectors, and the silver merchants' profit was likely to have been less than 12.5% due to transport and minting expenses (Hoberman 1991, pp. 75–76). Higher goods prices in frontier regions posed a risk to silver merchants, as there was a greater likelihood for the currency (*reales*) to reach melting point than in areas where goods prices were lower. Exchange rate fluctuations were a further source of risk. As evident in the *groot* per *maravedi* series for the relevant period, the additional 1–3 *reales* per mark charged by silver merchants fell within the range of exchange rate fluctuations at international currency fairs. There were other types of risk involved in the silver merchants' trade, such as the risk of theft during transport, for which quantification is perhaps not possible.

Silver merchants faced transport costs that may have been high for long distances, especially given the unsafe journey across Chichimec territory, on top of seigniorage and minting costs (3 *reales* per mark). The 6-month minting process implied an additional cost, given that the legal annual interest rate due to *lucrum cessans* was 5%. Cash payments for imported goods in Veracruz carried a 15% exchange charge. This may be linked to the 6–7% charge for issuing bills of exchange to transfer money from Seville to European fairs, where the goods in question would have been acquired.

Using goods and silver as collateral to engage in further trade activities helped to increase merchant profits, which seem moderate after examining the risks and expenses involved in the silver trade. Those activities also served as ways to reduce the silver merchants' risk exposure, as well as to finance the opportunity costs incurred by the slow transformation process of silver ores into coins. Mining in sixteenth-century and seventeenth-century Mexico was a key element of a wide and complex network of economic activities that shaped silver merchants' portfolios on both sides of the Atlantic. This defined the concept of mining held up by silver merchants, which contrasted with the narrower view of individual mine owners, whose priority was to cover their costs when silver production declined and to maximise their profits in boom times. Unlike silver merchants, mine owners limited possibilities to manage the risk that accompanied their business activities. This limitation put them in a position of disadvantage which manifests itself as resentment against silver merchants.

Bibliography

Primary Sources

Archival and Manuscript Sources
"Carta de Alonso de Peralta, Sidonia," October 10, 1603: Archivo General de Indias, México 258, fs. 559r–564v.
"Carta del factor Francisco de Ibarra," Caval, May 4, 1604": Archivo General de Indias, Mexico 325fs. 1r–3v.
"Petición de Baltasar de Bañuelos a la Corona," Zacatecas, April 2, 1587: Archivo General de Indias, Guadalajara 30, f. 331r.

Source Editions and Old Prints
Elhuyar, Fausto, 1825. *Memoria sobre el influjo de la minería*. Madrid: Imprenta de Amarita.
Humboldt, Alexander von, 1811. *Essaie Politique sûr le Royaume de la Nouvelle Espagne*. Paris: F. Schonel.
Mercado, Tomás de, 1571. *Summa de tratos y contratos*. Seville: H. Díaz.
Orozco y Berra, Manuel, 1855. "Moneda en México." In: *Diccionario Universal de Historia y Geografía*, vol. 5. México: Andrade.

Secondary Sources

Bakewell, Peter, 1971. *Silver Mining and Society in Colonial Mexico Zacatecas 1546–1700*. Cambridge: Cambridge University Press.

Borah, Woodrow, 1951. *New Spain's Century of Depression*. Berkeley: University of California Press.

Borah, Woodrow, 1992. *Price Trends of Royal Tribute Commodities in Nueva Galicia 1557–1598*. Berkeley: University of California Press.

Borah, Woodrow, 1994. *Tendencias de precios de bienes de tributo real en la Nueva Galicia, 1557–1598*. Michoacán: El Colegio de Michoacán.

Brading, D. A., 1970 "Mexican Silver Mining in the Eighteenth Century: The Revival of Zacatecas." In: *Hispanic American Historical Review* 50, 4: 665–681.

Brading, D. A., 1971. *Miners and Merchants in Bourbon Mexico: 1763–1810*. New York: Cambridge University Press.

Brading, D. A.; Harry E. Cross, 1972. "Colonial Silver Mining: Mexico and Peru." In: *The Hispanic American Historical Review* 52, 4: 545–579.

Braudel, Fernand; Spooner, Frank, 1967. "Prices in Europe from 1450 to 1750." In: *The Economy of Expanding Europe in the Sixteenth and Seventeenth Centuries*, edited by. E. E. Rich and C. H. Wilson. Cambridge: Cambridge University Press: 374–486.

Denzel, Markus A., 1995. *Europäische Wechselkurse von 1383 bis 1620*. Stuttgart: Steiner.

García Ruiz, Alfonso, 1954. "La moneda y otros medios de cambio en la Zacatecas colonial." In: *Historia Mexicana* 13: 20–46.

Garner, Richard L., 1988. "Long-Term Silver Mining Trends in Spanish America: A Comparative Analysis of Peru and Mexico." In: *American Historical Review* 93, 4: 889–914.

Giráldez, Arturo, 2012. "Cacao Beans in Colonial Mexico: Small Change in a Global Economy." In: *Money in the Pre-industrial World*, edited by John Munro. London: Pickering and Chatto: 147–161.

Hamilton, Earl, 1934. *American Treasure and the Price Revolution in Spain, 1501–1650*. Cambridge, MA: Harvard University Press.

Hoberman, Louisa S., 1991. *Mexico's Merchant Elite, 1590–1660: Silver, State, and Society*. Durham: Duke University Press.

Hoberman, Louisa S., 1998. "Aportación del mercader de plata a la economía nacional." In: *Crédito en la Nueva España en el siglo XVIII*, edited by M. P. López Cano and G. Del Valle Pavón. México: Instituto Mora: 61–82.

Israel, J. I., 1974. "Mexico and the 'General Crisis' of the Seventeenth Century." In: *Past and Present* 63: 33–57.

Pérez Herrero, Pedro, 1988. *Plata y libranzas. La articulación comercial del México borbónico*. Mexico: El Colegio de México.

Sargent, Thomas J.; Velde, François R., 2003. *The Big Problem of Small Change*. Princeton: Princeton University Press.

TePaske, John, 2007. *New World of Gold and Silver (Atlantic World)*, edited by Kendall Brown. Leiden: Brill.

Vilar, Pierre, 1974. *Or et monnaie dans l'Histoire: 1450–1920*. Paris: Flammarion.

10

Copper Money in Mexico: The Transition from the Eighteenth to the Nineteenth Century

José Enrique Covarrubias

In the last months of 1841, the central region of Mexico, including the capital, experienced an economic and social crisis caused by an excess of copper currency, including a large amount of counterfeit coins that were sometimes indistinguishable from the true ones. People tried to get rid of their coins as fast as possible: some shopkeepers refused to accept them and others accepted them only at a fraction of their value. Riots took place, principally in the poorest quarters, with men and women throwing away the coins in the streets. A classic record of this episode appears in the book *Life in Mexico* (1842), by Fanny Erskine Inglis, the Scottish wife of the Spanish ambassador in Mexico at that time, Marquis Calderón de la Barca, who refers specifically to a mutiny in Toluca, near Mexico City (Calderón de la Barca 1984, p. 384). Between 1829 and 1836, the Mexico City mint produced more than four million pesos in copper coins, and it is not unreasonable to estimate a similar amount of counterfeit coins according to the calculations of contemporaries. The number of pieces in circulation must have

J. E. Covarrubias (✉)
Universidad Nacional Autónoma de México, Mexico City, Mexico

© The Author(s) 2019
R. Pieper et al. (eds.), *Mining, Money and Markets*
in the Early Modern Atlantic, Palgrave Studies in Economic History,
https://doi.org/10.1007/978-3-030-23894-0_10

231

been enormous: their values were 1/4, 1/8 and 1/16 of a real and one real was equal to 1/8 of a peso.

Discontent with copper coins had begun five or six years earlier. In March 1837, the government had dictated a 50% depreciation of the coins. In January of the same year, a national bank had been founded in order to collect and replace the copper currency with a new and better one. Sadly, the task was beyond the capabilities of the bank, whose closure in 1841 was followed by spartan governmental orders and decrees enforcing the recollection of coins. The public obeyed and the replacement took place in 1842.

The story of copper currency in the first decades of the nineteenth century illustrates some of the principal obstacles faced by Mexico in its attempts to establish a modern monetary system. One of them was the financial and technical decay of the old Mexico City mint. A second cause of the failure was the fact that the establishment of monetary order did not figure among the priorities of the Mexican politicians and intellectuals at that time, at least not to the same degree as the endeavour to reorganise and strengthen the treasury. Before considering these two points, it is necessary to speak about the copper production in Mexico and the appearance of copper currency in this country at the end of the eighteenth century.

Tokens and Copper Money in Colonial Mexico

Colonial Mexico (New Spain) had been an important producer of copper in the Spanish Empire. Mining centres in Michoacán and other regions provided enough copper to satisfy local needs and, from the eighteenth century on, to supply the military and sugar industries of Cuba and Spain (Barrett 1981, pp. 73–96). In line with the mercantilist and absolutist spirit of the Spanish Empire, the Crown established a copper distribution monopoly during the second half of the eighteenth century to serve its own needs related to the coinage of silver and the manufacture of cannons and other matériel. Treasury officials in Mexico City fixed a price for the quintal of copper and used their offices in the city as warehouses for that metal. Much of the stored copper was exported; some was also sent to the *Casa de Moneda de México* (Mexico City mint) to be used in the alloying of metals. The local need for copper in New Spain could not be entirely satisfied by the officially distributed metal, and smuggling became commonplace until approximately 1800, when the monopoly ended and copper price controls relaxed.

After an initial precedent in the sixteenth century, copper minting resumed in New Spain at the end of the colonial period. Small purchases were important drivers of this change. For almost three hundred years, the only means to purchase small quantities of items in shops had been the tokens fabricated by local merchants and shopkeepers. Tokens made of copper and other materials were used principally in the so-called *tiendas mestizas*, also known as *cacahuaterías* or *pulperías*, retail shops in Mexico City and other urban centres in New Spain. This kind of trade was practiced by low-level merchants, some of them immigrants from Spain, some natives of Mexico. These shops supplied the lower, middle and upper classes of the population with everyday goods. Wholesale trade, on the other hand, was facilitated with silver money and through wide and flexible credit in the *almacenes* (warehouses) owned by Spanish immigrants from the Peninsula who had ascended through the ranks of traders in New Spain. An intermediate level between the *mestizo* and the *almacenero* trade was represented by the *cajoneros*, who rented stores of the principal marketplace in Mexico City and also used silver money for their deals. The last level of the commercial hierarchy was represented by the so-called *buhoneros* or *mercachifles*, vendors whose transactions took place at stands on plazas or simply went through the streets and alleys while announcing and showing their items. This last form of trade seems to have been carried out originally with cocoa beans or with the tokens made by the owners of *tiendas mestizas*.

The hierarchy of the New Spanish traders has been clearly identified by the historical research. Nevertheless, recent research has thrown significant light upon the social significance and economic scope of the so-called *baratillo* trade, which involved the buying and selling of second-hand merchandise. In the eighteenth century, the *baratillo* trade also included the buying and selling of imported luxurious items, such as glasses, mirrors and screens from China and European countries.

Contrary to the usual narrative, Andrew Konove has recently shown the wide range of classes and interests that became part of the *baratillo* trade (Konove 2018, p. 63) and implicitly confirms the assumption that small-trade tokens were being used in a wider circle than the simple toing and froing of the *mestizo* trade. It is no coincidence that the Mexican Merchant Consulate opposed any proposal of copper money in New Spain with the argument that it would endanger the exclusivity of silver coins as a means of exchange in foreign trade (Covarrubias 2000, pp. 44–46). Copper coins would inundate foreign trade circuits and expel silver ones, or increase their exchange value, or both at the same time, disrupting the carefully established links between New Spain and the markets of Spain and other countries.

From 1760 on, reflecting the reformist spirit which moved statesmen like Campomanes, Floridablanca and Gálvez, subjects and officials demanded the minting of copper currency in New Spain to facilitate purchases in shops and to avoid frauds and abuses by shopkeepers when giving change to customers or accepting pawns. Fraud and usury were also a common problem when using cocoa beans as coins in shops.[1] The copper coinage issued by the Mexico City mint in 1814, 1815, 1816 and 1821 solved these problems in a large part of the country. In some towns, municipal authorities had already minted coins or made tokens for retail trade (Covarrubias 2000, pp. 57–60). But, plagued by civil war (1810–1821) and the consequent scarcity of money, New Spain became the scene of an exotic monetary scenario with a profusion of counterfeit or spontaneously fabricated coins of all kinds and values, including the numerous faulty pieces produced in recently founded provincial mints, established during the Mexican War of Independence in Chihuahua, Durango, Guadalajara, Guanajuato, Sombrerete, Zacatecas, Real de Catorce, Oaxaca, Valladolid and Sierra de Pinos (Ortiz Peralta

---

[1]Ruggiero Romano called the tokens and cocoa beans "pseudocoins" and took them as evidence of an extended natural economy—as opposed to a strict silver coin economy—in many regions of colonial Mexico (Romano 1998, pp. 150–180). Far from assuming this natural economy in New Spain, Vornefeld (1992), Pietschmann (1996), and Covarrubias (2000) put a stronger emphasis on the administrative dimension of monetary problems in Spanish America, principally in the frame of the Bourbon monarchy of the eighteenth century.

1998, pp. 131–154; Velasco Herrera 2017, p. 377). This arrangement continued almost unchanged from the colonial era on to the early years of the Mexican Republic. During this period, the profusion of illegal currency and continuous counterfeiting constantly harmed the poorer classes.

Though the quality of the late minted copper coins was good, the coins were nevertheless criticised and there were public demands to replace them. They were seen as an intolerable remnant of Spanish dominion in Mexico (Bustamante 1980, pp. 222–328) Simultaneously, politicians and writers worried about the popular practice of using soaps or small pieces of metal, leather and wood to make purchases in shops or to barter among particulars, also a vestige of colonial times. As a response, the government decided in 1829 to mint a new national copper coin. The *Casa de Moneda* of Mexico City started the minting. How is it possible that the government did not stop the excessive minting of copper until a law on the subject was passed in July of 1836? What were the circumstances that led to such an unpropitious situation as the superabundance of copper coins? To answer this, we must turn our attention to the coexistence of different mints in independent Mexico and to the federal model that framed their activities at that time.

Mints Administration During the Federal and Central Republic

During the first federal Mexican Republic (1824–1835), coinage revenues belonged to the states. The decree of August 4th, 1824, does not assign the coinage revenues to the "general federal revenues", i.e. to the income of the Mexican federal government (Velasco Herrera 2017, p. 378). The federal congress would determine and make uniform the weight, "law" [purity], value, kind and denomination of the coins in all the states of the federation. The federal treasury was committed to guaranteeing the quality of provincial mints via local inspections. Nevertheless, a certain confusion and ambiguity prevailed about the administration of the different mints since the constitution did not mandate that mints be managed by the federal treasury, nor even by the treasury of the respective state. Consequently, the possibility of leasing the mints to private interests, who would pay a toll to the government, was considered. This republican model of rents gave each federal state the right to lease the mint to a financier or company which would also enjoy the profits of minting.

Guanajuato, Jalisco and Zacatecas illustrate the administrative models of the mints at that time. Since colonial times, these states were important mining districts and of great interest to businessmen as well as national and local officials. In 1828, a ten-year concession to the Guanajuato mint was awarded to an Anglo-Mexican Company. This decision was the result of several discussions and proposals by the Guanajuato state congressional committees and associated dialogues, after a similar but interrupted concession to the same company in 1825. In 1830 a new agreement extended the concession four years, making 1842 the final year of the *contrata*, the name of this kind of contract. At the end of this period, the Anglo-Mexican Company would give the mint to the government. Due to the high profits of this mint, in 1839, under the recently created central republic (1835–1846), the company presented another request to extend the validity of the *contrata*. The result was splendid: the government extended the *contrata* for 14 years after 1841, and the Anglo-Mexican Company also received the *contrata* for the Zacatecas mint in 1845. The company would give 8% of the production of both mints to the government, of which 10,000 pesos would be for the department of Zacatecas and the rest for the department of Guanajuato. The re-establishment of the federal constitution in 1846 brought a reorganisation of the public rents. All the profits from minting were distributed to the federal government.

We also have the case of Jalisco (Guadalajara), where the English businessman Richard Exter received a similar concession for ten years in 1825. It was the same year as Guanajuato's first *contrata*, only a few months after the promulgation of the federal constitution (1824). Nevertheless, the agreement was not concluded and the Guadalajara mint remained under the direct management of the state treasury until 1849.[2]

Zacatecas also decided to subordinate the mint under the direct administration of its own treasury. Inclined to exert its sovereignty to the full in this matter, the case of Zacatecas is nevertheless an example of the gradual loss of coinage revenues by a state in the Mexican Republic owing to an increasing administrative centralisation which began in 1835.

After ten years of receiving all the income of their mint, Zacatecas was in danger of being affected by an 1835 decree that destined any rent or revenue of a state to the central treasury, with the corresponding interference of the bureaucracy in Mexico City. As a result of tense negotiations, the mint revenues remained a rent of that "department", as it was now called by the centralist regime. By January 1836, however, a new decree stated that one half of the department revenues would go to the federal treasury to finance the cost of the war against rebellious Texas. One year later, a new central government, now under President Anastasio Bustamante, assigned all the rents, contributions and properties formerly conceded to the states by the federal constitution to the central government. Once again, negotiations between Zacatecas and the central government resulted in a redistribution of the mint revenues, half and half. The authorities at Zacatecas were allowed to designate their half of the revenues to increase the funds of the mint and to promote increased coining in that jurisdiction.

This system of mixed revenue distribution lasted until 1842, when the central government (the so-called provisional government or *administración provisional* of Santa Anna) temporarily conceded the Zacatecas mint to the British merchant house of Manning and Marshall, a typical example of *agio* practitioners in Mexico during the first three decades in the nineteenth century.[3] But, as we have seen, the definitive *contrata*

---

[2]A relatively exceptional situation for such an important provincial mint, as shown by Velasco (2016, p. 53).

[3]Tenenbaum (1986) exposes some of the still dominant ideas among historians about the *agio* practices in Mexico at that time.

(1845) was for the Anglo-Mexican Company, which enabled it to manage both the mint of Zacatecas and the one of Guanajuato.

The rivalry between the local and central authorities over the distribution of profits for gold and silver coining stimulated a discussion about the degree of power, or *potestad*, of each entity. There was simultaneous power struggle over gold and silver located in mines, as the provincial mints sought to obtain these precious metals to the detriment of the Mexico City mint. As shown in Tables 10.1 and 10.2, the revenues of the old central mint at Mexico City had been conspicuously declining in the first decades of the nineteenth century, due to the emergence of the provincial mints. Maintained under the direct administration of the government and far from regaining its old pre-eminence, the Mexico City mint had been seriously affected by the aggressive newcomers and resented the preference given by many gold and silver producers to the mint at Zacatecas and other provincial towns when it came to transform their metal into money. The good old colonial times when the *Casa de Moneda de México* (the only mint in

Table 10.1 Silver currency produced in Mexico City mint 1824–1842

Year	Pesos	Reales	Granos
1824	3,267,000	2	6
January–August 1825	2,112,758	4	0
September 1825–June 1826	2,733,201	4	6
July 1826–June 1827	2,884,892	4	0
July 1827–June 1828	2,113,487	6	0
July 1828–June 1829	975,652	4	0
July 1829–June 1830	97,358	5	6
July 1830–June 1831	934,142	2	0
July 1831–June 1832	1,103,114	3	0
July 1832–June 1833	1,164,358	6	0
July 1833–June 1834	977,267	4	6
July 1834–June 1835	448,282	1	6
July 1835–June 1836	90,544	0	0
July 1836–June 1837	380,579	1	6
July 1837–December 1838	1,557,845	4	0
1839	1,742,915	6	0
1840	1,917,617	4	0
1841	2,151,496	7	6
1842	1,964,537	0	9

Source Orozco y Berra, *Moneda* (1993, p. 116)

Table 10.2 Zacatecas and Guanajuato minting as a percentage in the national total (1821–1845)

Year	Minting in Zacatecas (pesos)	Minting in Guanajuato (pesos)	National total (pesos)	Zacatecas percentage in national total (pesos)	Guanajuato percentage in national total (pesos)
1821	2,468,577	29,145	9,127,956	27.00	3.19
1822	2,468,577	390,228	9,942,370	24.80	3.92
1823	4,029,031	502,358	9,641,995	41.80	5.21
1824	4,018,062	587,312	9,820,428	40.90	5.98
1825	3,213,356	401,673	9,251,441	34.70	4.34
1826	3,233,266	540,046	8,493,812	38.10	6.36
1827	4,010,820	933,011	9,715,990	41.30	9.60
1828	3,880,630	1,409,644	10,695,963	36.30	13.18
1829	4,009,201	1,902,084	10,261,306	39.10	18.54
1830	4,965,045	2,602,788	11,632,507	42.70	22.38
1831	4,469,450	2,282,882	9,781,047	45.70	23.34
1832	5,012,000	2,752,528	12,218,457	41.00	22.53
1833	5,372,000	3,206,256	12,660,944	42.40	25.32
1834	5,526,600	2,732,948	12,989,525	42.50	21.04
1835	6,154,690	2,407,976	11,832,188	52.00	20.35
1836	5,459,579	2,511,972	12,055,493	45.30	20.84
1837	5,238,253	3,008,024	11,616,302	45.10	25.89
1838	5,115,930	3,028,520	12,179,713	42.00	24.87
1839	4,490,935	3,360,256	12,264,929	36.60	27.40
1840	4,066,310	3,896,668	13,162,565	30.90	29.60
1841	4,386,641	3,736,540	13,544,026	32.40	27.59
1842	5,034,145	3,476,820	13,965,009	36.00	24.90
1843	4,605,862	3,346,664	12,149,169	37.90	27.55
1844	4,429,353	4,635,740	13,691,631	32.40	33.86
1845	4,435,576	4,385,702	15,141,794	29.30	28.96

Source Ibarra Bellon (1998, pp. 188–189)

those years) could process gold and silver in a few days had ended. This decline was something of an embarrassment that damaged the name and prestige of a mint that had been eagerly visited, described and even eulogised by Alexander von Humboldt for its productive capacity in 1803 (Humboldt 1966, pp. 457–461).[4] The cause of this decline was

---

[4]Soria (1994) shows the financial and technical improvements that took place in the mint in that period and were not matched again in the first half of the nineteenth century.

not only the sinking but also the shrinking state of its *fondo dotal*, i.e. the fund which provided money to pay for the entered gold and silver. The mint still had some opportunities to regain its competitiveness. It is true that in 1830, the treasury secretary in the so-called Alamán Administration, Rafael Mangino, considered closing the establishment, but he also considered merely reducing the number of employees and updating the minting machinery. Three years later, the officials of the mint proposed: (1) to reduce or retain the payment of wages for employees and to designate the surplus money to the fund and (2) to facilitate the quick compensation of those who entered their gold or silver in the mint for coining with some percentage of copper money (Velasco Herrera 2016, pp. 111–112).

Though the available documents from the Mexico City mint from the 1830s show that these two measures were taken, the last one had greater consequences (Covarrubias 2000, pp. 149–150). Clearly, minting copper coins in order to pay 1/10 of the introduced gold and silver must have appeared to be the easiest and most bearable way out of the mint's predicament, since the *Casa de Moneda* had been minting copper for some years and faced no unsurmountable impediments to expanding or intensifying the material and human resources available for the task.

Research about the copper money from Mexico City has piled up evidence of the priority given by the authorities to the repair and survival of the treasury over monetary reform. I have previously exposed the problems and dilemmas faced by the Mexican capital mint during the first years of the 1830s (Covarrubias 2000, pp. 147–159), adding to the earlier investigation of Torres (1994), who recently picked up on the subject in a new book (Torres Medina 2013, pp. 221–261). Velasco has increased our knowledge of the Mexico City mint and its relatively low profile in the competition with the provincial mints of Zacatecas and Guanajuato at that time (Velasco Herrera 2016, pp. 113–142).

Scholarly conclusions regarding the "copper crisis" of the 1830s, mentioned at the beginning of this article, can be summarised in the following manner. The drastic reduction of gold and silver coinage in the Mexico City mint was the principal argument for the federal and (from 1835 on) central government to consider the copper coinage as a means to restore, even in a modest degree, the finances of that establishment and make it more competitive. Already in 1829, the activities of the other mints had affected Mexico City's silver coinage: from 2,113,487 in 1827–1828 to 975,652 pesos in 1828–1829. A similar drop was registered between 1833–1834 and 1834–1835, from 977,267 to 448,282 pesos, as shown in Table 10.1. Besides the attraction to the other mints, the depletion of mines in the surroundings of the capital—traditional sources of metal for the mint—accentuated this scarcity. Grants for exporters of uncoined silver, the so-called *plata en pasta*, also contributed to this painful financial decline. The first copper emission by the Mexico City mint in the independent period (1829) was intended to maintain an approximate correspondence between the real and the nominal value of the coins. Due to the size of the coins of highest denominations (*cuartillas* of 1/4 of real) and also to technical difficulties in minting them, the copper emissions in the following years did not include pieces of this value and were confined to coins of 1/8 and 1/16 real. The purity and weight of the coins were reduced to a half and the public recognised this. Unsurprisingly, the copper coins were not accepted in their nominal value and suffered a discount.

The minting of coins with diminished real value worked and induced the government to find a way to repair its own finances. Copper coinage was not only to maintain the prestige and status of the old mint but also to aid the national treasury, which was the ultimate manager of that establishment. In the fiscal year of 1832–1833, copper minting reached enormous proportions: 491,300 pesos instead of the 180,000 of the previous year, and this increased production would not finish until the government ordered it stopped in 1836. The Banco Nacional de Amortización de la Moneda de Cobre was created to collect the copper coins and finance their replacement with new and more reliable pieces, less susceptible to counterfeiting. At the beginning of this paper, I have already explained the prevailing situation in the years between 1837 and 1841.

The Form of Considering Copper Money

But why did the government create that strange national bank, which seems to be a very original, unparalleled design to solve the already described monetary problem?

The perplexity is exacerbated by the fact that the usual approach to solving monetary problems of that sort, i.e. replacing a faulty fiduciary currency, was to create a fund to provide the necessary money and credit, as well as to financially support the mint in its technical operations. A tentative answer to this question can be that this bank followed the example of the Banco de San Carlos, whose objective was to finance the national debt and by extension to fund the wars of the Spanish Empire through emissions of *vales reales* at the beginning of the nineteenth century. A similar imperial bank had been projected immediately after Mexican independence to amortise a faulty money paper printed by the government of Agustín de Iturbide in 1822 (Covarrubias 2000, pp. 111–114). The Mexican "copper-bank" of 1837 was at one time charged with the administration of national goods and the collection of funds for the war against secessionist Texas. But this "copper-bank" was also in charge of managing the tobacco monopoly, a business at the heart of a heated and politicised dispute between a group of greedy businessmen that craved a *contrata* and the corporate subjects that also had part in the branch: the sowers (*cosecheros*) and the manufacturing workers (Walker 1984, pp. 675–705; Covarrubias 1993, pp. 384–400). The bank seems to have been intended to channel and disarm this dispute through a new administrative solution that would simultaneously facilitate monetary amortisation. Moved by the idea of very profitable rents, each side in the struggle claimed a different managerial model (entrepreneurs as concessionaries of the rent against ministerial administration of this one), ignoring that in fact the tobacco monopoly suffered from many impasses with historical antecedents.

The bank, therefore, was more a political instrument to channel and dismantle this struggle than a real economic and administrative solution to the monetary problem. A great deal of ink was expended in Mexican newspapers and pamphlets to express proposals and monetary arguments and to discuss "Copper, Texas and Tobacco" (Torres Medina

2013, pp. 250–261). Most of them remained ineffectual and were swiftly forgotten. The truth is that from 1836, when governmental efforts to stop the copper monetary crisis began, the authorities' real priority was to restore and reorganise the treasury and not to correct the whole monetary chaos. This last problem was understood as an effect of the disarray of public finances.

It is pertinent to mention the reception of a Spanish economist in Mexico who turned out to be influential in the strategy regarding copper coinage. Mexican politicians and pundits of the 1820s and 1830s had read some of the most famous economists of their age. Adam Smith, Jean Baptiste Say, Jean C. L. Simonde de Sismondi, Thomas R. Malthus, David Ricardo, James Mill, etc. were not unfamiliar to them. They also knew the works of more ancient and old-fashioned writers, such as Montesquieu, Baron von Bielfeld, Jacques Necker, David Hume and F. L. A. Ferrier. More influential in Mexico, however, were the contemporary Spanish economists José Canga Argüelles and Álvaro Flórez Estrada. Now, if those Mexican politicians knew about the inflationary effects and other inconveniencies of excessive currency, how could they, in 1836, urged and convinced the congress to decree the validity of counterfeited coins as if they were real (Orozco y Berra 1993, p. 25)?

The centralising impetus during the Mexican Central Republic was fundamentally oriented at enforcing the administrative and financial powers sitting in the capital. As in Guanajuato, where a group of entrepreneurs received the *contrata* for minting and the state government benefited from cash and ready money for emergencies and local needs, the central government launched itself through copper minting in a similar way, though on a larger scale. In the 1830s, copper for the Mexico City mint came not only from the traditional mines in Michoacán but also from the now intensively exploited districts of Durango (Tepezalá), Chihuahua (Santa Rita) and other places (Covarrubias 2000, p. 149). The strategy was obvious: high consumption of copper by the Mexico City mint guaranteed high prices for the metal, which in turn meant good returns for investors in copper mines, many of whom had the same entrepreneurial profile as the Anglo-Mexican Company of Guanajuato. The excessive and unprecedented flood of copper from the Mexico City mint between 1829 and 1836

also had the effect of expelling the copper coins which had been pro-
duced for small circumscriptions in some parts of central Mexico (see
Table 10.3). Some of these coins were of better quality than the Mexico
City copper, according to an Austrian traveller in Mexico in 1838,
Isidore Löwenstern, who wrote on the subject (Löwenstern 1843, pp.
315–317). Table 10.4 shows that copper coins from the Mexico City
mint circulated principally in the states or departments of México,
Puebla, Oaxaca and Veracruz and, to a much lesser degree, in parts of
Jalisco (Junta 1841, pp. 41–42). The copper flood also had the effect of
raising the price of copper across the whole country, even in its crudest
exploitation or production. The increase in price was deliberately pursued by the government, otherwise the massive production of copper
coins could not be profitable for the Mexico City mint, the national
treasury and the copper suppliers.

The crisis of 1836–1837 meant the end of this ambitious system.
The devaluation was implemented in order to formalise the real value

Table 10.3 Copper coining in Mexico City mint and in provincial mints, 1824–1842

Mint	Year	Pesos	Reales	Granos
Chihuahua	1833	18.069	0	0
	1834	15.858	3	0
	1835	16.501	2	0
Guadalajara	1831	730	6	0
	1832	7.066	3	0
	1833	10.692	4	0
	1834	20.461	0	0
	1835	14.102	6	6
	1836	8.164	1	0
San Luis Potosí	October 1827–December 1828	2.45	0	0
	1829	6.501	3	0
	1830	9.05	0	0
	1831–1832	3.996	0	0
	August 1833–February 1835	1.52	0	0
Zacatecas	1824–1827	30.2	0	0
	1828–1829	77.749	4	0
Mexico City		4,950,871	5	9
Total		5,193,984	6	3

Source Orozco y Berra (1993, pp. 118–124)

Table 10.4 Copper money in the interior of Mexico about 1840

Regions, places and owners	Pesos
Department of Puebla	1,100,000
City of Puebla	100,000
Private and corporative owners	400,000
Jurisdictions of Cuernavaca, Cuautla and Izúcar	200,000
Valleys of Toluca, Ixtlahuaca, Temascaltepec, Tenancingo y Zitácuaro	200,000
States and small towns in the north of Mexico	200,000
Department of Oaxaca	400,000
Villages and mountains	300,000
Towns, villages and cities in other departments (Jalisco and some others)	100,000
Total	**3,000,000**

Source Junta Directiva (1841, pp. 41–42)

of copper money and, by diminishing the profits, to dissuade coin counterfeiters. The aim of establishing a national bank was to regard the interest of businessmen close to the central government and send them a message of confidence and even to invite them to have a seat in the executive board of the bank. Additionally, the bank was intended to revitalise the national tobacco monopoly with better outlets for its inventories and stock. In terms of political expedience, the bank appeared as a new institutional instrument to reconcile the aforementioned commercial interests with the interests of miners, land proprietors and clerics, who were also "classes" to be represented on the executive board. Such a pluralistic composition of the executive organ was justified by the idea of a common interest of all these groups in credit (Torres Medina 2013, pp. 243–244). While leasing the tobacco monopoly guaranteed the government a regular income to amortise copper coins and consequently reduce the public debt, the wide field of transactions and exchanges opened to the tobacco company and other businessmen was an important means of enlarging the scope of credit. This schema follows the principles of the Spanish economist José Canga Argüelles who, dismissing the principle of a careful balance between coined money and credit instruments (i.e. between metallic monetary support and fiduciary papers), championed credit as the great booster of circulation in modern economies and recommended the extensive use of warrants, bills of exchanges, orders of payment, paper money and other instruments which he considered substitutes for money (Canga Argüelles 1833, p. 25). To support his statement, Canga Argüelles cited the old Spanish economist or *arbitrista* Luis Valle de la Cerda, who in the seventeenth century considered credit "feigned money" (*dinero fingido*) and emphasised how it increased the frequency and range of economic exchanges and operations. Just when the Mexican national bank began its operations, the government newspaper *Diario del gobierno* cited Canga Argüelles as the great authority to hear.

However, Canga Argüelles's statements were not confirmed by the experiences of many merchants in Mexico at that time, as evidenced in a book by Luis Manuel del Rivero (Rivero 1844, pp. 256–259), a Spanish writer and philosopher who lived in Mexico between 1839 and 1842 and spoke for the resident Spanish businessmen. According to him, people who experienced the final colonial decades and lived to see the first two of independence could testify about the sad decline of credit and money supply in Mexico, a clear consequence of the prevalent *agio* scenario and the death struggle among merchants trying to become the favoured lenders of the government. Mining had also decayed and failed to supply silver in a way that kept pace with the amount of money being extracted by foreign traders. Old-fashioned trade in the Spanish style—that is, in a family net frame and with a heavy reliance on long-term flexible credit (Kicza 1986)—was significantly reduced. That left space for the forms of credit, association and concurrence preferred by the other foreign traders, from Britain, France, Germany and the United States. The activities of these businessmen were conducive to a large money market sustained by a national debt (Tenenbaum 1995, pp. 257–292), while the Spanish merchants preferred, according to Rivero, investments in productive branches of industry (Rivero 1844, pp. 256–259). According to Rivero, a historically proven way to reconcile the public and private financial interests in Mexico had been shown by the mining activities in the eighteenth century, which contributed to a productive investment in the mines and at the same time provided an effective means of exchange among businessmen. Tokens, warrants, bills of exchange and fiduciary money should be in correspondence with a productive basis, so that there could be no real divergence of economic interests between the Crown and the vassals (in the colonial times) or the government and the citizens (after independence) without promoting productivity. Promotion of circulation did not make great sense without this promotion of production. Humboldt had argued in the same vein, though he put more emphasis on the desirability of producing goods which satisfied elemental needs rather than precious metals (Humboldt 1966, pp. 236–239).[5]

It is worthwhile to note that a certain ambiguity already existed in the eighteenth century about the principal reasons for minting copper money in Mexico and its advantages for circulation and production. In a 1766 proposal addressed to the Crown by a neighbour of New Spain (Agustín de Coronas y Paredes), which in turn was followed by consultations with officials and the Merchant Consulate of Mexico City, emphasis fell on the strict monetary advantages of copper currency and the consequent elimination of usury and similarly undesirable practices of shopkeepers. According to the enlightened reformist spirit of the high Spanish bureaucracy in the 1760s and 1770s, discussions of copper minting hinged heavily upon the resulting currency's suitability for trade, saving and transportation. At the end of the century, however, the emphasis shifted to other aspects of the issue: the profits for the treasury.

Mariano Briones Larriqueta wrote a *Memorial* in 1805 to convince the metropolitan bureaucracy about the advantages of copper minting in New Spain. He mentioned usuries, abuses and the other inconveniencies of shop tokens, unsurprising after such a long history of public complaint about those issues. Briones, however, advanced a new argument trying to show a wider dimension of the issue: the potential benefits to the royal treasury. His point was that the interests of the public and the Crown converged in this subject, which concerned both the

---

[5]The reader will find a general history of projectism and economic ideas in México between the eighteenth century and the nineteenth century in Altable et al. (2015). Economic and administrative discussions on money and credit were particularly intense in the 1820s, 1830s and 1840s, focusing principally on the harms or benefits of exporting silver in a *laissez faire, laissez passer* manner (Altable et al. 2015, pp. 119–122, 140–141, 151, 167–169).

availability of sound fractionary money (a benefit for the vassals) and the profitable taxation of copper mining production (a benefit for the Crown). A massive copper coinage would mean a continuous, increased exploitation of that metal and also the opportunity of new incomes for the royal treasury through taxes on the produce of the mines. Briones's proposal appeared during the years of massive monopolistic copper consumption by the Crown, when the price of this metal would not fall under a minimum level that guaranteed the intrinsic value of the coin (Covarrubias 2000, pp. 73–75). It was also the time when the mining authorities in New Spain were pondering the establishment of a melting centre for the copper coming from Michoacán to Mexico City (Covarrubias 2000, p. 32).

Briones's proposal supposed a parallelism between copper and silver as mining and minting products in New Spain. If silver mining and minting had been one of the principal revenues for the royal treasury for almost three centuries, why could copper not be the same? The highest official and scientific authority in mining affairs in New Spain at the time, Fausto de Elhuyar, denied Briones' parallel between the silver and copper minting processes. Elhuyar saw copper minting as a governmental duty and found no reason to draw public revenue through it. Any coin was an eminently useful public object, and, further, copper did not have the same relevance of gold and silver as a merchandise or a means to accumulate capital and wealth. In fact, he reasoned in the opposite way: in an ideal scenario, every minting process (even that of gold and silver) should be free of any fixed right or taxation. This principle was repeated by the first secretaries of the treasury in independent Mexico, who, however, did not eliminate all the charges on gold and silver minting due to the precarious situation of the treasury. State sovereignty was the only justification for any charge on metal trade or minting (Covarrubias 2000, pp. 75–76).

Liberal economic thought reached Mexico during the last colonial and first independent decades. Spanish economists such as José Alonso Ortiz (1796) and Álvaro Flórez Estrada (1812, 1958) wrote about money and credit in order to ban any "bullionism" (a critical identification of national wealth with metal currency) and to recall that no national welfare or prosperity could be achieved without industriousness and hard work.[6] Consequently, a copper currency enthusiast like Briones could only resort to the effects of such a measure in employment (mining and metal industry), not to mention the impulse it would provide for the purchase of certain popular products in small quantities, according to the small fraction values of copper money.

Evidence of the increasing *agio* character of the financial activities and the money market in Mexico in the 1840s appear in the sequence of events after the amortisation of faulty copper money at the end of 1841. Charles Lempriere, an Englishman who travelled in Mexico in 1861 and 1862 and informed the British public about the Mexican affairs, offered a good summary of these events (Lempriere 1862, pp. 258–260, 265–266). In compliance with what had been prescribed by the decree of March 1841 and other government provisions, copper coins were collected by the government at the end of that year and the holders of the copper money received so-called copper certificates of deposit. According to the law, the holders would be reimbursed in new coins within six months. This reimbursement would consist in new, properly minted copper coins. However, problems arose when the government sold the melted metal of the recollected coins instead of minting it again, contrary to what had been announced. Clearly, the holders were not to be paid in the prescribed manner.

After a new decree in 1842, the Mexican government ordered a new way of paying the holders of the copper certificates, resorting to stamped paper. One year later, the provision of payment was reinforced by the mortgage upon the wastelands of the nation in the interior of the country. It also resorted some credits officially before the national independence took place. As a new step in its efforts to pay, in 1843 the authorities decreed the suspension of the circulation of the old depositors' certificates, which would be now substituted with bonds payable to the bearer, who could rely on the aforementioned government pledges

---

[6]Alonso Ortiz criticised Montesquieu's formulation of the quantitative theory while Flórez Estrada combined this theory, in its Humean version, with Smith's "real analysis" of money. Flórez Estrada became more influential in the Spanish-speaking world through the different editions of his *Curso de economía política* (first published in London in 1828), into which he incorporated gradually the principal ideas of the British and French economists. In monetary issues, he always stuck to the mentioned theories of money.

in support of this credit. These, now known as copper bonds, were subject to an interest rate of 6% per annum from 1 January 1844.

Another measure to inspire confidence in the copper holders had been the creation of a special board of creditors, who were in charge of administering the products of the stamped paper, which they in fact did until the end of 1850. At this point, the Mexican Minister of Finance ordered the interruption of this procedure and eliminated the stamped paper as a source of payment for the copper creditors. The copper bonds would now be part of the Consolidated National Debt. This put an end to an arrangement that had been well received by the copper bondholders, who to a great extent were foreign merchants, invariably well connected with the earlier governments and some of their European embassies. Among them seems to have been a large number of French people, whose interests were defended by Mr. Levasseur, their minister plenipotentiary in Mexico. Levasseur reached an agreement or "convention" with the Mexican government in order to redress the losses of the copper holders. This convention was concluded in 1853 and arranged the ongoing payment of the creditors for the later years.

This was not the only example of replacing money with certificates, bonds and other fiduciary means under the system of *agio* procedures with the compliance of the government. A similar story unfolded with regard to the so-called tobacco bonds, issued to pay the farmers of the tobacco monopoly that had been leased to them by the Banco de Amortización. After a financially disastrous experience, the entrepreneurs gave up farming and received from the government a compensation of several millions of pesos in paper. To pay the tobacco bonds, soon consolidated with other public debts, the creditors obtained 25 or 26% of the revenues of the maritime custom houses. As was the case with the copper bonds, the tobacco bonds were progressively concentrated by powerful foreign businessmen, principally the house Martínez del Río Hermanos, originally from Panama but opportunely affiliated to British interests. The house had the right to freely endorse, transfer and give as guarantee the bonds, and anyone who held the bonds would enjoy the same right and the share in the tobacco enterprise.

The tobacco negotiations also ended in an international convention, this time between the Mexican government and the English minister, who acceded to the demand of Martínez del Río and included the debt as an issue of English property. The question was settled in 1855 and as precedent had a sentence of the Supreme Court on the case that was—as Lempriere points out—"against the law of the land, which ordained that a certain class of credits should have assigned for their payment a certain specified fund, with such augmentations as might indemnify the holders for loss" (Lempriere 1862, p. 258). Here, we see again the story of a big business that failed to adjust to the supposedly good scheme of an *agio* circulation, which would in turn in benefit of the treasury by gradually enabling it to consolidate debts and bring order to the public credit market while stimulating the basis of commercial exchanges. Some of the most powerful financial houses remained focused on a principal credit client (the government) and would not submit themselves to national credit systems, funds or institutions (a national bank, for example) because they wished to retain the possibility of resorting to foreign, embassy-supported demands.

Conclusions

This article exposes Mexican copper money troubles and dilemmas in the first half of the nineteenth century in a very general manner. Monetary problems in Mexico at that time were not confined to copper money. In fact, the whole century was full of monetary problems and irregularities in relation to silver money and the issuance of paper money. It is also important to point out that Mexican troubles with currency had two sides, internal and external, of which only the first has been examined here. Preserved documents from the 1830s mention the illegal coinage of Mexican silver pesos outside the country, and there is evidence of counterfeit copper money which came from the United States across the border to Mexico (Torres Medina 1998, pp. 109–110).

It is difficult to understand all these issues without considering the utilitarian way of thinking that prevailed at the end of the eighteenth century in New Spain. After a radical reformist phase in the 1760s and 1770s, a period when statesmen and officials clearly echoed the monetary theories of Montesquieu and Hume, the emphasis shifted to the usefulness of currency for national and international trade and priority was already given to the interest identity or "interest combination" (the favourite expression at that time) between the Crown and its subjects and especially between the royal treasury and the miners and the Royal Mint and its metal suppliers. Briones's *Memorial* exemplifies this at its best. As we have seen, the schema persisted after independence in different states or departments across the country, as we see in the examples of mint-leasing in Guanajuato and Zacatecas.

Canga Argüelles and the ideas about credit and currency he articulated at the end of the 1830s and beginning of the 1840s contrasted with the monetary principles of Flórez Estrada and influential Mexican writers from a couple of years before. Flórez Estrada and these Mexican writers had been staunch opponents of public loans and *agio* practices. Resorting to bonds and similar fiduciary values took the lead among the possible government policies during the years of the Mexican Central Republic and the Santa Anna administration from 1841 to 1845, even during the loss of Mexican territory after the war with the United States and the payment period for the lost territories. It was only after the chaotic and bloody years of civil war and French military intervention between 1857 and 1867 that the old *agio* practices were discredited and dismissed, and progressive control from the government in monetary issues led to the creation of the first national bank in 1925, as well as the establishment of a true modern monetary system in the twentieth century.

Bibliography

Primary Sources

Alonso Ortiz, José, 1796. *Ensayo económico sobre el sistema de la moneda-papel y sobre el crédito público*. Madrid: Imprenta Real.

Canga Argüelles, José, 1833. *Elementos de la ciencia de Hacienda (los publica D. Felipe Canga Argüelles)*. Madrid: J. Palacios.

Flórez Estrada, Álvaro, 1812. *Examen imparcial de las disensiones de la América con la España, de los medios de su reconciliación y de la prosperidad de todas las naciones*. Sevilla: Manuel Ximénez Carreño.

Junta Directiva del Banco Nacional de Amortización de la Moneda de Cobre, 1841. *Informe de la Junta…sobre los diversos proyectos que se han presentado para ella, dirigido a la Comisión de Hacienda de la Cámara de Diputados.* México: Imp. del Águila.

Secondary Sources

Altable, Francisco, et al., 2015. *El mito de una riqueza proverbial: ideas, utopías y proyectos económicos en torno a México en los siglos XVIII y XIX.* México: Instituto de Investigaciones Históricas de la UNAM.

Barrett, Ellinore M., 1981. "Copper in New Spain's Eighteenth Century Economy: Crisis and Resolution." In: *Jahrbuch für Geschichte von Staat, Wirtschaft und Gesellschaft Lateinamerikas,* 18: 73–96.

Barrett, Ellinore M., 1987. *The Mexican Colonial Copper Industry.* Albuquerque: University of New Mexico.

Bustamante, Carlos María, 1980. *La Abispa de Chilpancingo (1821–1823).* México: Miguel Ángel Porrúa.

Calderón de la Barca, M., 1984. *La vida en México, durante una residencia de dos años en ese país.* México: Porrúa.

Covarrubias, José Enrique, 1993. "El Banco Nacional de Amortización de la Moneda de Cobre y la pugna por la renta del tabaco." In: *Los negocios y las ganancias, de la Colonia al México moderno,* edited by Leonor Ludlow and Jorge Silva Riquer. México: Instituto de Investigaciones Dr. José María Luis Mora/Instituto de Investigaciones Históricas de la UNAM: 384–400.

Covarrubias, José Enrique, 2000. *La moneda de cobre en México (1760–1842). Un problema administrativo.* México: Instituto de Investigaciones Dr. José María Luis Mora/Instituto de Investigaciones Históricas de la UNAM.

Flórez Estrada, Álvaro, 1958. "Curso de economía política." In: Álvaro Flórez Estrada, *Obras* I. Madrid: Ediciones Atlas: 1–332.

Humboldt, Alejandro de, 1966. *Ensayo político sobre el reino de la Nueva España.* México: Porrúa.

Ibarra, José Antonio, 1998. "Reforma y fiscalidad republicana en Jalisco: ingresos estatales, contribuciones directas y pacto federal." In: *Hacienda y política, las finanzas públicas y los grupos de poder en la primera república federal mexicana,* edited by José Antonio Serrano and Luis Jáuregui. México: Instituto de Investigaciones Dr. José María Luis Mora/El Colegio de Michoacán: 133–174.

Ibarra Bellon, Araceli, 1998. *El comercio y el poder en México, 1821–1864: la lucha por las fuentes financieras entre el Estado Central y las regiones.* México: Fondo de Cultura Económica/Universidad de Guadalajara.

Kicza, John, 1986. *Empresarios coloniales: familias y negocios en la ciudad de México durante los borbones.* México: Fondo de Cultura Económica.

Konove, Andrew, 2018. *Black Market Capital: Urban Politics and the Shadow Economy in Mexico City.* Oakland: University of California Press.

Lempriere, Charles, 1862. *Notes in Mexico in 1861 and 1862, Politically and Socially Considered.* London: Longman.

Löwenstern, Isidore, 1843. *Le Mexique: souvernirs d'un voyageur.* Paris: Arthus Bertrand.

Matamala, Juan Fernando, 1998. "La casa de moneda de Zacatecas (1810–1842)." In: *La moneda en México, 1750–1920*, edited by José Antonio Bátiz Vázquez and José Enrique Covarrubias. México: El Colegio de México/Instituto de Investigaciones Históricas de la UNAM/El Colegio de Michoacán: 169–185.

Matamala, Juan Fernando, 2008. "Las casas de moneda foráneas (1810–1905)." In: *Historias*, 71: 61–85.

Muñoz, Miguel L., 1976. *Tlacos y pilones. La moneda del pueblo de México.* México: Fomento Cultural Banamex.

Orozco y Berra, Manuel, 1993. *Moneda en México.* México: Banco de México.

Ortiz Peralta, Rina, 1998. "Las casas de moneda provinciales en México en el siglo XIX." In: *La moneda en México, 1750–1920*, edited by José Antonio Bátiz Vázquez and José Enrique Covarrubias. México: El Colegio de México/Instituto de Investigaciones Históricas de la UNAM/El Colegio de Michoacán: 131–154.

Pietschmann, Horst, 1996. "Dinero y crédito en la economía mexicana (1750–1810). Reflexiones sobre el estado actual de las investigaciones." In: *Históricas. Boletín del Instituto de Investigaciones Históricas de la UNAM* 47: 27–52.

Rivero, Luis Manuel del, 1844. *Méjico en 1842.* Madrid: Imprenta de D. Eusebio Aguado.

Romano, Ruggiero, 1998. *Monedas, seudomonedas y circulación monetaria en las economías de México.* México: Fideicomiso Historia de las Américas/Fondo de Cultura Económica.

Soria Murillo, Vítor M., 1994. *La Casa de Moneda de México bajo la administración borbónica, 1733–1821.* México: Universidad Metropolitana de Iztapalapa, División de Ciencias Sociales y Humanidades.

Tenenbaum, Barbara A., 1986. *The Politics of Penury: Debts and Taxes in Mexico, 1821–1856*. Albuquerque: University of New Mexico.

Tenenbaum, Barbara A., 1995. "Mexico´s Money Market and the Internal Debt, 1821–1855." In: *La deuda pública en América Latina en perspectiva histórica*, edited by Reinhard Liehr. Madrid: Vervuert Iberoamericana: 257–292.

Torres Medina, Javier, 1994. *De monedas y motines: los problemas del cobre durante la primera república central de México, 1835–1842*. M.D. thesis, Universidad Nacional Autónoma de México.

Torres Medina, Javier, 1998. "La ronda de los monederos falsos. Falsificadores de moneda de cobre (1835–1842)." In: *La moneda en México, 1750–1920*, edited by José Antonio Bátiz Vázquez and José Enrique Covarrubias. México: El Colegio de México/Instituto de Investigaciones Históricas de la UNAM/El Colegio de Michoacán: 107–130.

Torres Medina, Javier, 2013. *Centralismo y reorganización. La Hacienda pública y la administración durante la primera república central de México, 1835–1842*. México: Instituto de Investigaciones Dr. José María Luis Mora.

Velasco Herrera, Omar, 2016. *Política, ingresos y negociación: el arrendamiento de las casas de moneda de Guanajuato, Zacatecas y la ciudad de México frente a la construcción de la Hacienda pública nacional, 1825–1857*. Ph.D. thesis, Instituto de Investigaciones Dr. José María Luis Mora.

Velasco Herrera, Omar, 2017. "Orden constitucional y repartición de rentas: la disputa por los ingresos de las casas de moneda de Guanajuato y Zacatecas, 1824–1845." In: *Procesos constitucionales mexicanos: la Constitución de 1824 y la antigua constitución*, edited by Beatriz Rojas. México: Instituto de Investigaciones Dr. José María Luis Mora: 376–395.

Vornefeld, Ruth A., 1992. *Spanische Geldpolitik in Hispanoamerika 1750–1808. Konzepte und Massnahmen im Rahmen der bourbonischen Reformpolitik*. Stuttgart: Steiner.

Walker, David W., 1984. "Business as Usual: The Empresa del Tabaco in Mexico, 1837–1844." In: *Hispanic American Historical Review*, 64, 4: 675–705.

11

Minting the Picture: Machines and Coinage in Transition from the Sixteenth to the Eighteenth Century

Harald Kleinberger-Pierer

Minting machines shaped the coins as well as the money market of the early modern era. The technology used to produce coins had a strong influence on their design. Progress in coinage technology, therefore, had the potential not only to affect money markets by reducing production costs and increasing output, but also to define the appearance of the coin itself.

Drawings and prints are a particularly good source with which to analyse and trace the development and usage of minting technology in history, and previous research has shown some interest. Volker Benad-Wagenhoff produced a comprehensive analysis of a wide range of pictures, examining the development of coining technology from the beginning of the "mechanisation" of the coining process in the Renaissance up to the nineteenth century, when industrialised production lines for coins became standard (Benad-Wagenhoff 2008). In addition, previous research has highlighted the crucial role technical drawings played in the nineteenth-century process of industrialisation,

H. Kleinberger-Pierer (✉)
Institute of History, University of Graz, Graz, Austria

© The Author(s) 2019
R. Pieper et al. (eds.), *Mining, Money and Markets in the Early Modern Atlantic*, Palgrave Studies in Economic History,
https://doi.org/10.1007/978-3-030-23894-0_11

acting as important vectors for the transfer and documentation of technology (cf. Brown 1999; König 1999).

Technical drawings from the Middle Ages up to 1700 have been comprehensively analysed (Lefèvre 2004). Technical drawings in the eighteenth century, on the other hand, and the transition from linear perspective to orthographic view have not been sufficiently studied until now. In order to provide insights into the different aspects of design and usage of technical drawings in early modern times, this study employs qualitative and quantitative analyses of drawings from before and after 1700. In particular, this approach highlights the shift in the use of perspective in art as well as engineering between 1500 and 1800.

This paper argues that the most important elements of technical drawings for the documentation and transfer of technology were developed and used in engineering long before the nineteenth century. By analysing selected drawings concerning minting technology, this article demonstrates the development and importance of technical documentation between approximately 1500 and 1800.

Machine technology is a good example of the importance of technical documentation in early modern times. Due to its critical importance for the economy, coinage technology was a princely affair (cf. Schmitz-Esser 2008). Coinage technology also, in general, used more metal than other machinery. Machines of the eighteenth century were predominantly made of wood (cf. Radkau 2008, p. 73ff.). In contrast, the construction of machines like the screw press, *Taschenwerke* (rocker press), piercing tools, drop press and other minting technologies employed a significant amount of metal. Machines used in coinage, especially those working with rolls, had to be precisely constructed. Due to the confluence of high quality and precision, as well as the use of large quantities of metal, coinage technology was particularly "high tech" in early modern times. Consequently, this paper focuses on drawings of machine technology related to coinage.

From the eighteenth century, drawings became important vectors for knowledge in the field of engineering. The eighteenth century also saw an enormous increase in the production of precious metals and the expansion of commerce with Asia, for which silver coins were indispensable. In addition, technical literature containing technical drawings became widespread during the eighteenth century in comparison with earlier centuries (cf. Kronick 1976; Ferguson 1977), illustrating the growth of demand for accessible recorded technical knowledge.

Drawings are subject to temporal change, showing not a *reflection* of a presumed reality but an *interpretation* or *idea* of it. Today, technical drawings in general are classified as scientific pictures, which can be visual aids for communication under specific circumstances, following certain rules and norms of depiction (Madsen and Madsen 2017; Heßler 2006; Boehm 2004, 2007). In contrast, early modern technical drawings cannot be classified solely as scientific images, but also as artistic ones. They combined aesthetics, artistic expression, representation, staging and technical documentation and focused technical communication in one frame. In fact, a separation between the artistic and scientific elements of the image is neither possible nor expedient (cf. Hensel 2011, p. 10ff; Lazardzig 2007; Bredekamp et al. 2008). Consequently, this article considers the expression, characterisation and methods of depiction, as well as the context and target audiences of the drawings.

The main sources of early modern technical drawings are printed technical literature. Although technical literature had some specific regional characteristics, in general it was largely similar across Europe and the Americas. Technical literature moved across borders and regional boundaries as well as being translated into other languages.[1] Drawing techniques were also comparable across different regions and were frequently based on similar instructions and textbooks (cf. Andersen 2007).

This paper analyses early modern drawing methods in two main regions: France (Félibien 1690, p. 359, pl. LIV) and the mining centre Clausthal-Zellerfeld in the German-speaking territories (Calvör 1763). France and the mining centres in German-speaking territories were key players in the realms of arts and technology in the early modern period. The first printed example discussed in this article dates to the seventeenth century and covers the architect and art theoretician

---

[1]For example, one of the most considerable textbooks for mining of the eighteenth century from the instructor (Delius 1773) which was translated 1778 in French.

André Félibien. It demonstrates the importance of the French art industry and the quality of prints and drawings in early modern Europe. The second printed example, from Calvör (1763), illustrates the use of technical drawings in textbooks for mining of the eighteenth century. Printed technical literature regarding mining technology appears primarily in the eighteenth century and was often related to the mining schools and academies emerging throughout Europe at this time (cf. Schleiff and Konečný 2013). Consequently, this study draws upon the first instructional text related to mining and its educational institutions: the textbook of the instructor Henning Calvör, focusing on mechanics in the mining centre Clausthal-Zellerfeld, Oberharz. Calvör's text also served as a role model for later mining publications, underlining its importance. Additionally, Calvör's drawings of minting technology were reused in the economics encyclopaedia of Johann Georg Krünitz (Flörke and Krünitz 1805).

Because minting technology was crucial for early modern economies, it was frequently the focus of the court and administration (cf. Schmitz-Esser 2008; Benad-Wagenhoff 2008). The first hand-drawn example presented in this paper is an illustration of a screw press from the early fifteenth century (Weimar Ingenieurkunst- und Wunderbuch 1520, p. 37). While the specific context of this drawing is lost, the design points to an important shift in depiction from the fifteenth to the sixteenth century. The second hand-drawn example (Architectural Drawing 1790s), on the other hand, was chosen not because of its design but because it represents a shift towards steam-powered minting technology at the end of the eighteenth century. This machine design, originally developed by Matthew Boulton, not only showcases the style employed for technical drawings in the eighteenth century, but also demonstrates the connection between arts and engineering in this period (Baynes and Pugh 1981).

In the first part of the paper, two early modern drawing methods, linear perspective and orthographic view, are presented and their construction is explained. In the second part, specific drawings in the two different perspectives are discussed in detail. Sources containing imagery of early modern coinage technology exist in both prints and hand-drawn pictures and span a period from approximately 1500 to the end of the eighteenth century. To underpin these specific examples, I will also present the findings of a comprehensive analysis of drawing methods of complex machinery in printed technical literature from 1600 to 1800. Based on a screening of library catalogues of engineering schools and other institutions related to early modern engineering, I will examine the design and perspective methods of the hundred most important books containing drawings of machine technology.

Perspective: The View on the Machine

Today, technical drawings are used to depict an object with its measurements. For this purpose, engineers can employ a broad range of techniques and depiction. In general, the techniques of depiction are defined by established norms, so the transfer of technical information via drawings is possible in an unambiguous, universally understandable way. It is important to note that the design of the drawing and the methods of depiction depend upon and simultaneously affect the type of application and possibilities for communication. General views, design drawings, freehand sketches, part drawings and blueprints all have different purposes and leverage different methods and norms (cf. Madsen and Madsen 2017; Ferguson 1994).

One of the main criteria for technical drawings is perceptivity, the *point of view* in the drawing. In the early modern period, linear perspective and orthographic projection (parallel projection) were the most commonly used perspectives. Other parallel projections used for technical purposes, such as axonometric and isometric projections, only played minor roles before the nineteenth century.[2]

Linear perspective was employed in artwork from the fifteenth century onwards because it enabled the artist to display objects in a perfectly geometrical view. The principle of the linear perspective is to

---

[2]It was the chemist William Farish who consistently defined and applied the isometric projection in the first half of the nineteenth century by an article in the Cambridge Philosophical Transactions (Farish 1822).

display the object using one or more vanishing points. This view is similar to the way we see: objects grow smaller with distance and parallel lines converge into a vanishing point. However, drawings in linear perspective are difficult to make true-to-scale (cf. Bärtschi 1976). The orthographic projection is similarly based on geometrical correlation. Orthographic projection is a mode of parallel projection which makes it possible to represent a three-dimensional object in two dimensions. Additionally, it is used to create drawings in which the principal axes or planes of the object are parallel with the projection plane (also known as multiview or combined orthographic view). This projection makes it very easy to render the object true-to-scale, and as a result, it is used for technical documentation even in the present day (Madsen and Madsen 2017; Maynard 2005).

Examples of the construction and application of these two views appear in an early seventeenth-century treatise on perspective by the mathematician, architect and engineer Andreas Albrecht. In his drawing (Fig. 11.1), a wayside cross is presented in both linear and orthographic views in order to demonstrate the construction of perspective. The left upper drawings are two different views of the cross in orthographic perspective, while the left lower one shows the cross in fully developed linear perspective. The three drawings on the right demonstrate how the linear perspective is constructed on paper. While in the linear perspective one view gives a good impression of the subject, the orthographic view requires two or more views. In contrast to the linear perspective, the object in the orthographic view appears true-to-scale.

The Oblique-Linear Perspective on the Minting Machine

Drawings of machines in manuscripts appear throughout the Middle Ages in different styles and designs, often without following any unifying concept (cf. McGee 2004; Vigevano and Ostuni 1993; Leng 2004). A significant shift in methods of depiction in arts and engineering, which were highly intertwined areas in the early modern period, occurred during the Renaissance: the rediscovery of the

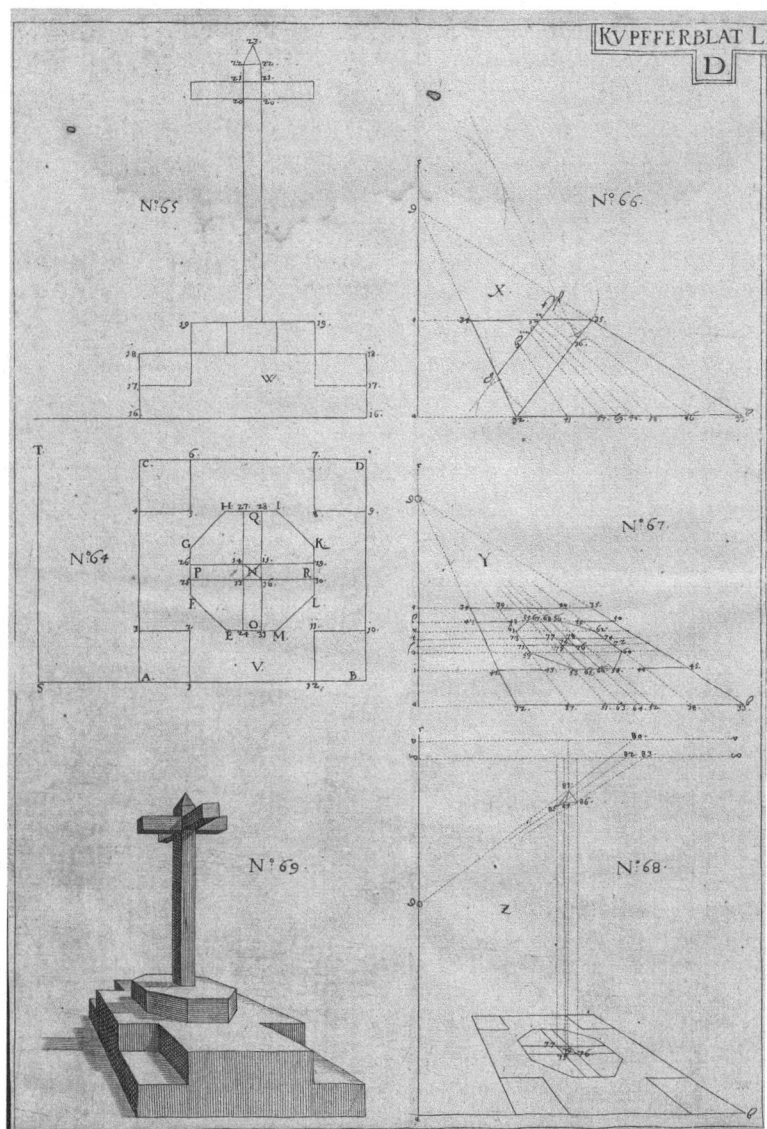

Fig. 11.1 Example of linear and orthographic perspective, wayside cross, left upper corner two views in orthographic view, other pictures in linear perspective (*Source* Albrecht [1623, Pl. D, No. 64–69])

linear perspective (Edgerton 1975), which became a standard in the arts for centuries. This section discusses two examples of the application of linear perspective for coinage technology from the sixteenth and seventeenth centuries. The first example concerns a drawing from a sixteenth-century manuscript; the second is a French print from the second half of the seventeenth century.

Early Engineering and Perspective: The Art of War

Essential sources for the depiction of technical devices, including machines for coinage, appear in a particular type of manuscript called "Kriegsbücher" or "books for the art of war". These manuscripts, mainly written and distributed in the fifteenth and sixteenth centuries, are collections of hand-drawn illustrations of devices and objects related to artillery, the arts of fireworks and fields related to engineering in general. Such commonly used manuscripts were often dedicated to an important ruler and used for administrative and, *inter alia*, representational purposes (Leng 2002, 2004).

The first example of early modern minting technology discussed here is a hand-drawn depiction of a screw press included in one of these "book of war" manuscripts (Weimar Ingenieurkunst- und Wunderbuch 1520, p. 37). This manuscript, produced by anonymous authors and illustrators around 1520, is part of the collection of the "Herzogin Anna Amalia Bibliothek" (Klassik Stiftung Weimar). The manuscript includes a range of drawings on 325 parchment folios, dealing with artillery, fortifications, weapons, machines,[3] wonders/magic and architecture in both practical and fictional ways. The drawings were produced by different illustrators and are often borrowed from other manuscripts, for example the fifteenth-century's most famous manuscript, Konrad Keyser's "Bellifortis" (for more details of this manuscript see Kratzsch 2017, p. 180ff).

---

[3]Before the nineteenth century, the term "machine" meant any mechanical object with moving parts and thus also includes what we would now understand as "tools" (scissors, pliers).

Although the exact origin and ultimate purpose of the screw press drawing are unknown, the depicted objects and design of the machine give us some hints about the circumstances in which the drawing was created. It is a replication of a drawing in the "Kriegsbuch" of Ludwig von Eyb, manufactured around 1500 (Eyb 1500, f. 116r).[4] The screw press in this manuscript appears in two full views from different angles, surrounded by depictions of the individual parts of the device. Compared to other early modern (pre-eighteenth century) depictions of machinery, the drawing is very detailed; according to David McGee, machine designs from the fifteenth to the eighteenth century usually only consisted of one full view of the device in linear perspective (McGee 2004, p. 53). In contrast, this presentation shows two full views, as well as details of individual parts.

Screw presses were used from the fifteenth century onwards and applied techniques that were formerly used in the extraction of oil or other fluids from seeds and fruits. This device represents one of the first uses of so-called machine tool technology[5] in the coinage process as the motion of the tool was determined by the design of the machine. While before the application of that technology, the outcome had been highly dependent on the skills of craftsmen who struck the coin by hand, and the screw press guided the force repeatedly in the same way.

The device is drawn in an oblique-linear perspective (cf. Scolari 2012). Some elements are presented in linear perspective, while others are drawn more arbitrarily. The application of linear perspectives by artists as well as engineers was a visual revolution around 1500 and permanently altered the methods of depiction in European artwork towards a precise geometrical construction (cf. Edgerton 1975). Linear perspective replaced the "flat" and non-perspective (or anti-perspective) view in drawings. As some elements are not depicted accurately, this drawing represents a transition from a non-perspective to a geometrical depiction of objects in the manner of a linear perspective (for other examples, see

---

[4]For the connection between the drawing presented here and the *Kriegsbuch*, see Leng (2002, p. 292ff.).

[5]For more details on machine tools and their significance for the industrialisation process of the nineteenth century, see Paulinyi (1991).

Leng 2004). The linear perspective remained an important method for artistic painting throughout the early modern period and was generally constructed on a more accurate, geometrical basis than this example. In the next section, we will discuss a drawing that illustrates a development up to the seventeenth century, when drawings of machinery began to draw upon a common concept and perfectly utilise the pictorial space.

Perfection and Arts: Linear Perspective in the Seventeenth Century

Figure 11.2 shows a symbolic collection by the architect and art theorist André Félibien, depicting machines used for minting and coinage. This drawing first appeared in his 1674 work (Félibien 1690, second edition), which translates into "Principles of architecture, sculpture, painting and other related arts, with a dictionary of terms specific to each of these arts". The term "other related arts" included manufactures and crafts like coinage.

Félibien's print (Fig. 11.2) depicts a number of different machines and tools used in the coinage process. A *Balancier* and a drop press appear in the foreground; the background shows a piercing machine and cutter and hand-operated pliers on the floor of the workshop. In comparison with the screw press from the sixteenth century, this drawing is much better designed and executed. The linear perspective is fully developed and precisely constructed, meeting the highest artistic standards.

Félibien held a key position in the administration for art and architecture in France during the seventeenth century. His drawing was printed in the same period that Jean-Baptiste Colbert consolidated a mercantilist programme for science and art, most famously including the foundation of the *Académie des sciences* in Paris in 1666 (cf. Hahn 1971). Similarly, the *Académie royale de peinture et de sculpture* was founded for the promotion and control of the arts. Félebien was closely involved in the *Académie royale de peinture et de sculpture*. This institution provided a broad education for artists, especially regarding perspective and geometry (Bettag 2009; Michel 2018). Besides his involvement in the *Académie royale d'architecture* and the *Académie*

Fig. 11.2 Collection of machines and tools for coinage in linear perspective, copperplate (*Source* Félibien [1690, p. 359, pl. LIV])

royale des Inscriptions, Félebien also held the position of "Historiographe du Roy et de ses bastiments, des arts et manufactures de France". Much of Félibien's activity was related to these positions, where he promoted arts and technology through his writing (cf. Germer 1997 for a comprehensive work on the life and work of André Félibien). His work and engagement in the scientific and cultural efforts of the French crown were completely intertwined with his work *Des principes* as well as the drawing presented here.

Félibien's work *Des principes* is marked, according to its title, as a "dictionary". However, it also cleaves to the programme of the eighteenth-century *Encyclopédie ou Dictionnaire raisonné des sciences, des arts et des métiers* and other early modern encyclopaedias: compilation, description and the dissemination of knowledge and skills (cf. Pannabecker 1994, 1998; Fröhner 1994; Birn 1985). Félibien's work can therefore be seen as a forerunner of the comprehensive eighteenth-century encyclopaedia, contributing to the dissemination of knowledge.

Félibien did not aim to show the machines and tools in detail or to give exact instructions for their reconstruction, but rather to give an overview of what was possible and used in practice for different applications. The different machines and tools, presented in one view (a linear perspective), characterise the drawing as a museum or a collection of wonders rather than as an actual workshop. Besides the accomplished use of the linear perspective, this picture illustrates the close connection between art, craftsmanship and engineering in the seventeenth century. In some ways, arts and engineering remained connected into the eighteenth century, but in terms of visual presentation, they developed in different directions. While the arts adhered to linear perspective, engineering oriented towards the orthographic view.

The Orthographic View on the Minting Machine

From 1700 on, the orthographic view began to challenge linear perspective as the primary form of depiction. With the change of view, the application of technical drawings also changed. Before 1700, drawings mainly served representational purposes, but now they gained a more practical, technical purpose. This is evident in the introductions of textbooks for machines and mechanics, which more explicitly highlighted engineers and mechanics as target groups. Furthermore, the publication of technical literature in the eighteenth century intertwined with the foundation of institutions related to technical teaching; technical education required instructional literature. The next section discusses one piece of this "institutional" literature, demonstrating the implementation of the orthographic view for technical purposes.

Mining and Minting: Textbook for Practitioners

Coinage technology was closely related to the mining industry, so it is no surprise that the textbooks and instructional manuals which appeared in the second half of the eighteenth century increasingly dealt with aspects of coinage. One of the early practically oriented textbooks was written by Henning Calvör, an instructor (from 1713) and later rector at the lyceum in Clausthal-Zellerfeld (Oberharz). At the lyceum, Calvör taught mathematics as well as mechanics and (machine) engineering, although his original profession was theology. His efforts to establish special courses for mining officers—which would also become the basis and inspiration for the academy of mining founded in Clausthal in 1775—were of outstanding importance (Scharlau 1990, p. 71ff).

In 1763, Calvör published a comprehensive work about the technology used in mining, surveying, blasting and smelting, which covered the whole process from ore to workable metal. His efforts led to other publications on mining technology in the eighteenth century, for instance the publications of Christoph Traugott Delius (1773), instructor for Cameralistics and Mining at the academy of mining in Schemnitz (today Banská Štiavnica, Slovakia), and Johann Friedrich Lempe, professor for Mathematics and Physics at the mining academy in Freiberg, Saxony (Lempe 1795–1797). All three books targeted practitioners and sought to instruct mining administrators and engineers in mining schools and academies (see Introductions in Calvör 1763; Delius 1773; Lempe 1795–1797). Calvör's publication, furthermore, had an impact beyond the realm of mining. His drawings and descriptions of coinage technology found their way into Krünitz's encyclopaedia around 1800 (Flörke and Krünitz 1805). In contrast to the previous example, this source contains a large number of drawings and displays the details of machines used in coinage, unlike any other piece of eighteenth-century technical literature. The focus on technical detail is highlighted in the title of Calvör's work: "von den Münzmaschinen"—"of coinage machines". Calvör provided the reader with detailed descriptions of machines and tools related to coinage, such as hammer presses, knurling machines, screw presses, machine parts for drive and gear ratio and tools for metalworking (for the drawings see Calvör 1763, Tab. XXIII, XXIV).

The drawings in Calvör's works were not exclusively presented in orthographic view, but also used a linear perspective for specific purposes and illustrations. Although drawings no longer relied exclusively on linear perspective anymore, it still appeared in eighteenth-century technical literature. In particular, full views of machines often used linear perspective, while detailed views were presented true-to-scale in orthographic view. Calvör also used linear perspective as a didactic tool for presenting the machine in an approachable way, while providing the details of parts and assembly of the machine in other illustrations with orthographic views.

Calvör's work demonstrates the use of orthographic view for technical documentation in printed literature, especially in material designed for other practitioners and as a textbook for institutions in technical education. Corresponding to the presence of orthographic, true-to-scale technical documentation in printed textbooks, this perspective was widely used in the practice of engineering in the eighteenth century. The last example presented in this paper, a hand-drawn illustration, illustrates the practical use of the orthographic projection and its connection with the representational use of technical drawings in the early modern period.

Acceleration of the Minting Process: Machines Under Pressure

The last example is one of the first industrial coinage machines, built by Mathew Boulton at the end of the eighteenth century. Boulton, a partner of James Watt, was an English engineer and medallist whose machine designs influenced not just the steam engine, but also the coinage technology of his time. In cooperation with Jean Pierre Droz, he developed a steam-driven minting machine for the efficient production of copper coins in the 1780s. The coins produced with this machine were difficult to counterfeit because they were embossed with complex patterns, but were still cost-effective to manufacture. Furthermore, the machine produced consistently designed coins at a much faster pace than other minting machines (Doty 1990).

Another version of Boulton's steam-driven machine was constructed for the production of silver coins. A drawing of the machine exists in the collection of the Museum of Applied Arts & Sciences in Sydney. The drawing was a proposal for the Russian mint in St. Petersburg, made in the second half of the 1790s by Alexey Nikolaevich Olenin and an architect named Baboshin, who took charge of minting from 1797 onwards. This drawing was one of several concerning the modernisation of the St. Petersburg mint with new machines for gold, silver and copper coins. Although this machine ultimately was not constructed in Russia, similar machines were built later (cf. details in Rudder 2009a, b; Doty 1998, pp. 74–123), and the drawing itself allows for comparison between different ways of recording technical knowledge in the early modern era.

The design of the drawing is very different from the previous examples: first, the drawing is in an orthographic view. It is measured and constructed true-to-scale and follows the design of contemporary technical documentation. Unusual for early modern technical drawings, the scale is given as an inscription ("two 10ths of an inch per foot", bottom centre) and not as a drawn scale directly on the paper with a division. Second, the target audience is similar to that of the screw press drawing from around 1500. The drawing is realised in a colourful way with the highest standard and precision on a large drawing paper. Thus, it would seem also to be intended for representational purposes and directly aimed at higher administration and the court. On the other hand, tools and perspectives are employed as in other technical drawings intended for practical use in engineering. We can assume, therefore, that this drawing had a dual purpose—technical documentation *and* representation.

Third, the dual use of the drawing is emphasised by its colourful design. While workshop drawings in general used monochromatic black or grey, drawings of this kind have more colours and show more artistic efforts, like shade.[6] On the other hand, the use of colour in this particular drawing had a practical purpose: it marks the different materials used in building the machine. This shows the ambiguous, unstandardised characteristics of early modern technical drawings.

Art and engineering in the early modern period in general were not considered separate, and this is particularly true of the eighteenth century. Art schools in the eighteenth century often dealt with artistic as well as engineering and surveying drawing. Drawing courses in technical schools also gave instructions in artistic and landscape painting (cf. Kleinberger-Pierer 2018, ch. 6; Plank 1999). The link between art and engineering is also evident in the drawing of the steam-driven coinage machine. On the one hand, the drawing is in orthographic view and true-to-scale, while on the other hand, the illustrators used colour, shading and other elements with primarily artistic representational purposes.

Framing the Transition of Perspective: Overall Evaluation

Between the creation of the exemplar drawings from the Renaissance and the eighteenth century, a revolutionary change occurred in European illustrators' methods of depiction. Before 1700, an artistic

---

[6]For the method of shading in historical technical drawings, see, for example, de Montesson (1790), a textbook for shading in drawings in the eighteenth century.

approach to depicting machines for coinage prevailed; after 1700, the technical aspects of machinery came to the fore. This change was not limited to drawings regarding coinage technology, but appeared on a large scale all around Europe and the Americas.

This is illustrated by a comprehensive analysis of printed technical literature spanning the period between 1600 and 1800 and drawing upon imagery of machine technology of different types and areas of application. My analysis included (a) printed technical literature, selected based on its dual educational and practical usage and (b) the personal belongings of engineers. Book inventories and catalogues from the Prague engineering school (Černá-Šlapáková 1971), the mining school/ academy of Schemnitz (Zsámboki 1978) and the polytechnic school of Vienna (Martin 1850) helped to determine which technical literature was utilised for the teaching and practice of engineering in the early modern period. In addition, I also examined bibliographies for technical machine science, handbooks for technical administration and the personal libraries of engineers.[7]

Altogether, my analysis considered the use of drawings in one hundred publications between 1600 and 1800. Effectively 1525 drawings of machines dating from 1600 to 1800 were analysed, particularly regarding their application, use of perspective, measurements and other characteristics. The findings from this quantitative analysis support the argument of a change in depiction described in this paper: before 1700, approximately 95% of drawings concerning complex machinery in printed literature were presented in linear perspective. Overall, 578 of 617 drawings from the period between 1600 and 1700 exclusively used linear perspective for depicting machinery. After 1700, in contrast, the majority used orthographic view (415 exclusively and 135 in combination with linear perspective, from a total of 908 drawings), and only around 25% were still exclusively presented in linear perspective.

The orthographic view enabled engineers to communicate and record technical devices true-to-scale. With linear perspective, in contrast, true-to-scale depictions of devices were difficult. However, true-to-scale

---

[7]For a detailed description of this analysis and results, see Kleinberger-Pierer (2018).

replication was not the purpose of depictions of machine technology in the early modern era. Drawings from before 1700 tended to be representational, while later drawings were intended also to serve as technical documentation. This shift is evidenced by the findings of my analysis: only about 5% of machine drawings in printed technical literature before 1700 used any kind of scale, while after 1700 ten times more drawings included a scale.

Additionally, machines were increasingly depicted in multiple views. Before 1700, multiple views of a single machine rarely appeared in printed technical literature (less than 1% of cases). After 1700, in contrast, more than a third of all the analysed depictions of machines in technical literature included additional full views of the machine. Further, illustrations of machines after 1700 were often supplemented with detailed drawings of individual parts (the majority of which were also in orthographic view), providing more information for the reader.

The change in drawing methods around 1700 was not unintentional. Orthographic and scaled drawings appear in some manuscripts and prints before 1700, but they are rare (e.g. Leng 2004, pp. 109–110). The method of orthographic view is discussed in drawing instructions like Albrecht's from 1623 (see Fig. 11.1), but the authors of technical literature after 1700 consciously pursued better technical documentation via visual representation of machines. For example, one of the first eighteenth-century authors who predominantly used orthographic views for machine drawings in the German-speaking territories, Leonard Sturm, explicitly mentioned the significance of drawings for technical practice. He criticised his predecessors for using an unsuitable (i.e. linear) perspective to document machinery and favoured the orthographic perspective. Needless to say, he and other authors of eighteenth-century technical literature addressed mechanics and engineers with their work (Sturm 1718, Introduction).

The new ways of depicting and recording knowledge also led to new forms of knowledge and technical practice. The revolutionary change to a true-to-scale orthographic view prepared the ground for the industrialisation process of the nineteenth century, where technical drawings played a crucial role.

Discussion and Conclusion: The Impact of the Picture

Historiography has highlighted the importance of the visual elements of coins as well as technical drawings of minting technology to illustrate the evolution of minting and its process of mechanisation, but the design of the drawings themselves has largely been neglected. A comparison of drawings of coinage technology from the sixteenth to the eighteenth centuries reveals a significant, and in some ways revolutionary, development of technical documentation in practical engineering.

In particular, there were two major changes in the early modern period. First, the development or rediscovery of the linear perspective, which made it possible to create an exact geometrical abstraction of an object and second, the replacement of linear perspective with the orthographic view for a better technical documentation of machine designs around 1700. This was an important step in making engineering knowledge visible and establishing technical drawings as a tool for communication in the field of engineering.

The early modern transition in representations was intertwined with a change of practice in engineering, especially in the eighteenth century. New institutions like engineering schools, engineer military academies, mining schools and mining academies can be seen as evidence of a developing (technical) expert culture (Ash 2010; Klein 2016) that used a common visual language for communication. Recent supplementary historical research highlights the growing influence of empiricism and practical knowledge on early modern science that led to the establishment of a "practical science" (Long 2011; Holenstein et al. 2013; Jacob and Stewart 2004).

Leibniz's experiments in improving technology in mining and other fields were widely recognised (Wellmer 2010). In addition, Leibniz tried to establish other academies related to practical science and to develop a comprehensive programme enhancing the connection between art, craftsmanship and science. With the foundation of the Prussian Academy of Sciences (*Kurfürstlich-Brandenburgische Societät der Wissenschaften*) in 1700, his vision was at least partially fulfilled

(Schneiders 1975; Knobloch 1987). Another connection between science and the arts of mechanics and machine engineering is less well known: the mathematician, solicitor and philosopher Christian Wolff was well connected in science and politics, as evidenced by his membership of the Royal Society, the Prussian Academy of Sciences, the Russian Academy of Science in St. Petersburg and the *Académie des sciences* in Paris. In a 1710 guideline for mathematical and mechanical publications, Wolff highlighted that earlier publications on machinery were not practically usable and suggested a new form of publication focusing on practical use in mechanics (Wolff 1710, pp. 452–454). His network, and the support he provided, was one of the reasons why the Leipzig mechanic Jacob Leupold could publish the first volumes of his "Theatrum Machinarum" (Leupold 1724–1727), which became the standard reference for practical machine technology and mechanics in the eighteenth century (Ferguson 1971). Of course, drawings in this anthology applied the orthographic view and, like other publications of the time, made this specific method of technical documentation more widely known. A connection between natural science and engineers, or fruitful cooperation as between Wolff and Leupold, was important for the development and dissemination of new methods of documenting technical knowledge. Further, the transitions and developments in early modern methods of depiction also reflected the efforts of the court and administration to turn technology and technical expertise to their advantage.

The first example presented in this paper, a screw press from the anonymous author of an "arts of war" manuscript from around 1500, reflects the transition to linear perspective in arts and engineering; it also helped to fulfil the need for military expertise in a period when black powder and artillery challenged fortifications and sieges (cf. Leng 2004, pp. 88–89). Princes and courts had historically monopolised military power and needed to maintain military dominance by adjusting to new techniques like black powder. Even though the provenance of this manuscript cannot be traced with complete certainty, its style and military content would have made it highly precious for courts and administration in the early modern period.

The work of André Félibien from the second half of the seventeenth century, similarly, is not *just* a superb collection of images of minting technology in linear perspective; it also reflects the French effort to improve art and crafts in an effort to better economic performance. Publications, especially drawings, were employed by the crown as a means to disseminate knowledge about these technical improvements.

The next example, the textbook Henning Calvör's textbook, symbolises the next step in achieving economic prosperity with the use of technical education in the eighteenth century. Experts in engineering were educated in engineering schools and academies and developed a common visual language for technical communication and documentation. Calvör's work, as well as the efforts to improve mining in the early modern period, can be interpreted from a different perspective: the arrival of American silver had a tremendous impact on the silver mining industry in Europe (Renate Pieper in this volume) and the establishment of mining schools, and academies could be understood as an attempt to compete against American silver as well as an effort to diversify risk (Domenic Hoffmann in this volume). Unsurprisingly, the eighteenth-century mining schools and academies were established directly at the centres of silver mining.

Finally, the technology developed *inter alia* by Matthew Boulton not only changed the production costs of copper coins, but was also exported to India, Mexico (Doty 1990, p. 184) and Russia. Copper coins changed the money market in the nineteenth century, sometimes with disastrous effects (José Enrique Covarrubias in this volume). The aforementioned plan for the St. Petersburg mint, based on Boulton's concept, was an early attempt to enhance the efficiency of coin production. In presenting these machines, a style of drawing was chosen which combined the requirements of technical documentation (i.e. orthographic true-to-scale view) with princely representational entitlements.

Minting was apparently a princely affair in early modern times, but technical documentation, expertise and literature were intimately connected to such affairs, and the two were mutually dependent. As the methods and scope of technical drawings changed, so did the requirements and demands for technical expertise by the court and administration.

Bibliography

Primary Sources

Archival and Manuscript Sources

"Architectural Drawing," Proposal for St Petersburg Mint, 1790s. Museum of Applied Arts & Sciences, Object Number 87/646.

Eyb, Ludwig von der Jüngere, 1500. *Kriegsbuch* [Franken]. Handschrift der Universitätsbibliothek Erlangen-Nürnberg, Signatur H62/MS.B 26.

Weimar Ingenieurkunst- und Wunderbuch, 1520 (ca.), s.l. Handschriften der HAAB Weimar Fol 328. https://haab-digital.klassik-stiftung.de/viewer/resolver?identifier=1669&field=MD_DIGIMOID. Accessed January 9, 2018.

Source Editions and Old Prints

Albrecht, Andreas, 1623. *Zwey Bücher. Das erste Von der Ohne und durch die Arithmetica gefundenen Perspectiva. Das andere Von dem dartzu gehörigen Schatten.* Nürnberg: bey Simon Halbmayrn 1623. ETH-Bibliothek Zürich, Rar 3111. http://doi.org/10.3931/e-rara-8283.

Calvör, Henning, 1763. *Acta historico-chronologico-mechanica circa metallurgiam in hercynia superiori: oder Historisch-chronologische Nachricht und theoretische und practische Beschreibung des Maschinenwesens, und der Hülfsmittel bey dem Bergbau auf dem [...].* Braunschweig: Verlag der Fürstl. Waysenhaus-Buchhandlung. ETH-Bibliothek Zürich, Rar 9003. http://doi.org/10.3931/e-rara-33356.

Delius, Christoph Traugott, 1773. *Anleitung zu der Bergbaukunst nach ihrer Theorie und Ausübung, nebst einer Abhandlung von den Grundsätzen der Berg - Kammeralwissenschaft, für die K. K. Schemnitzer Bergakademie entworfen.* Wien: Johann Thomas Edlen v. Trattnern.

Dupain de Montesson, Louis C., 1790. *Die zum Zeichnen und Mahlen unentbehrliche Wissenschaft des Schattens oder sogenannte Schattir-Kunst. Wie man auf allerhand Flächen d. Schatten nach richtigen u. gewissen Gründen bestimmen, u. dadurch schöne Risse sowol in d. bürgerlichen als Kriegs-Bau-Kunst verfertigen soll; nebst e. Anhang, d. Zeichner im Cabinet u. bey d. Armee.* Nürnberg: Christoph Weigel- und A. G. Schneiderschen-Kunsthandlung.

Farish, William, 1822. "On Isometrical Perspective." In: *Cambridge Philosophical Transactions*, 1: 1–19.

Félibien, André, 1690. *Des principes de l'architecture, de la sculpture, de la peinture, et des autres arts qui en dependent: avec un dictionnaire des termes propres à chacun de ces arts.* Paris: Veuve de Jean Baptiste Coignard. ETH-Bibliothek Zürich, RAR. http://doi.org/10.3931/e-rara-1162/.

Flörke, Heinrich G.; Krünitz, Johann, 1805. "Münze und Münzwissenschaft." In: *Johann Krünitz's ökonomisch-technologische Encyklopädie*, 97: 759ff.

Lempe, Johann Friedrich, 1795–1797. *Lehrbegrif der Maschinenlehre, mit Rücksicht auf den Bergbau.* Leipzig: Crusius.

Leupold, Jacob, 1724–1727. *Theatrum machinarum.* Leipzig: Zunkel.

Martin, Anton, 1850. *Katalog der Bibliothek des K.k. polytechnischen Institutes in Wien.* Wien: Ueberreuter.

Monge, Gaspard, 1799. *Géométrie descriptive.* Paris: Baudouin.

Sturm, Leonhard Christoph, 1718. *Vollständige Mühlen Baukunst.* Augsburg: Jeremias Wolff.

Vigevano, Guido; Ostuni, Giustina, 1993. *Le macchine del re. Il Texaurus Regis Francie.* Vigevano: Diakronia.

Wolff, Christian von, 1710. *Der Anfangs-Gründe Aller Mathematischen Wissenschafften. Letzter Theil. Welcher so wol die gemeine Algebra, als die Differential- und Jntegral-Rechnung, und einen Anhang Von den vornehmsten Mathematischen-Schrieften Jn sich begreifet.* Halle (Saale): Regner.

Secondary Sources

Andersen, Kirsti, 2007. *The Geometry of An Art: The History of the Mathematical Theory of Perspective from Alberti to Monge.* New York: Springer.

Ash, Eric H., 2010. "Introduction: Expertise: Practical Knowledge and the Early Modern State." In: *Osiris*, 25: 1–24.

Bärtschi, Willi, 1976. *Linear Perspective: Its History, Directions for Construction, and in the Fine Arts.* New York: Van Nostrand Reinhold Company.

Baynes, Ken; Pugh, Francis, 1981. *The Art of the Engineer.* Guildford, Surrey: Overlook Press.

Benad-Wagenhoff, Volker, 2008. "Die Maschinisierung der Münzfertigung – Entwicklung und technikhistorische Stellung der Prägetechnik zwischen 1450 und 1850." In: *Interdisziplinäre Tagung zur Geschichte der neuzeitlichen Metallgeldproduktion – Projektberichte und Forschungsergebnisse – Beiträge zur Tagung in Stolberg (Harz) im April 2006 (Abhandlungen der Braunschweigischen Wissenschaftlichen Gesellschaft, Bd. LX u. LXI),* edited by Reiner Cunz, Ulf Dräger, and Monika Lücke. Braunschweig: VDS-Verlagsdruckerei Schmidt: 213–283.

Bettag, Alexandra, 2009. "Die Académie de Peinture et de Sculpture als kunstpolitisches Instrument Colberts – Anspruch und Praxis." In: *Akademie und/oder Autonomie. Akademische Diskurse vom 16. bis 18. Jahrhundert*, edited by Barbara Marx and Christoph Oliver Mayer. Frankfurt am Main: Peter Lang: 237–260.

Birn, Raymond, 1985. "Words and Pictures: Didierot's Vision and Publishers' Perceptions of Popular and Learned Culture in the Encyclopédie." In: *Popular Traditions and Learned Culture in France: From the Sixteenth to the Twentieth Century*, edited by Marc Bertrand. Saratoga: Anma Libri: 73–92.

Boehm, Gottfried, 2004. "Jenseits der Sprache? Anmerkungen zur Logik der Bilder." In: *Iconic turn. Die neue Macht der Bilder*, edited by Christa Maar and Hubert Burda. Köln: DuMont: 28–43.

Boehm, Gottfried, 2007. *Wie Bilder Sinn erzeugen. Die Macht des Zeigens*. Berlin: Berlin University Press.

Bredekamp, Horst; Schneider, Birgit; Dünkel, Vera, eds., 2008. *Das technische Bild. Kompendium zu einer Stilgeschichte wissenschaftlicher Bilder*. Berlin: De Gruyter.

Brown, John K., 1999. "When Machines Became Gray and Drawings Black and White: William Sellers and the Rationalization of Mechanical Engineering." In: *IA. The Journal of the Society for Industrial Archeology*, 25, 2: 29–54.

Černá-Šlapáková, Marie L., 1971. *Vzácné staré knihy ve Státni technické Knihovně v Praze*. Praha: Státní pedagogické nakladatstvi.

Doty, Richard G., 1990. "The World Coin: Matthew Boulton and His Industrialisation of Coinage." In: *Interdisciplinary Science Reviews*, 15, 2: 177–186.

Doty, Richard G., 1998. *The Soho Mint and the Industrialization of Money*. London: Smithsonian Institution.

Edgerton, Samuel Y., 1975. *The Renaissance Rediscovery of Linear Perspective*. New York: Basic Books.

Ferguson, Eugene S., 1971. "Leupold, Jacob, Leupold's Theatrum Machinarum." In: *Technology and Culture*, 1: 64–68.

Ferguson, Eugene S., 1977. "The Mind's Eye: Nonverbal Thought in Technology." In: *Science*, 197, 4306 (August 26): 827–836.

Ferguson, Eugene S., 1994. *Engineering and the Mind's Eye*. Cambridge: MIT Press.

Fröhner, Annette, 1994. *Technologie und Enzyklopädismus im Übergang vom 18. zum 19. Jahrhundert. Johann Georg Krünitz (1728–1796) und seine Oeconomisch-technologische Encyklopädie*. Mannheim: Palatium.

Germer, Stefan, 1997. *Kunst, Macht, Diskurs: die intellektuelle Karriere des André Félibien im Frankreich von Louis XIV*. München: Fink.

Hahn, Roger, 1971. *The Anatomy of a Scientific Institution: The Paris Academy of Sciences, 1666–1803*. Berkeley, Los Angeles, and London: University of California Press.

Heßler, Martina, 2006. "Einleitung. Annäherungen an Wissenschaftsbilder." In: *Konstruierte Sichtbarkeiten. Wissenschafts- und Technikbilder seit der Frühen Neuzeit*, edited by Martina Heßler. München: Wilhelm Fink: 11–37.

Hensel, Thomas, 2011. *Wie aus der Kunstgeschichte eine Bildwissenschaft wurde. Aby Warburgs Graphien*. Berlin: DeGruyter.

Holenstein, André; Steinke, Hubert; Stuber, Martin, eds., 2013. *Scholars in Action: The Practice of Knowledge and the Figure of the Savant in the 18th Century*, 2 vols. Leiden: Brill.

Jacob, Margaret C.; Stewart, Larry, 2004. *Practical Matter: Newton's Science in the Service of Industry and Empire, 1687–1851*. Cambridge: Harvard University Press.

Klein, Ursula, 2016. *Nützliches Wissen: die Erfindung der Technikwissenschaften*. Göttingen: Wallstein.

Kleinberger-Pierer, Harald, 2018. *Die visuelle Revolution des Maschinenbildes. Mediale Repräsentation von Technik im 18. Jahrhundert*. Unpublished thesis, University of Graz.

Knobloch, Eberhard, 1987. "Theoria cum praxi. Leibniz und die Folgen für Wissenschaft und Technik." In: *Studia Leibnitiana*, 19, 2: 129–147.

König, Wolfgang, 1999. *Künstler und Strichezieher. Konstruktions- und Technikkulturen im deutschen, britischen, amerikanischen und französischen Maschinenbau zwischen 1850 und 1930*. Frankfurt am Main: Suhrkamp.

Kratzsch, Konrad, 2017. *Von Büchern und Menschen: Arbeiten aus drei Jahrzehnten als Bibliothekar an der Herzogin Anna Amalia Bibliothek in Weimar*. Hamburg: tredition.

Kronick, David A., 1976. *A History of Scientific and Technical Periodicals: The Origins and Development of the Scientific and Technical Press 1665–1790*. Metuchen, NJ: The Scarecrow Press.

Lazardzig, Jan, 2007. *Theatermaschine und Festungsbau. Paradoxien der Wissensproduktion im 17. Jahrhundert*. Berlin: Akademie Verlag.

Lefèvre, Wolfgang, ed., 2004. *Picturing Machines 1400–1700*. Cambridge: MIT Press.

Leng, Rainer, 2002. *Ars Belli. Deutsche taktische und kriegstechnische Bilderhandschriften und Traktate im 15. und 16. Jahrhundert. Bd. 2: Beschreibung der Handschriften.* Wiesbaden: Reichert L.

Leng, Rainer, 2004. "Social Character, Pictorial Style, and the Grammar of Technical Illustration in Craftsmen's Manuscripts in the Late Middle Ages." In: *Picturing Machines 1400–1700*, edited by Wolfgang Lefèvre. Cambridge: MIT Press: 85–111.

Long, Pamela O., 2011. *Artisan/Practitioners and the Rise of the New Sciences, 1400–1600.* Corvallis: Oregon State University Press.

Madsen, David A.; Madsen, David P., 2017. *Engineering Drawing and Design.* Boston: Cengage Learning.

Maynard, Patrick, 2005. *Drawing Distinctions: The Varieties of Graphic Expression.* New York: Cornell University Press.

McGee, David, 2004. "The Origins of Early Modern Machine Design." In: *Picturing Machines 1400–1700*, edited by Wolfgang Lefèvre. Cambridge: MIT Press: 53–84.

Michel, Christian, 2018. *The Académie royale de Peinture et de Sculpture: The Birth of the French School, 1648–1793.* Los Angeles: Getty Research Institute.

Pannabecker, John R., 1994. "Diderot, the Mechanical Arts, and the Encyclopedie." In: *Journal of Technology Education*, 1: 45–57.

Pannabecker, John R., 1998. "Representing Mechanical Arts in Diderot's Encyclopedie." In: *Technology and Culture*, 1: 33–73.

Paulinyi, Akos, 1991. "Bemerkungen zu Bedeutung, Begriff und industrielle Vorgeschichte der Werkzeugmaschinen." In: *Technikgeschichte*, 58, 4: 263–277.

Plank, Angelika, 1999. *Akademischer und schulischer Elementarzeichenunterricht im 18. Jahrhundert.* Frankfurt am Main and Wien: Peter Lang.

Radkau, Joachim, 2008. *Technik in Deutschland. Vom 18. Jahrhundert bis heute.* Frankfurt am Main: Campus Verlag.

Rae, John; Volti, Rudi, 1999. *The Engineer in History.* New York and Berlin: Peter Lang.

Rudder, Debbie, 2009a. "Architectural Drawing, Proposal for St Petersburg Mint." https://collection.maas.museum/object/76923#&gid=1&pid=1. Accessed January 9, 2018.

Rudder, Debbie, 2009b. "Matthew Boulton and the Imperial Bank Mint in St Petersburg." https://maas.museum/inside-the-collection/2009/06/03/matthew-boulton-and-the-imperial-bank-mint-in-st-petersburg/. Accessed January 9, 2018.

Scharlau, Winfried, 1990. *Mathematische Institute in Deutschland 1800–1945.* Wiesbaden: Vieweg.

Schleiff, Hartmut; Konečný, Peter, eds., 2013. *Staat, Bergbau und Bergakademie. Montanexperten im 18. und frühen 19. Jahrhundert.* Stuttgart: Steiner.

Schmitz-Esser, Romedio, 2008. "Die Walzenprägung der Münze Hall in Tirol Innovation – innerhabsburgischer Technologietransfer – Rekonstruktion." In: *Interdisziplinäre Tagung zur Geschichte der neuzeitlichen Metallgeldproduktion – Projektberichte und Forschungsergebnisse – Beiträge zur Tagung in Stolberg (Harz) im April 2006 (Abhandlungen der Braunschweigischen Wissenschaftlichen Gesellschaft, Bd. LX u. LXI)*, edited by Reiner Cunz, Ulf Dräger, and Monika Lücke. Braunschweig: VDS-Verlagsdruckerei Schmidt: 285–314.

Schneiders, Werner, 1975. "Sozietätspläne und Sozialutopie bei Leibniz." In: *Studia Leibnitiana*, 7, 1: 58–80.

Scolari, Massimo, 2012. *Oblique Drawing: A History of Anti-perspective.* Cambridge: MIT Press.

Wellmer, Friedrich-Wilhelm; Gottschalk, Jürgen, 2010. "Leibniz' Scheitern im Oberharzer Silberbergbau – neu betrachtet, insbesondere unter klimatischen Gesichtspunkten." In: *Studia Leibnitiana*, 42, 2: 186–207.

Zsámboki, László, 1978. *Die Schemnitzer Gedenkbibliothek von Miskolc in Ungarn.* Miskolc: Zentralbibliothek der TU Schwerindustrie.

Part V

Markets

12

Reloading the Price Revolution in Seville: Four Stages of High Inflation with Different Causes

Manuel González-Mariscal

Introduction

New estimates about the evolution of prices in Seville during the sixteenth century (González-Mariscal 2015, 2017) challenge the conclusions of Earl J. Hamilton's *American Treasure and the Price Revolution in Spain, 1501–1650* (Hamilton 1934). While Hamilton's calculations suggest that prices increased by a factor of 3.7 between 1514 and 1603—dates which, as we shall see shortly, mark the beginning and the end of the "price revolution"—new estimates indicate that prices in fact increased by a factor of 7.1. Both price curves begin to separate in 1547 and can be largely explained by the behaviour of house rentals, which Hamilton did not take into consideration.

In order to address the main differences between the two price curves, an in-depth analysis of the distinct trajectories and stages needs to be undertaken. The aim of this paper is, therefore, to review the

M. González-Mariscal (✉)
Universidad de Sevilla, Seville, Spain
e-mail: mgmariscal@us.es

© The Author(s) 2019
R. Pieper et al. (eds.), *Mining, Money and Markets in the Early Modern Atlantic*, Palgrave Studies in Economic History,
https://doi.org/10.1007/978-3-030-23894-0_12

economic phenomenon known as the "price revolution" and examine the causes that have been advanced to explain it, while incorporating new possible interpretations. The article is organised as follows: section one examines the characteristics of the new price curve, outlines the chronological limits of the "price revolution" and identifies the periods in which inflation was greatest, namely 1515–1526, 1540–1560, 1575–1587 and 1598–1602; section two calculates the impact that the different products making up the "basket of goods" have on the general increase in prices and shows the possible factors responsible for the inflation in each of the four periods outlined above. Among them are the effects of agrarian crises, population growth, increases in money supply, division of labour and growth in urbanisation, the relationship between supply and demand in the house rental market, changes in the income velocity of money, coinage debasement, and the impact of taxes. These factors will be compared and contrasted with the theories put forward by other specialists. My premise is that each of the inflationary episodes occurred in response to a different combination of factors. In the near future, I intend to build an econometric model incorporating the factors that underlay the inflationary waves, helping to determine the impact of each factor on the increase in prices. In order to achieve this, I still have to quantify some of the factors that I have identified as causing inflation and gain a better understanding of certain variables, especially money supply.

The New Shape of the Price Revolution: Main Features, Duration and Stages

The causes of the rise in prices in Europe during the sixteenth century comprise one of the most controversial debates among economic historians. Although the debate can be traced back to the sixteenth century—when Jean Bodin argued that prices rose as a consequence of the arrival of American silver, while Jean de Malestroit believed that inflation was caused by the debasement of coinage (Mauro 1976)—it was not until 1934 that Hamilton established the modern foundations of the discussion, when he presented a vast quantity of statistical data concerning prices, salaries and precious metal imports, and provided a precise quantification of the phenomenon (Hamilton 1928, 1929a, b, 1934). This was the beginning of a debate that, with ebbs and flows, has continued until the present day. Other pioneers in the creation of price databases for the sixteenth century were Hoszowski (1928) for the Ukraine; Pelc (1935, 1937) and Adamczyk (1938) for Poland; Elsas (1936/1940) for Germany; Hauser (1936) for France; Pribram (1938) for Austria; Parenti (1939) and Coniglio (1952) for Italy; Posthumus (1946, 1964) for the Netherlands; Phelps Brown and Hopkins (1956) for Britain; and Verlinden (1965) and Van der Wee (1975) for Belgium.[1] Their work led to the elaboration of the so-called *first-generation price indexes*.

Although the history of prices has been revived over the last twenty years, most studies have relied on the statistical data compiled by those "pioneers".[2] This has resulted in the publication of *second-generation price indexes*, which have substantially improved the elaboration of index numbers, including those concerning Hamilton's Spanish data (Martín Aceña 1992; Reher and Ballesteros 1993; Llopis et al. 2000). On the other hand, very little new information about sixteenth-century prices has been published in recent years, with the exception of the work of Feliù (1991a, b) for Catalonia, González Agudo (2015) for Toledo and González-Mariscal (2015, 2017) for Seville.

Concerning Seville, a close examination reveals that Hamilton's data is very problematic, especially for the period between 1501 and 1580 (Hamilton 1983, pp. 337, 353–355). In the period 1501–1550, of the 24 commodities for which he provides price information, no information whatsoever exists for 70% of the years and, overall, 79% of the data is missing. For the period 1551–1580, of the 56 commodities for which he provides price information, we have no information for 47% of the years and, overall, 61% of the data is missing. It is worth noting

---

[1]Bibliographic references taken from Allen (2001).
[2]See the databases of International Institute of Social History: http://www.iisg.nl/hpw/data.php; and the databases of Global Price and Income History Group: http://gpih.ucdavis.edu.

that Hamilton's commodity list does not include some products which were very important for domestic economies, such as bread, house rentals and finished textiles. Finally, Hamilton decided not to weight his price index, assigning equal importance to such products as wheat, gunpowder, beef and lead, whose importance in the household basket of goods was very different.

The identification of these problems led me to a search for alternative primary sources to try and rectify the deficiencies of Hamilton's price index (González-Mariscal 2013). Substantial differences in the results suggest that, in order to improve our understanding of prices and standards of living in the *Ancien Régime*, scholars must go beyond the statistical data compiled by the "pioneers", expanding their databases as well as investigating changes in the spending structure of ordinary families.

The use of new primary sources (for a complete list cf. González-Mariscal 2015, pp. 360–362) and improved calculation methods allowed me to obtain a new estimate of the evolution of prices in Seville between 1501 and 1603 (González-Mariscal 2013, 2015, 2017). Improvements include the incorporation of house rentals into the index; the use of four different baskets of goods, inspired by changes detected in consumption patterns over time; a more careful selection of the products that compose these baskets; and a more precise weighting of each product. The new Consumer Price Index (CPI) is illustrated in Fig. 12.1, alongside Hamilton's. If we take the central years of the 11-year moving averages—that is, if we focus on the trend of the variable—we may say that, between 1506 and 1603, prices in Seville increased by a factor of 6.1. This corresponds to a Compound Annual Growth Rate (CAGR) of 1.89%. The most inflationary decades were the 1520s, 1550s and 1540s, with CAGR of 4.29, 3.53 and 2.52%, respectively.

Concerning the price of discrete categories, those which rose the most between the 1500s and 1590s concerned *house rentals*, the price of which increased by a factor of 19.9, followed by *fuel* (×4.6), *food* (×4.2) and *alcoholic beverages* (×3.5); overall, prices increased by a factor of 5.1. If we compare the 1520s and the 1590s—for which I have data pertaining to all the defined categories—the prices that rose the most were *house rentals* (×12.7), *domestic furnishings* (×3.9), *food* (×3.3) and *fuel* (×3.2), while the categories which saw the least growth

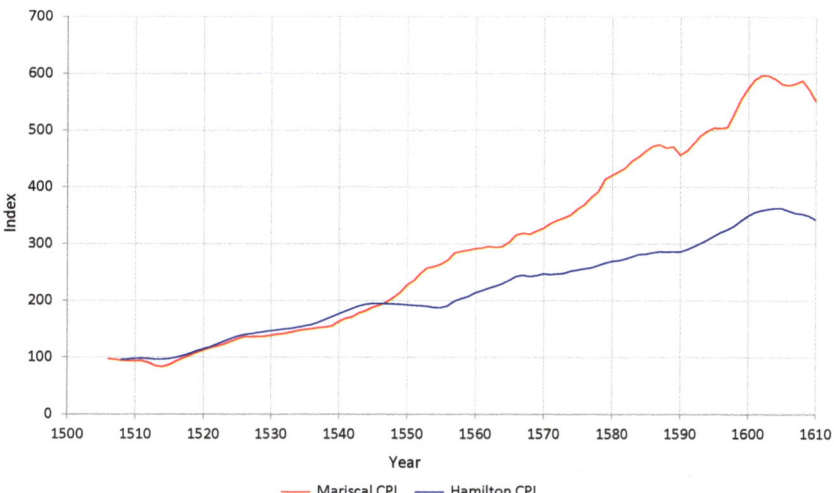

Fig. 12.1 Consumer Price Index in Seville, 1501–1610 (in vellón [In Castile, the sixteenth century was characterised by monetary stability, so that silver prices were equal to vellón prices between 1501 and 1598. Monetary debasement started to take place in 1599 and still in 1610 the differential between vellón prices and silver prices was not very relevant—less than 2% (González-Mariscal 2017, p. 287, Fig. 3)], 11-year moving averages, base 100 = 1501–1510 average) (*Source* González-Mariscal [2015, 2017] and Hamilton [1983])

were *clothing and shoes* (×3.0), *other goods* (×2.8) and *alcoholic beverages* (×2.0). Overall, prices increased by a factor of 3.8. To understand these different price increases, we must consider the elasticities of supply and demand of each product.

The substantial differences between the evolution of house-rental prices and the other categories could have been caused by the inability of the housing supply to meet the demand posed by the rapidly growing Sevillian population between 1530 and 1580. Speculation during the second half of the century may have further increased house rental prices (Carmona 1984; Collantes de Terán 1989). Population censuses and baptism records suggest that between the 1530s and 1580s the population of Seville increased by around 150% (González-Mariscal 2015), while residential districts barely expanded outside the limits of the twelfth-century city wall (Morales Padrón 1989, pp. 26–29).

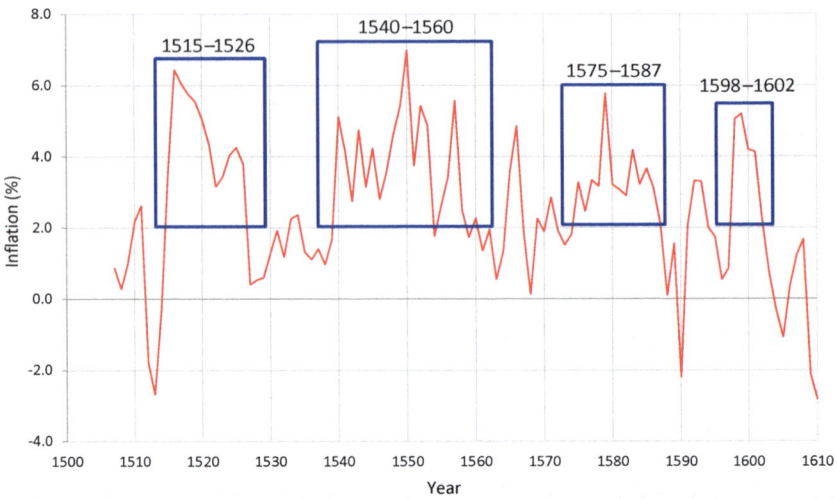

Fig. 12.2 Inflation in Seville, 1501–1610 (11-year moving averages). Highlighted: high inflationary periods (*Source* González-Mariscal [2015, 2017])

As shown in Fig. 12.1, the curves run close to one another between 1508 and 1547. From 1547 onwards, however, the new curve begins to grow at a much faster pace, and the difference between the two curves by 1603 is 65%. According to Hamilton, prices in Seville grew by 275% between 1508 and 1603, while the new index suggests an increase of 529%. The periods in which the greatest differences exist are 1542–1557 and 1570–1588. The reasons for these differences are the consideration of house rental prices (González-Mariscal 2015, p. 377), but also the inclusion of other basic goods at the household level, such as bread and textile manufactures; improvements introduced to the database; greater accuracy in the design of baskets of goods, which reflect changes in consumption patterns; and, a more precise weighting of products included in said baskets of goods.

The 11-year moving averages of inflation rates of the new CPI from 1501 to 1603 (Fig. 12.2) suggest that the entire sixteenth century was characterised by price rises, except for the 3-year period 1512–1514 and the year 1590. *Grosso modo*, four waves of rising inflation are evident—1508–1516, 1527–1550, 1568–1579 and 1590–1599—with

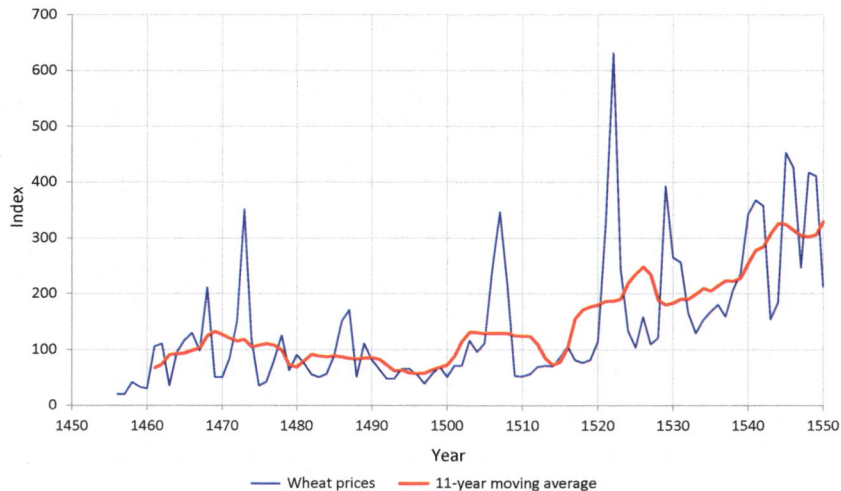

Fig. 12.3 Wheat prices in Carmona and Seville, 1456–1550. Index numbers, base 100 = 1461–1470 average (*Source* González Jiménez [1976] and González-Mariscal [2013])

moving averages of around 6.5, 7, 6 and 5%. The second of these waves is the most significant, both in terms of duration (23 years) and intensity (the highest values in the whole series are attested during this period). Conversely, three periods during which inflationary pressure abated are also evident: 1516–1527, 1550–1568 and 1579–1590.

With the main characteristics of the new price index examined and the most important differences with Hamilton's series outlined, we can attempt to establish the chronological limits of the so-called price revolution and define its different stages. Although most specialists agree that the price revolution began in the period 1516–1520 (Munro 2007, p. 5), this narrative does not tally with Sevillian wheat price data between 1450 and 1550 (Fig. 12.3). Wheat is the only commodity for which I have continuous price data for the second half of the fifteenth century. The purpose of this exercise is to ascertain whether inflationary tendencies may have begun before the sixteenth century. As Fig. 12.3 illustrates, wheat prices tended to decrease between 1469 and 1496. Between 1496 and 1504 wheat prices rose sharply, coincident with the

Table 12.1 Classification of inflation In Seville, 1501–1800, based on 11-year moving averages (by percentage)

Values	Description	Percentage of total data (%)
$-1.00 \geq X$	Very low	9
$-1.00 < X \leq 0.00$	Low	10
$0.00 < X \leq 1.00$	Medium-low	15
$1.00 < X \leq 2.00$	Medium	26
$2.00 < X \leq 3.00$	Medium-high	17
$3.00 < X \leq 4.00$	High	11
$X > 4.00$	Very high	12

Source González-Mariscal (2013)

agrarian crisis of 1502–1508 (González Jiménez 1976, p. 14), but it is not until 1514 that the series shows a clear upward trend. This date coincides with the new price index (Fig. 12.1), as such, 1514 seems like the most appropriate date to consider as the beginning of the price revolution.

The inflationary process that characterised the sixteenth century can be considered to havwe ended by 1603 (Fig. 12.1), when the prices began decreasing until 1615—the average inflation rate during this period was −0.5%. In 1615, the *vellón inflation* began in Seville. Overall, prices in Seville increased between 1514 and 1603 by a factor of 7.13, rather than Hamilton's 3.74. Therefore, prices increased almost twice as much as we have previously believed. On the other hand, CAGR during this period was 2.23%, while during the so-called *vellón inflation* (1615–1679), it was 1.01% (González-Mariscal 2017).

Once the chronological limits of the sixteenth-century inflationary process have been delimited, the next step is to define its different stages. The evolution of prices in Seville between 1501 and 1800 has been analysed previously (González-Mariscal 2013, pp. 163–165), and the annual inflation rates and their 11-year moving averages calculated. The latter oscillate between 7.33% in 1645 and −6.91% in 1684, averaging 1.59%. Following this, the 289 data points have been grouped according to the classification of Table 12.1. I have established that the notion of high inflation—and, therefore, "price revolution"—applies when the moving average is in excess of 2% in three consecutive years. In this way, the price revolution can be placed in the periods 1514–1526, 1540–1560, 1575–1587 and 1598–1602 (Fig. 12.2). Concerning the seventeenth and eighteenth centuries, and based on the classification illustrated in Table 12.1, the most acutely inflationary periods were 1637–1650, 1660–1664 and 1697–1700. In the next section, we will analyse the factors which may be regarded as responsible for each of the inflationary processes during the "price revolution".

The Four Inflationary Waves and Their Causes

The First Inflationary Wave, 1515–1526: The Combined Effects of a Deep Agrarian Crisis and the Impact of German Silver

Since American silver did not arrive in large quantities until the 1540s, the advocates of monetarist theories have explained the inflationary process of 1516–1520 as a combined effect of: (1) the sharp increase in silver and copper production in southern Germany and Central Europe between the 1460s and the 1540s (Munro 1994, 2003, 2007; Hatcher 1996; Nightingale 1997); and (2) changes in private and public finance in the early sixteenth century (Munro 1991; Van de Wee 1963, 2000). Those economic historians who do not concur with these arguments, on the other hand, have blamed the increase in prices largely on demographic growth (Nadal 1959; Brenner 1961; Vilar 1969; Ramsay 1971). This section examines how these theories tally with the new price curve in Seville.

If we analyse the relationship between silver production in southern Germany and Central Europe between 1471 and 1540 (Munro 2003, pp. 43–44) and the evolution of wheat prices in Seville or the new CPI, the following conclusions may be drawn (Table 12.2): first, between 1471 and 1500 the increase in silver production ran parallel to a sharp drop in the price of wheat; and second, between 1501 and 1540, there is a hard correlation between prices and increases in silver production ($R^2 = 0.86$). Based on these conclusions, we can

Table 12.2 Precious metals (in million of pesos of 272 maravedíes) and Seville prices (index numbers, base 100=1501–1510 average), 1471–1610

Decade	Precious metals				Seville prices	
	I	II	III	IV	V	VI
1471–80	–	–	–	1.4	–	83
1481–90	–	–	–	1.6	–	65
1491–00	44.6	0.7	–	2.0	–	41
1501–10	55.7	8.2	2.0	2.5	100	100
1511–20	55.7	7.2	3.6	2.7	88	59
1521–30	81.3	4.3	1.9	3.3	134	181
1531–40	81.3	18.7	9.2	4.1	148	145
1541–50	161.4	36.9	17.3	3.4	190	236
1551–60	176.4	53.4	29.6	–	269	224
1561–70	160.9	64.9	41.9	–	309	249
1571–80	160.9	84.5	48.2	–	365	282
1581–90	211.0	110.4	88.0	–	467	467
1591–00	211.0	125.3	115.2	–	506	569
1601–10	219.9	134.6	92.3	–	586	564

Key I. World Silver and Gold Production; II. American Silver and Gold Output; III. Imports of Gold and Silver to Seville; IV. Silver Output from the Major South German and Central European Mines; V. Consumer Price Index in Seville (Mariscal); and VI. Wheat prices in Seville and Carmona
Sources Tepaske (2010), Hamilton (1983), Munro (2003), González-Mariscal (2015, 2017); and González Jiménez (1976)

argue that the abundance of Central European silver may have had an effect on Sevillian prices from 1500 onwards. This idea finds support if we analyse the relationship between the production of precious metals in America (TePaske 2010) and prices in Seville between 1501 and 1550, or the relationship between the arrival of precious metal in Spain (Hamilton 1983) and price levels in Seville. The correlation is considerably lower in both cases ($R^2 = 0.29$ and 0.35, respectively). This seems to indicate that the main background factors in the inflationary process of the period 1515–1526 in Seville were chiefly to do with the growth in silver production in southern Germany and Central Europe.

At this point, a problem that underlies this whole section is encountered for the first time: namely that the only feasible way of approaching money supply is to use production and import data for precious metals. In order to get a more accurate view, we would also need data concerning the export and existing stocks of the same metals. This would give us some indication of the changes in net stock and thus its influence on the price index. Even if this information were available, however, it would still have to be taken into consideration that silver and gold stock is not the same as *coined money*, and that coined money is not the same as *money supply*. Therefore, using the import or production of precious metals as a source of information regarding money supply can be a problem when we try to understand the causes of the price revolution.

Concerning the European "financial revolution" of the early sixteenth century, it must be stressed that the changes were largely linked to an improvement in the "negotiability" of private financial products and a better organisation of markets for public debt instruments. These changes led to an expansion of credit. The first changes in private financial transactions can be traced back to Antwerp in 1507 (Munro 2007, p. 9). It is unclear whether those changes affected the operation of public and private finances in Spain during the early sixteenth century and what effect they had on the expansion of credit and the increase in prices in Seville.[3]

With respect to the argument that population growth may have been the cause of the sharp increase in prices, it must be stressed that during the first three decades of the sixteenth century, Seville and western Andalusia in general *lost* population, owing to recurrent and severe agrarian crises, epidemics and emigration to America (Carmona 2000, 2004). Fig. 12.4 illustrates the index of baptisms in Seville and western Andalusia between 1501 and 1610. The curve seems to confirm this demographic trend, that is, that significant population growth did not begin until the 1530s, which complicates the idea that inflation was caused by population growth.

---

[3]In order to corroborate, qualify or rule out this possibility, I am currently working alongside other researchers from the University of Seville, on a project which aims to reconstruct interest rates in the city during the sixteenth century, based on protests attached to bills of exchange. The aims of the project are twofold: on the one hand, to develop a more accurate approach to money supply than has hitherto been possible; on the other hand, to gain a better understanding of changes in the Spanish financial system.

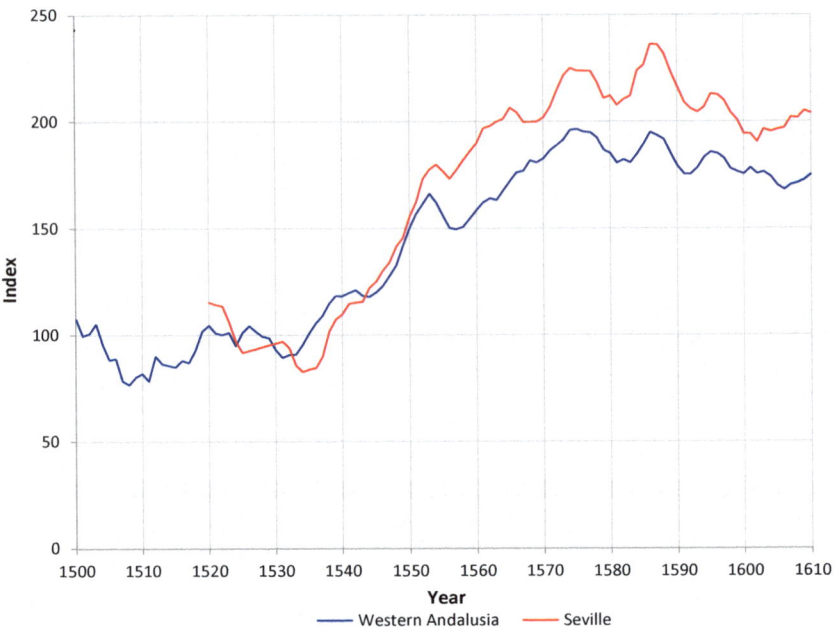

Fig. 12.4 Index of baptisms in Seville and Western Andalusia, 1501–1610 (5-year moving averages, base 100=1521–1530 average) (*Key* Western Andalusia includes the current provinces of Seville, Cádiz and Huelva. *Source* González-Mariscal [2013])

If we examine the first inflationary wave in the sixteenth century, as described by the new price index (1514–1526), we can see that during this period prices increased by 63%—a 4.1% annual rate. The products whose prices increased the most were bread (171%), fat (85%), and wine, salted cod and sugar (60% each). Bread was responsible for 56% of the total increase in prices, wine for 10%, beef for 9%, fat for 6% and salted cod for 5%. That is, these five products were collectively responsible for 86% of the overall increase in prices between 1514 and 1526. The importance of the increase in the price of bread and its impact on the growth of inflation cannot be emphasised enough.

The increase in the price of bread in the period 1514–1526 can easily be explained by the severe agrarian crisis in western Andalusia between

1520 and 1522 (González Jiménez 1976). 1521 was especially critical, so much that the Seville Cathedral council, responsible for managing the Archbishop's tithes,[4] was forced to suspend the collection of the tax in most parts of its jurisdiction because of the serious shortage of grain.[5] While the tithes collected in 1521 amounted to 73.6 tons of wheat and 28.9 tons of barley, in 1524 the council collected 891.3 tons of wheat and 338.2 tons of barley, and in 1525, 1629.4 tons of wheat and 634.9 tons of barley.[6] There are full tithe records of the Archbishopric of Seville for the series 1520–1800,[7] a 280-year period, and at no other time were such extreme measures taken. This period witnessed one of the most severe agrarian crises of the early modern period in western Andalusia.

The fact that the most inflationary products, alongside bread, were fat, wine and salted cod could suggest that, in addition to the agrarian crisis, high inflation was also been triggered by American demand (mostly for wheat and wine, as Hamilton pointed out) and the need to supply ships bound for America (biscuit, fat, wine and salted cod). We should point out that the cultivation of wheat in New Spain did not begin until 1521, and that significant crops are only attested from 1524 onwards (Del Río Moreno and López 1996). It is also worth stressing that outbound ships began forming convoys in 1522 and this may have affected the pricing of products which held a prominent place among ships' supplies.

In conclusion, this first inflationary period was chiefly driven by two factors: the arrival of silver from southern Germany and Central Europe and the agrarian crisis of 1520–1522. The former was a catalyst of inflation from 1500 onwards, but the peak of the price index between 1515 and 1526 was largely caused by the scarcity of cereal brought about by the latter.

---

[4]The Archbishopric of Seville included the modern provinces of Seville and Huelva and the northern half of the province of Cádiz.

[5]Archivo de la Catedral de Sevilla (ACS), sección II, serie 1ª, libro 1A.

[6]ACS, sección II, serie 1ª, libros 1A, 1B and 2.

[7]ACS, sección II, serie 1ª, libros 1A-241C.

The Second Inflationary Wave, 1540–1560: Demographic Growth, American Silver and Real Estate Boom

During our second inflationary period between 1539 and 1560, prices increased by 83%—an annual rate of 2.92%. The products whose prices increased more than average are house rentals (176%), chickpeas (106%), walnuts (93%), bread (91%), beef (87%) and honey (85%). Bread was responsible for 18% of the overall price increase, house rentals for 17%, beef for 14%, chickpeas for 10%, wine for 9% and fat and canvas textiles for 4% each. That is, the increase in prices of these seven products accounted for 76% of overall price increase. The responsibility for this second inflationary process is borne by a larger number of products than the first, in which bread was the main factor by a wide margin.

The factors which have traditionally been related to the price revolution converge in this second inflationary episode, namely population growth, the beginning of production of large quantities of silver in America, and the impact of both on the income velocity of money. To these general factors, in the specific case of Seville, we must add the operation of the house rental market. The considerable increase of demand for real estate, driven by population growth and the increase in money supply, met an extremely rigid real estate supply that could not and did not want to absorb the demand—which, due to technical limitations and to the structure of real estate property, was strongly concentrated in the hands of religious institutions (Carmona García 2015). From 1536 onwards, there was a real estate boom in Seville. Given the structure of the real estate market in the city, this could only be followed by a sharp growth in house rental prices. Therefore, between the 1530s and the 1570s, house rental prices increased by 692% (González-Mariscal 2013).

The first of these factors, population growth, began in Seville in 1534—when the population was approximately 45,000—and accelerated from 1538 onward (Fig. 12.4). Between 1539 and 1560, the curve of baptisms in Seville grew by 77%. Between 1560 and 1586, when the number of baptisms peaked—the population was approximately

120,000—the growth rate slowed to 23%. Finally, from 1586 until 1603, the curve took a downward turn, with the baptism rate decreasing by 17% to the end of the period. As such, it is difficult to argue that from 1560 onwards the increase in prices can be related to demographic growth.

Second, the arrival of American precious metal in Spain significantly increased during the middle decades of the sixteenth century. While between the 1500s and the 1520s the quantity of precious metals brought from America was between 1.9 and 3.6 million pesos (Table 12.2), it increased to 9.2 million in the 1530s, 17.3 million in the 1540s and 29.6 million in the 1550s, a total increase, between the 1520s and the 1550s, of 1558%. Up until 1530, the main precious metal arriving in Seville was gold, thereafter replaced by silver (Hamilton 1983, p. 54) following the discovery of the mines of Potosí in 1545 and Zacatecas in 1546.

The beginning of this inflationary period coincided with sharp population growth and the massive influx of American silver. In the near future, I will try to determine, by econometric means, the relative importance of these factors in the inflationary process. The linear regression models concerning the baptisms index and the new CPI for the period 1530–1570 on the one hand and the arrival of precious metals and the new CPI on the other hand only reveal that both variables were closely intertwined with the evolution of prices ($R^2 = 0.98$ and 0.99, respectively).

In contrast with the postulates of Keynesian economists, who argue that an increase in money supply is followed by a decrease in the income velocity of money, V—that is by an increase of the so-called Cambridge k—some authors (Miskimin 1975; Goldstone 1984, 1991a, b; Lindert 1985; Mayhew 1995) have argued that inflation in the 1550s and 1560s was mainly caused by an increase in V. According to these authors, population growth in the sixteenth century introduced structural changes in the economy, which in turn resulted in the increase of V. According to Goldstone, population growth was accompanied by a disproportionate growth in urbanisation and the rapid development of commercialised agriculture, urban markets and use of credit instruments. All of these factors contributed to an increase of V. Lindert argue that demographic growth was accompanied by two factors: changes in relative prices (the price of agricultural produce grew more than that of industrial goods and nominal wages, which caused important changes in household budgets) and changes in age structure and, consequently, in productive/dependent population ratios. These changes led to a decrease of k and, therefore, an increase of V.

I expect that my ongoing research will soon provide additional information concerning the income velocity of money and its effect on inflation. For now, it is worth stressing that demographic growth in western Andalusia during the sixteenth century (Fig. 12.4) was accompanied by an increase in urbanisation. In 1535, there were 16 towns or cities in western Andalusia with more than 5000 inhabitants, of which four had populations in excess of 10,000. By 1587, the former group included 35 towns or cities, and the latter 9 (González-Mariscal 2013). On the other hand, the Sevillian data seems to confirm Lindert's argument: while wheat prices increased by a factor of 1.9 between 1530 and 1570, Anjou linen fabrics increased by 1.4, and the average salary of bricklayers by 1.7. However, I do not have data concerning Sevillian population structure.

In conclusion, this second stage represents the apex of the "price revolution". Not only because of its duration—twenty years, the longest of the four stages identified—and because it witnesses the highest inflation rates of the whole series—7%—but also because it is the period that sees the causes traditionally used to explain the inflation of the sixteenth century converge: demographic growth and the arrival of large quantities of silver from America. In the specific case of Seville, we must add additional explanatory factors: the operation of the real estate market, characterised by an explosion of demand driven by increasing population and money supply, and a rigid supply, unable to meet the increase in population—baptisms in Seville increased by 77% between 1539 and 1560—both for technical reasons and because of the very structure of real estate property, heavily concentrated in the hands of religious institutions. Finally, one factor with a still-uncertain effect is the income velocity of money.

The Third Inflationary Wave, 1575–1587: Agrarian Crisis, Amalgamation, Real Estate Speculation and *Alcabalas*

In the third inflationary wave between 1574 and 1587, prices grew by 39%, that is, at an annual rate of 2.56%. The products whose price increased above average were house rentals (86%), sardines (60%) and bread (59%). Bread was responsible for 45% of the overall price increase, house rentals for 32% and chickpeas for 8%. That is, these products accounted for 85% of the overall price increase. Therefore, as in the first stage, most of the inflation was caused by a few products, in this case bread and house rentals.

As we have seen, these years saw population growth slow down substantially, but the production of American silver and its arrival in Spain reached new heights with the introduction of amalgamation in Zacatecas in 1554–1557 and, more importantly, in Potosí in 1572. Silver production in America grew from 29.6 million pesos in the 1550s to 88.0 million in the 1580s, nearly a 200% increase (Table 12.2).

Two other factors which contributed to the inflationary process were the severe agrarian crisis suffered by western Andalusia from 1575 onwards and, despite the slower population growth, the continued increase in house rental prices, largely brought about by the generalisation of speculative strategies in the real estate market. Concerning the agrarian crisis, cereal production dropped by 32% in western Andalusia between 1575 and 1587 (Fig. 12.5) leading to a substantial increase in the price of bread. In addition, house rental prices continued to grow—rising 40% between 1575 and 1587—despite the fact that population barely grew during this period, which implies that the increase in house rental prices was caused by speculation. Finally, the increase of the *alcabalas* tax by the Crown of Castile in 1575 may have had a considerable impact on prices. In Seville, the surcharge grew from 10.2 to 17.5% of the total amount (Álvarez Nogal and Chamley 2012) and the Crown managed to double its revenue. This must have had a substantial impact on prices.

In conclusion, in the third phase as in the first, we find a long-term factor—the increase in the arrival of silver from America, boosted by

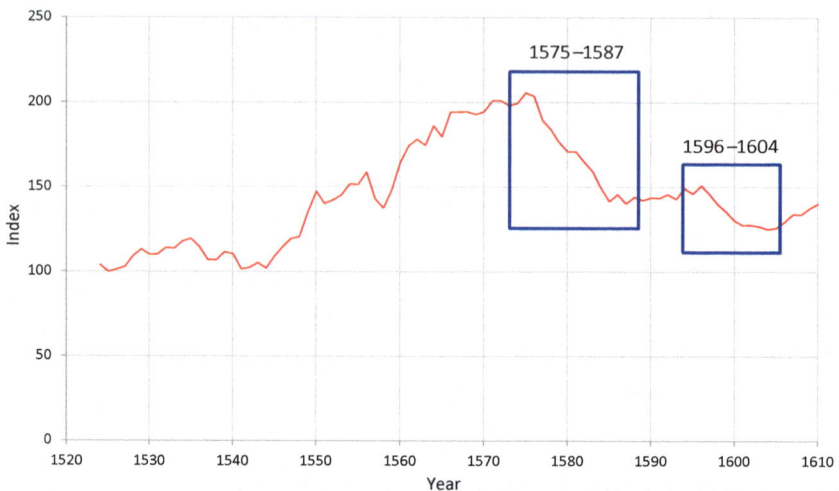

Fig. 12.5 Cereal production in the Archbishopric of Seville, 1521–1610 (9-year moving averages; Index number, base 100 = 1521–1530 average) (*Source* Llopis and González Mariscal [2010])

the implementation of the amalgamation method—and three circumstantial factors: the agrarian crisis, real estate speculation and the increase of the *alcabalas* tax.

The Last Inflationary Wave, 1598–1602: The Early Effects of Monetary Debasement? Agrarian Crisis and *Millones* Tax

In the final inflationary period under consideration, between 1597 and 1602, prices increased by 17%, an annual rate of 3.25%. Products whose price increased more than average were chickpeas (49%), wine (40%), bread (26%) and fat (18%). Bread was responsible for 48% of the price increase, chickpeas for 20%, wine for 10% and house rentals for 8%. These products accounted for 86% of the overall price increase.

The reasons behind this last inflationary episode could be the coinage debasement (in 1599, Philip III accepted for the first time the issue of pure copper money), the agrarian crisis and the imposition of the tax of *millones*. After the production of cereal recovered by 8% between 1587 and 1596 (Fig. 12.5), the period 1596–1604 saw another crisis, with cereal production dropping by 17% (Llopis and González Mariscal 2010). The crisis affected agricultural products other than cereal; the production of wine and oil, especially, had been decreasing since 1587, and by 1605 the drop in production amounted to 62%. Finally, the creation of the tax of *millones* in 1590, and its extension in 1596, could also have contributed to the price increase (Andrés Ucendo 1998).

Conclusions

1. In Seville, the economic phenomenon known as the "price revolution" began in 1514 and ended in 1603. According to new calculations, during this period prices increased nearly twice as much as hitherto believed. The difference can be explained by the inclusion of house rental prices and, to a lesser extent, by the use of an improved dataset, a better selection of the products to constitute the basket of goods and a more precise weighting of those goods.
2. Inflationary tensions took place in four waves: the most important was the wave between 1540 and 1560. It was not only the longest period of inflation but also the most severe. In second place we find the waves of 1515–1526 and 1575–1587, which were briefer and less severe; finally, 1598–1602 was merely a residual aftershock. Each episode was caused by a different set of factors.
3. The main causes of inflation during the period 1515–1526 were the boom of silver production in southern Germany and Central Europe and the agrarian crisis that devastated the western Andalusian countryside between 1520 and 1522. The first of these processes was driving prices up from the beginning of the century, while the second, although it was caused by circumstantial economic conditions, caused a sharp and sudden increase of prices on the side of supply, which placed the price of bread at the centre of the inflationary dynamic. Other factors also contributed, albeit in a minor way, to

inflation: the arrival of American gold, the need to meet American demand for food and goods and the beginning of convoy navigation to America. It is also possible that changes in private and public finance in Europe from the beginning of the sixteenth century affected prices by expanding credit, but this is still pending confirmation for Seville.

4. The main causes of inflation during the period 1540–1560 were the enormous demographic growth witnessed by Seville—during this period the population grew by approximately 80%; the arrival of huge quantities of American silver, which increased the money supply to unknown levels; and the operation of the real estate market, which combined an explosively growing demand (following the demographic growth) with a rigid supply, caused by the market's inability, owing to technical difficulties, to meet the increase in population, and a concentration of properties in the hands of religious institutions. It is possible that an additional factor also contributed to inflation: the income velocity of money.

5. During the third inflationary wave (1575–1587), price increases were, again, driven by factors on both sides supply and demand. On the one hand, the agrarian crisis of 1575–1587 drove cereal production down by 32%. On the other hand, money supply enjoyed another boost following the implementation of silver amalgamation in America. In addition, the characteristics of the real estate market led prices to continue growing, despite the fact that population growth had decelerated. Finally, the substantial increase in the *alcabalas* tax in 1575 also could have contributed to the increase in prices.

6. The causes for the 1598–1602 aftershock may be found in the 1596–1602 agrarian crisis, the creation of a new tax—the *millones*—and the early effects of coinage debasement policies.

7. Finally, if we want to continue improving our knowledge about the factors that caused the price revolution, we need to make a better approach to money supply, which probably involves the calculation of interest rates; to quantify some variables such as velocity of money and taxes; and to elaborate an econometric model with all the factors pointed out throughout this text. These will be the next steps of my research.

Acknowledgements Work funded by the Ministerio de Economía, Industria y Competitividad (AEI/FEDER, UE), HAR-2016-78026-P. I wish to express my gratitude for comments made about previous versions of this work—presented in Graz, London, Salamanca and Boston—by Renate Pieper, Claudia Jefferies, Markus Denzel and Rafael Dobado. Having greatly profited from their ideas, the errors that the work may still contain are mine alone.

Bibliography

Adamczyk, Wadyslaw, 1938. *Ceny w Warszawie w Latach w XVI i XVII Wieku.* Lwow: Instytut Popierania Polskiej Tworczosci Naukowej Warszawa.

Allen, Robert Carson, 2001. "The Great Divergence in European Wages and Prices from the Middle Ages to the First World War." In: *Explorations in Economic History* 38: 411–447.

Álvarez Nogal, Carlos; Chamley, Cristophe, 2012. "La Crisis Financiera en Castilla en 1575–1577: Fiscalidad y Estrategia." In: *Revista de la Historia de la Economía y la Empresa* VII: 187–211.

Andrés Ucendo, José Ignacio, 1998. "Una Herencia de Felipe II: los Servicios de Millones en Castilla durante el Siglo XVII." In: *Felipe II (1527–1598): Europa y la Monarquía Católica. Congreso Internacional "Felipe II (1598–1998), Europa Dividida, la Monarquía Católica de Felipe II*, edited by José Martínez Millán, vol. 2. Madrid: Universidad Autónoma de Madrid.

Brenner, Yehojachim Simon, 1961. "The Inflation of Prices in Sixteenth Century England." In: *Economic History Review* XIV, 2 (2nd ser.): 225–239.

Brown Phelps, Henry Ernest; Hopkins, Sheila V., 1956. "Seven Centuries of the Prices of Consumables, Compared with Builders' Wage Rates." In: *Economica* 23 (nueva serie): 196–314.

Carmona García, Juan Ignacio, 1984. "Valor, Rentabilidad y Formas de Cesión de la Propiedad Inmobiliaria en la Sevilla de Finales del Siglo XVI." In: *Archivo Hispalense* 205: 3–38.

Carmona García, Juan Ignacio, 2000. *Crónica Urbana del Malvivir (s. XIV–XVII): Insalubridad, Desamparo y Hambre en Sevilla.* Sevilla: Universidad de Sevilla.

Carmona García, Juan Ignacio, 2004. *La Peste en Sevilla.* Sevilla: Ayuntamiento de Sevilla.

Carmona García, Juan Ignacio, 2015. *Mercado Inmobiliario, Población, Realidad Social. Sevilla en los Tiempos de la Edad Moderna.* Sevilla: Universidad de Sevilla.

Collantes de Terán, Antonio, 1989. "El Mercado Inmobiliario en Sevilla (Siglos XIII–XVI)." In: *D'une Ville à L'autre, Structures Matérielles et Organisation de L'espace dans les Villes Européennes (XIIIe–XVie Siècle). Actes du Colloque de Rome.* Roma: Publications de l'école Française de Rome: 227–242.

Coniglio, Giuseppe, 1952. "La Rivoluzione dei Prezzi nella Città di Napoli nei Secoli XVI e XVII." In: *Atti della IX Riunione Scientifica della Società Italiana di Statistica.* Spoleto: 204–240.

Del Río Moreno, Justo Luis; López y Sebastián, Lorenzo Eladio, 1996. "El Trigo en la Ciudad de México. Industria y Comercio de un Cultivo Importado (1521–1564)." In: *Revista Complutense de Historia de América* 22: 33–51.

Elsas, M. J., 1936/1940. *Umriss einer Geschichte der Preise und Löhne in Deutschland.* Leiden: Sijthoff.

Feliu i Montfort, Gaspar, 1991a. *Precios y Salarios en la Cataluña Moderna. Volumen I: Alimentos.* Madrid: Servicio de Estudios del Banco de España.

Feliu i Montfort, Gaspar, 1991b. *Precios y Salarios en la Cataluña Moderna. Volumen II: Combustibles, Productos Manufacturados y Salarios.* Madrid: Servicio de Estudios del Banco de España.

Goldstone, Jack A., 1984. "Urbanization and Inflation: Lessons from the English Price Revolution of the Sixteenth and Seventeenth Centuries." In: *American Journal of Sociology* 89: 1122–1260.

Goldstone, Jack A., 1991a. "Monetary Versus Velocity Interpretations of the 'Price Revolution': A Comment." In: *Journal of Economic History* 51: 176–181.

Goldstone, Jack A., 1991b. "The Causes of Long Waves in Early Modern Economic History." In: *The Vital One: Essays in Honor of Jonathan R. T. Hughes,* edited by Joel Mokyr. Greenwich, CT: JAI Press: 51–92.

González Agudo, David, 2015. *Población, Precios y Renta de la Tierra en Toledo, Siglos XVI–XVII.* PhD thesis, Universidad Complutense de Madrid.

González Jiménez, Manuel, 1976. "Las Crisis Cerealistas en Carmona a Fines de la Edad Media." In: *Historia. Instituciones. Documentos* 3: 283–308.

González-Mariscal, Manuel, 2013. *"Población, Coste de la Vida, Producción Agraria y Renta de la Tierra en Andalucía Occidental, 1521–1800."* PhD thesis, Universidad Complutense de Madrid.

González-Mariscal, Manuel, 2015. "Inflación y Niveles de Vida en Sevilla durante la Revolución de los Precios." In: *Revista de Historia Económica* 33: 353–386.

González-Mariscal, Manuel. 2017. "Precios y Niveles de Vida en Sevilla durante la Inflación del Vellón." In: *The Prices of Things in Pre-Industrial Times*. Firenze: Firenze University Press.

Hamilton, Earl J., 1928. "American Treasure and Andalusian Prices, 1503–1660: A Study in the Spanish Price Revolution." In: *Journal of Economic and Business History* 1: 1–35.

Hamilton, Earl J., 1929a. "American Treasure and the Rise of Capitalism, 1500–1700." In: *Economica* 27: 338–357.

Hamilton, Earl J., 1929b. "Imports of American Gold and Silver into Spain, 1503–1660." In: *Quarterly Journal of Economics* 43: 436–472.

Hamilton, Earl J., 1934. *American Treasure and the Price Revolution in Spain, 1501–1650*. Cambridge: Harvard University Press.

Hamilton, Earl J., 1983. *El Tesoro Americano y la Revolución de los Precios en España, 1501–1650*. Barcelona: Ariel.

Hatcher, John, 1996. "The Great Slump of the Mid-Fifteenth Century." In: *Progress and Problems in Medieval England*, edited by Richard Britnell and John Hatcher. Cambridge and New York: Cambridge University Press: 237–272.

Hauser, H., 1936. *Recherches et Documents sur l'histoire des prix en France de 1500 a 1800*. Genève: Slatkine Reprints, 1985.

Hoszowski, Stanilaw, 1928. *Ceny we Lwowie w XVI i XVII Wieku*. Lwow: Instytut Popierania Polskiej Tworczosci Naukowej Warszawa.

Lindert, Peter, 1985. "English Population, Wages and Prices: 1541–1913." In: *Journal of Interdisciplinary History* 15: 609–634.

Llopis Agelán, Enrique et al., 2000. "Indices de Precios de la Zona Noroccidental de Castilla y León, 1518–1650." In: *Revista de Historia Económica* XVIII, 3: 665–684.

Llopis, Enrique; González Mariscal, Manuel, 2010. "Un crecimiento tempranamente quebrado: el producto agrario en Andalucía occidental en la Edad Moderna." In: *Historia Agraria. Revista de Agricultura e Historia Rural* 50: 30–42.

Martín Aceña, Pablo, 1992. "Los Precios en Europa durante los Siglos XVI y XVII: Estudio Comparativo." In: *Revista de Historia Económica* X, 3: 359–395.

Mauro, Frédéric, 1976. *Europa en el Siglo XVI: Aspectos Económicos*. Barcelona: Barcelona Labor.

Mayhew, Nicholas J., 1995. "Population, Money Supply and the Velocity of Circulation in England, 1300–1700." In: *Economic History Review* 4, 2 (2nd ser.): 238–257.

Miskimin, Harry, 1975. "Population Growth and the Price Revolution in England." In: *Journal of European Economic History* 4: 179–185.

Morales Padrón, Francisco, 1989. *Historia de Sevilla. la Ciudad del Quinientos*. Sevilla: Universidad de Sevilla.

Munro, John, 1991. "The Central European Mining Boom, Mint Outputs, and Prices in the Low Countries and England, 1450–1550." In: *Money, Coins, and Commerce: Essays in the Monetary History of Asia and Europe (from Antiquity to Modern Times)*, edited by Eddy H. G. Van Cauwenberghe. Leuven: Leuven University Press: 119–183.

Munro, John, 1994. "Patterns of Trade, Money, and Credit." In: *Handbook of European History, 1400–1600: Late Middle Ages, Renaissance and Reformation, vol. I: Structures and Assertions*, edited by James Tracy, Thomas Brady Jr., and Heiko O. Oberman. Leiden, New York and Köln: E. J. Brill: 170–179.

Munro, John, 2003. "The Monetary Origins of the 'Price Revolution': South German Silver Mining, Merchant-Banking, and Venetian Commerce, 1470–1540." In: *Global Connections and Monetary History, 1470–1800*, edited by Dennis Flynn, Arturo Giráldez, and Richard Von Glahn. Aldershot and Brookfield, VT: Ashgate Publishing: 1–34.

Munro, John, 2007. "Hamilton and the Price Revolution: A Revindication of His Tarnished Reputation and of a Modified Quantity Theory." *EH.net*. http://eh.net/book_reviews/american-treasure-and-the-price-revolution-in-spain-1501-1650/. Accessed October 21, 2018.

Nadal, Jordi, 1959. "La Revolución de los Precios Españoles en el Siglo XVI: Estado Actual de la Cuestión." In: *Hispania* 19: 511–514.

Nightingale, Pamela, 1997. "England and the European Depression of the Mid-Fifteenth Century." In: *Journal of European Economic History* 26, 3: 631–656.

Parenti, G., 1939. *Prime Ricerche sulla Rivoluzione dei Prezzi in Firenze*. Firenze: Casa Editrice del Dott. Carlo Cya.

Pelc, J., 1935. *Ceny w Krakowie w Latach 1369–1600*. Lwow: Instytut Popierania Polskiej Tworczosci Naukowej Warszawa.

Pelc, J., 1937. *Ceny w Gdansk w XVI i XVII Wieku*. Lwow: Instytut Popierania Polskiej Tworczosci Naukowej Warszawa.

Posthumus, Nicolaas W., 1946. *Inquiry into the History of Prices in Holland*, vol. i. Leiden: E. J. Brill.

Posthumus, Nicolaas W., 1964. *Inquiry into the History of Prices in Holland*, vol. ii. Leiden: E. J. Brill.

Pribram, A. F., 1938. *Materialien zur Geschichte der Preise und Löhne in Osterreich. Band i*. Wien: Carl Ueberreuter.

Ramsay, P. H., ed., 1971. *The Price Revolution in Sixteenth-Century England*. London: Methuen & Co.

Reher, David; Ballesteros, Esmeralda, 1993. "Precios y Salarios en Castilla la Nueva: la Construcción de un Índice de Salarios Reales, 1501–1991." In: *Revista de Historia Económica* XI, 1: 101–151.

TePaske, John J., 2010. *A New World of Gold and Silver*. Leiden and Boston: Brill.

Van der Wee, Herman, 1963. *Growth of the Antwerp Market and the European Economy, 14th to 16th Centuries*, 3 vols. Den Haag: [s.d.].

Van der Wee, Herman, 1975. "Prijzen en Lonen als Ontwikkelingsvariabelen, een Vergelijkend Onderzoek Tussen Engeland en de Zuidelijke Nederlanden, 1400–1700." In: *Album Offert a Charles Verlinden a L'occasion de ses Trente ans de Profesorat*. Gent: Universia: 413–447.

Van der Wee, Herman, 2000. "European Banking in the Middle Ages and Early Modern Period (476–1789)." In: *A History of European Banking*, edited by Herman van der Wee and Ginette Kurgan-van Hentenryk. Antwerpen: Mercatorfont: 152–180.

Verlinden, C., 1965. *Dokumenten voor de Geschiedenis van Prijzen en Lonen in Vlaanderen en Brabant (XIVe – XIXe eeuw)*, vol ii: 3–70. Brugge: De Tempel.

Vilar, Pierre, 1969. *Oro y Moneda en la Historia (1450–1920)*. Barcelona: Ariel.

13

Interest Rates and Silver Production: Credit in Mexico City Between Market and Spirituality (1770–1779 and 1819–1828)

Andrés Calderón Fernández, Rafael Dobado González and Alfredo Garcia-Hiernaux

Introduction

The history of credit in Mexico before modern banks emerged in the second half of the nineteenth century was largely unknown before the 1980s, and what had been published presented a very sombre image of credit in viceregal times (Sánchez Cuen 1958; Lobato López 1945). The history of credit and financial markets has developed over the past thirty years. It has been studied from its origins in New Spain (Martínez López-Cano 2001) through its development in the eighteenth century

A. Calderón Fernández (✉)
Facultad de Economía, Universidad Nacional Autónoma de México,
Mexico City, Mexico

R. Dobado González · A. Garcia-Hiernaux
Universidad Complutense de Madrid, Madrid, Spain
e-mail: rdobado@ccee.ucm.es

A. Garcia-Hiernaux
e-mail: agarciah@ucm.es

© The Author(s) 2019
R. Pieper et al. (eds.), *Mining, Money and Markets
in the Early Modern Atlantic*, Palgrave Studies in Economic History,
https://doi.org/10.1007/978-3-030-23894-0_13

(Von Wobeser 1994; del Valle Pavón 2003, pp. 649–675) and up to its crisis during late viceregal times (Von Wobeser 2003; del Valle Pavón 2012); panoramic views of the viceregal period are also available (Chamoux et al. 1993; Martínez López-Cano and del Valle Pavón 1998). There are good pioneering works dealing with early independent Mexico (Tenenbaum 1985; Bernecker 1992) as well as general overviews (Ludlow and Marichal 1998). Nevertheless, there has been no attempt to quantify either the number of loans or the evolution of the interest rate.

To build our series of interest rates, we have used, and will continue to use as part of a larger project, a source that has not been systematically explored for this purpose: the Archivo Histórico de Notarías de la Ciudad de México (AHNCM), the Archive of the Notaries of Mexico City. Notaries in the Spanish world were officials appointed by the State, some of whom received public money to perform certain duties, but most of whom lived off the fees they charged for registering important civil—dowries, testaments and marriage contracts—and commercial—loans, property sales and mercantile companies—transactions.

The series for the first decade that we analyse, 1770–1779, was built with a sample of eleven out of 89 notaries taken directly from the register books and comprises more than 1100 transactions. The series for the second decade, covering the years 1819–1828, was constructed using data from the detailed catalogue of all transactions registered on the archive by the team of the Colegio de México, directed by Pilar Gonzalbo (http://notarias.colmex.mx/), and revised and completed with archival research. It is, therefore, not a sample but the totality of more than 1400 loans registered at the archive during those years.

The database that we have created has enormous potential; it not only allows us to determine the interest rate, but also provides a very rich picture of the economic life of Mexico City during the two periods, since it has the data of creditors, debtors, mortgaged property, purpose of the loan, etc. Nevertheless, our data has limitations as we lack insight into many private loans that were not made official with a notary; further research with other sources may fill these gaps. At the moment, we have only been able to check some of the books of the Yraeta-Ycaza-Yturbe family kept at the Universidad Iberoamericana. This was an important family of merchants that was active during roughly the same period that we have studied. What we have found does not point to a story other than the one told by the registers of the AHNCM.

The Evolution of Silver Production and Outflow in New Spain/Mexico

If we assume the quantity theory of money, in which prices are proportional to the amount of money in circulation, since interest rates *are* the price of money, it is necessary to track the evolution of money supply in order to understand the evolution of borrowing rates and credit volumes. New Spain/Mexico was by far the largest producer of silver in the world at the time. If all this precious metal had poured into the economy, the country would have experienced high inflation rates, but silver coins were also New Spain's main export, exchanged for textiles, porcelain, machines, etc. and, therefore, the much less silver remained in the country than was produced. A larger supply of silver as money would lead to higher inflation and lower real interest rates. This was the case in the first decade of our analysis. A tighter money supply would drive deflation and higher interests. This is what happened in the second decade of our study.

Thus, considerate is important to offer a new, adjusted series of silver production and silver outflow. They can still be improved, but they provide a better picture of actual figures than other estimations. For silver production, we have used two sources: Miguel Lerdo de Tejada's *Comercio exterior de México* (Lerdo de Tejada 1967, doc. 54) and a briefing from the Finance Ministry in 1857 (Siliceo 1857, pp. 17–34) (see Fig. 13.1). For silver outflow, the reconstruction was far more complicated.[1] Also, data from before 1810, the year of the beginning of the war of independence, is generally of better quality than the data

---

[1]Lerdo (1967), Garner (1982, pp. 558–559), Marichal (1999, p. 305), Marichal and Grafenstein (2012, pp. 92–93), Marichal and Souto Mantecón (1994, pp. 612–613), Ortiz de la Tabla Ducasse (1978, pp. 151, 153, 253, 257), Romero Sotelo (1997, p. 206), and data from the Archivo General de la Nación de México (AGNM) generously provided by Ernest Sánchez Santiró for the period between 1802 and 1820.

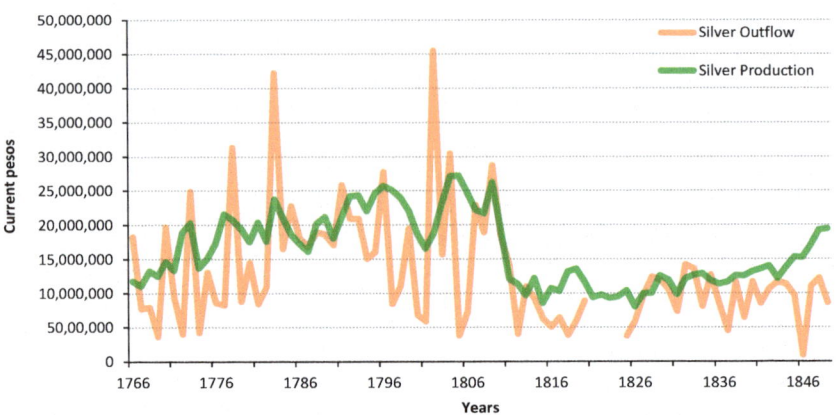

Fig. 13.1 Silver production and outflow, 1764–1850 (*Source* see Footnote 1)

available afterwards; the period between 1811 and 1824 is still subject to revisions of some importance.

These precautions do not prevent us from making some general remarks about the series: at the end of the viceregal period, we see a "high pressure system", with high levels of silver production and high levels of silver exports; after 1821, we have a "low pressure system", but with a similar balance between production and outflows (see Fig. 13.1).

As a percentage of registered silver production, the outflow in 1766–1820 (80.4%) is similar to that of 1825–1850 (75%). Interestingly, the highest percentage of outflows—dramatically exceeding a hundred per cent—is recorded in the 1780s rather than in the 1800s. Public transfers (*remesas* and *situados*) were far from unimportant (44.2% in 1781–1800), but a larger (55.8% in the same period) outflow of silver was the result of private transactions (imports of goods and services, private transfers and capital repatriation).

Mexico City's Money Market: Economic, Moral and Spiritual Aspects

Mexico City had developed a functional credit system in the sixteenth century (Martínez López-Cano 2001), with silver production and commerce as its axis. Nevertheless, beyond economic interests, a "spiritual economy" emerged quite early on (Martínez López-Cano et al. 1998, 2004; Von Wobeser 2005). In the counterreformation society of the Spanish monarchy, the salvation of the souls was no minor issue. Contradicting Luther and other Protestant reformers that defended predestination, the Council of Trento emphasised the importance of good deeds and prayers in saving souls. Therefore, chaplaincies and pious works were created by the inhabitants of the Kingdom of New Spain. Wealthy settlers left munificent legacies worth thousands of pesos, but medium and small fortunes also funded chaplaincies or pious works (obras pías) whose holder would pray to reduce the time that the soul of the donor would spend in purgatory. Confraternities, Spanish, Indian and black also tried to guarantee souls' passage to heaven by helping both their own members and other, less favoured, groups of society (Carrera 2011; Mejía Torres 2014).

Due to the accumulation of capital over more than two centuries, by the end of the eighteenth century, the Church and its institutions dominated the money market of Mexico City. The Juzgado de Capellanías y Obras Pías (Court of Chaplaincies and Pious Works) controlled the largest volume of money available for loans (Ludlow and Marichal 1998, p. 7) and settled all issues concerning loans granted from the funds of chaplaincies and pious works. Chaplaincies were assigned to a particular priest—if they were very large, they could be granted to more than one person. Priests were paid for praying masses for the rest of the souls of the donors, usually 1 peso if the mass was rezada (prayed) and 2 if it was cantada (sung). Chaplaincies had a principal, the trust itself, but this money was not given directly to the priest: it was lent to economic agents that paid réditos (interest) on the loan, 5% per annum in most cases. The interest paid for the masses that were celebrated on a monthly, weekly or daily basis, except for Sundays and the major festivities of the Church. The idea was to never exhaust the funds of the chaplaincy, so that a faster passage from purgatory to heaven was assured. Since a priest could live modestly on 250 pesos per year, one of the most common figures for chaplaincies was 5000 pesos; for smaller fortunes, the figure of 1000 pesos is also common, since that would secure a weekly mass, albeit not the maintenance of the chaplain. Pious works functioned like chaplaincies in the sense that the money spent on them came from interest paid by debtors and not from the trust itself. In this

318 A. Calderón Fernández et al.

case, the money was given to nuns, orphans, defenceless women, the sick, etc., who would pray for the soul of the donor.

Newly founded chaplaincies and pious works almost always had a patron—different from the donor but appointed by him or her with the approval of the *Juzgado*—who ensured that the money was used wisely. It was not uncommon for the beneficiary of the *réditos,* that is the holder of the chaplaincy or pious work, to be a member of the donor family or someone close to it. For instance, in 1781 Agustina Arias Favila founded a chaplaincy worth 4000 pesos that was immediately given to his brother, Juan José Arias Favila.[2] The money of the *principal* was not given immediately to the *Juzgado*, but was secured by a 9-year loan using the haciendas of San Nicolás and Santa Teresa Tilostoc on the Valley of Temascaltepec as warrants. Why would a family proceed in this way? The money given to the *Juzgado* and serving as the trust was untouchable and guaranteed by the *Juzgado* itself. On the other hand, using real estate or other valuable goods as mortgage, the family could establish chaplaincies or pious works without decapitalising their haciendas or companies by funding the whole trust all at once. In this way, the family could secure both their spiritual well-being *and* the material support of other members of the family. The Church allowed for these proceedings since it was not uncommon that older chaplaincies and pious works were, after four or five generations, left with no patron or beneficiary related to the donors. Thus, the *Juzgado* could decide which businesses would be favoured with loans or which chaplains would benefit from the endowments. The power bestowed upon bishops in New Spain, and in particular, the Archbishop of Mexico by this kind of financial instrument has probably been understated by the literature: in our sample for 1770–1779, the *Juzgado* was the sole creditor in 36.8% of all loans (2,457,690.31 pesos) and intervened alongside other creditors in another 16.9% of all financial operations (1,128,364.88 pesos).

How were these resources put on loan? The Church had a long tradition of condemning usury, from at least the fourth century: since time belonged only to God, someone charging interest on loans would

---

[2]AHNCM, Notary Andrés Delgado Camargo, vol. 1374, ff. 167 v–188 r.

be considered a thief of God's property (Von Wobeser 1993, pp. 124, 131). Thomas Aquinas did not accept usury as just, but tolerated some forms of usury, considering that it benefited many people, including the debtors who otherwise would not have access to money in times of hardship (Von Wobeser 1993, p. 133). To circumvent the prohibitions on usury, from medieval times the Church and Crown authorised the use of the *censo consignativo*, a contract in which one person sold to another person the right to receive yields from real estate in exchange for a certain amount of money. *Censos* were thus real loans (Von Wobeser 1998, p. 183) transmitted along with the property to new owners and would therefore perish if the mortgaged property suffered major losses. Some *censos* were perpetual, while others could be redeemable, but without a specific expiration term. *Censos* were the favourite form of credit used by the Church in New Spain in the sixteenth and seventeenth centuries, but the flooding of Mexico City between 1629 and 1633, which seriously damaged most buildings of the city, turned the eyes of the ecclesiastical institutions to another instrument: the *depósito irregular* (irregular deposit) (Berthe 1993, p. 30).

Irregular deposits were personal loans with fixed expiration terms; guarantors and movable property were also admitted as collateral. They were exempt from paying the *alcabala* (Martínez and del Valle Pavón 1998, p. 28) since they were not considered a sales contract. Theoretically, they transmitted the enjoyment of an object, in this case, money, in exchange for a rent. Due to their flexibility and lower transaction costs, the merchant elite of New Spain resorted to irregular deposits over *censos* more and more, and by the late eighteenth century, they were the dominant form of credit operation.

Fifteenth-century treatises began to justify charging interest, or equivalent mechanisms, based on a series of just titles: *stipendium laboris*, compensation of work, since lending money implied investigating the possible debtor and his goals; *damnum emergens* and *periculum sortis*, the damage that the lender would suffer in case of delinquency; *lucrum cessans*, loss of earning, because the lender could have obtained profits by investing the money on a different venture and therefore should be paid for that sacrifice; and *ratio incertitudinis*, the uncertainty suffered by the lender which should somehow be compensated (Von

Wobeser 1993, p. 135). Finally in 1745, Pope Benedict XIV published his encyclical *Vix pervenit*, in which he limited the concept of usury to the operations in which charging interests was considered to be unfair (Von Wobeser 1993, p. 140); irregular deposits were officially approved by the Fourth Mexican Council in 1771 (Von Wobeser 1994, p. 46).

Nunneries and confraternities also played a major role as creditors: in our sample for 1770–1779, nunneries accounted for 12.1% of all loans (809,680 pesos), while confraternities accounted for 1.9% of them (128,750).[3] Altogether, these religious institutions accounted for more than two-thirds of all the credit in our sample.

Individuals make up about one quarter of the sources of credit in our 1770s sample. The other instruments registered by the notaries of Mexico City before independence were *reconocimientos de deuda* and *obligaciones de pago,* in which the creditor was usually one of the *almaceneros* of Mexico City (wholesale merchants). The *almaceneros* controlled the circulation of silver in New Spain, since they provided credit for high risk, and potentially high profit, operations in mining and trade. The *Juzgado* usually lent money for lower risk ventures such as agriculture, livestock raising or urban workshops, although by the late eighteenth century it was also lending large sums of money to the wholesale merchants of Mexico City.

In the cases of loans presented by the *almaceneros* to notaries, the operation usually officially registered an expired loan that had been previously made privately, quite often by providing goods to a smaller merchant or miner that would pay their value after making profit. Many credits of this kind mention no interest rate, either because the debtor was someone close to the creditor or because the interest rate was implicit in the amount of the debt registered. If the debtor could not repay the loan on the promised date, the creditors tried to recoup their money by obtaining a mortgage or by demanding the presentation of

---

[3]Our sample may be exaggerating the role of the *Juzgado*, since we captured the data of the notary that registered most of its operations, Andrés Delgado Camargo. Nevertheless, the amounts it lent are very significant, especially when compared to other loans. Also, our sample may be understating the role of brotherhoods, since we have very few operations of some very important *cofradías* (Santísimo Sacramento, Aranzazu) or none at all (Rosario).

guarantors that would eventually pay in the case that the debtor again failed to pay the debt.[4] These loans were usually short-term,[5] as opposed to the ones registered with the *Juzgado,* which lasted for five years usually. Some *almaceneros* seemed to be quite active on the money market, for example the Gutiérrez de Terán brothers who lent 88,496.78 pesos in our sample for the 1770s, 1.3% of the total.

The *Juzgado* lost a lot of money during the process of *Consolidación de Vales Reales*. In 1804, the Spanish Crown, allied with Napoleon, was at war with Britain. Public debt titles, the *Vales Reales,* were losing significant value due to widespread distrust of the Crown's ability to repay the invested money. To back the *Vales* with fresh money, the Crown extended the consolidation decree to the American kingdoms by the end of 1804, immediately confiscating all debts due and liquid funds available; debts reaching maturity could not be extended—a common practice with ecclesiastical endowments, since they were more interested in the yields than in repayment of the principal. As a result, the *Juzgado* of the Archbishopric of Mexico paid 1,557,673 pesos to the *Caja de Consolidación* between 1805 and 1809 (Von Wobeser 2003, pp. 136–143), a significant amount by any measure—for instance *all* institutions affected in the Viceroyalty of Peru paid a total of 1,487,093 pesos to the *Caja de Consolidación*. The destruction caused by the civil war of independence (1810–1821), as well as the disruption of the commercial system which provided resources to support mining and other activities (Romero Sotelo 1997, pp. 15–17) depleted the coffers of the *Juzgado* even further. Thus, in the universe of loans for the decade

---

[4]For example, on 31 March 1781, the Marquis of Selvanevada registered a debt owed to him by José Zamora and Francisco Javier Lozano, neighbours in the town of Tepeapulco. He had lent them 1893 pesos and had only received 293 pesos back. Therefore, after waiting two months past the limit payment date, he secured the refinanced credit by means of the intervention of guarantors Francisco García Carrasco and Casimiro Cortés and established an interest rate of 5% on the new loan—seemingly, the previous one was paying no interest, since most of the money was given on merchandise and the profit of the creditor was included in the price of the goods provided to the debtors. AHNCM, Notary Andrés Delgado Camargo, vol. 1374, ff. 251 v–253 v.

[5]For instance, Gabriel y Damián Gutiérrez de Terán, *almaceneros* of Mexico City, usually gave 6 months to delinquent debtors after registering the operation to pay the money owned from previous operations. AHNCM, Notary Andrés Delgado Camargo, vol. 1374, fs. 160 v–161v and 164–165 v.

between 1819 and 1828 the Juzgado accounted only for 10.6% of new loans (920,699 pesos). Nunneries also saw a fall in their disposable funds, lending 631,516 pesos (7.3% of the total). Confraternities and *congregaciones* appear with 7% (602,605 pesos); their economy had also been affected (Calderón Fernández 2009, p. 47), but they seem to have gained importance at this moment since they are underrepresented in our sample for 1770–1779.

Almaceneros were still important during the 1820s, although they were facing difficulties due to the losses suffered during the war of independence and from the loss of access to a truly world market in the Spanish Empire. The Count of Ágreda, several times elected Prior of the *Consulado* of merchants, was the leading private lender, putting 177,899 pesos on loan during these years (1.4% of new loans) and refinancing further 55,000 pesos (4% of renegotiated debts). Among the persons lending more than 50,000 pesos, we found several merchants (José María Pérez, Luis de Escobar), as well as the director of the Pawnshop, Antonio Manuel Couto, and important landowners (the Garay family, owners of the Hacienda de los Morales close to Mexico City, and the heirs of the Counts of Valparaíso). Foreigners appeared in the credit market of the capital of the new country quite early on. For example, the American ambassador Joel Poinsett made a loan worth 2500 pesos and the British consul, George O'Gorman, one worth 8000. Nevertheless, neither merchants nor foreigners could make up for the diminished funds of the ecclesiastical institutions, leading to a severe fall in the volume of credit.

Interest Rates in Mexico City, 1770–1779 and 1819–1828

Average Rate

The most striking feature of our data is the omnipresence of the 5% interest rate before 1821, which dominates all *depósitos irregulares* contracted with the *Juzgado*, but is also frequent among private loans—see Fig. 13.2. This level of interest had been established by an *ordenanza* of

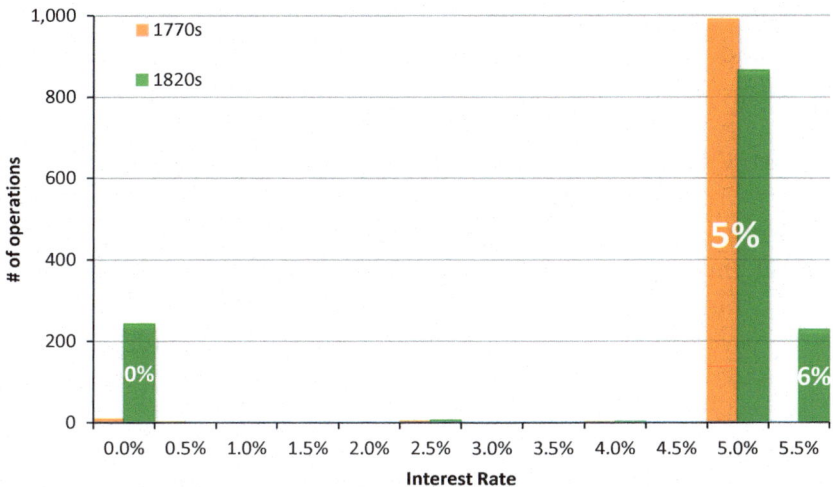

Fig. 13.2 Interest rate by decade (*Source* http://notarias.colmex.mx/; Archivo Histórico de Notarías de la Ciudad de México)

Philip III in 1608 (Berthe 1993, p. 28). Its persistence for more than two centuries could be explained by the fact that when interest reductions occurred in Spain, affecting the *censos* and *juros*, the money market in New Spain had developed to the point that it was autonomous of the metropolitan market. The existence of many chaplaincies and pious works by the mid-seventeenth century may have created a logic that was "spiritual" rather than economical: further reductions of interest rates would have been opposed in the Viceroyalty, despite the increasing production of silver, since a "reduction" in the prayers and masses expected by the donors and their descendants would have been unacceptable.

We have tried to compare interest rates with the profitability of other financial operations. In the books of the Yraeta-Yturbe-Ycaza family for the period 1797–1817,[6] we have found the rates charged for orders of payment (3.75%), cashing checks (2%) or the stocks of the Royal Company of the Philippines (5 − 1%, that is 4%). Lending money certainly seemed a more profitable activity than these

---

[6]Archivo Histórico de Comerciantes, Universidad Iberoamericana, vol. 2.6.6.

alternatives, but its profitability was clearly lower than that of commerce and mining. For instance, the company formed by Francisco Ignacio de Yraeta and Ana Gómez de Valencia, which existed from August 1770 through February 1777, was created with an investment of 99,012.53 pesos and yielded a profit of 136,419.13 (Torales Pacheco 1985, p. 120) in 6.5 years, that is 20,987.56 pesos per year, a striking return of 21% *per annum*. Certainly, Yraeta was one of the most intrepid and successful *almaceneros* of the second half of the eighteenth century, but the point here is that such large profits were possible in the economy of New Spain, for two reasons: (1) mining, which was risky but could yield enormous profits if an important ore was discovered and (2) the New Spain's position in the network of commerce of the Spanish Empire, which gave it privileged access to both Asian and European markets. Within this context, a 5% rate seemed reasonable and New Spaniards would not expect lower yields even if rates in Europe, even in Spain, were going down before the revolutionary and Napoleonic wars.

A 6% interest rate became more common after independence, although the 5% did not disappear and was still the most common interest rate for several years, showing that the availability of money was diminishing. Nevertheless, we see more 0% credits than 6% during the 1820s (see Fig. 13.2). What does this tell us about the market? We have no doubt that the *depósitos irregulares* made at the 0% rate were new credits given with no interest due to the hardship of the moment, in which a moral economy would operate among members of society. E. P. Thompson (2000, p. 213ff.) has shown how this mechanism worked among the poor, but the moral obligation to help fellow men, deriving from Christian ethics, would also be present among the middle and upper strata: the more prosperous would help their relatives and friends facing difficulties by lending them money without interest. Besides these new loans, most *obligaciones de pago* and *reconocimientos de deuda* of the 1820s with 0% interest rates (1,388,268 pesos or 13.8% of all loans) seem to be old debts, many of them mercantile, that could not be paid on time and were thus formalised with notaries in an effort to secure the repayment of the capital. While the maturity of irregular deposits in both decades is almost identical (see Table 13.1),

Table 13.1 Average maturity of loans by type of instrument (in days)

Depósito irregular			
1770	1425.24	1819	1557.86
1771	1625.07	1820	1546.61
1772	1324.9	1821	1273.13
1773	1542.32	1822	1523.78
1774	1558.12	1823	1919.13
1775	1491.26	1824	1453.04
1776	1612.14	1825	1825
1777	1565.34	1826	1376.48
1778	1470.46	1827	1091.87
1779	1358.88	1828	1456.24
Average	1497.37		1502.32
Reconocimiento de deuda			
1770	438.33	1819	1488.19
1771	1264	1820	1133.02
1772	668.75	1821	1317.24
1773	1095	1822	1213.03
1774	1496	1823	1465.42
1775	1825	1824	1603.49
1776	2068.33	1825	1460
1777		1826	1310.19
1778		1827	1683.41
1779	1594.55	1828	1921.36
Average	1306.25		1459.54
Obligación de pago			
1770	–	1819	884.94
1771	–	1820	1064.02
1772	529.29	1821	523.67
1773	228.57	1822	795.07
1774	703.5	1823	1092.52
1775	463.46	1824	1319.52
1776	499.24	1825	1460
1777	907.22	1826	737.1
1778	348.43	1827	841.76
1779	755	1828	1019.33
Average	554.34		973.79

Source http://notarias.colmex.mx/; Archivo Histórico de Notarías de la Ciudad de México

the maturity of *reconocimientos de deuda* went up by 12% and of *obligaciones de pago* more than 75%. We do not think that this indicates a deepening of the long-term loans market, but rather the difficulties of reimbursing outstanding debts: without a longer maturity, debtors could not possibly pay back the money they had received. The balances of businessmen in Mexico City were probably in a worse shape than their counterparts in other parts of the country, since the commercial and fiscal system that concentrated smelting and tax collection in the capital was totally dislocated by the war of independence: roads became unsafe, local authorities and merchants gained power, and the situation prior to 1810 was never restored (Sánchez Santiró 2014, p. 215).

Market Concentration and Large Credits

As evident in Table 13.2, loans under 2000 pesos accounted for 40% of the operations during the 1770s, but only represented 8.7% of the money lent in our sample. Credits over 15,000 pesos made up for a mere 8% of the operations, but comprised 41.7% of the total amount put on loan. Forty years later, concentration in the credit market had grown further. The most common amount lent remained 2000 pesos, but the median and average loans grew by 17 and 29%, respectively. This means that, besides seeing a credit crunch,[7] the resources that available to put on loan were held by a smaller group of individuals and institutions, leaving middle strata with limited access to financing through formal channels (Table 13.3).

A qualitative analysis of the largest credit of each year sheds more light on the changes in the money market between these two periods. For six years in the 1770s, the *Juzgado de Capellanías* had been, alone or in concurrence with other institutions, the largest lender in Mexico City.

---

[7]We have sampled 1/8 of the notaries active in Mexico City in the 1770s. The amount of money lent in our sample comes to approximately 6.7 million pesos. Since we have used the data of the notary preferred by the largest creditor of the kingdom, the *Juzgado de Capellanías* of the Archbishopric of Mexico—Andrés Delgado Camargo—we expect to obtain a total amount of loans between two and three times that figure.

Table 13.2 Size of credits

| | 1770–1779 | | | 1819–1828 | | | | | |
| | All operations of the sample | | | All operations | | | Subtracting presumably refinancing operations | | |
	# of op.	Amount	%	# of op.	Amount	%	# of op.	Amount	%
Less than 1000 pesos	136	71,070.25	1.07	186	94,025.66	0.94	100	50,774.66	0.59
From 1000 to 2000	326	505,853.44	7.58	367	561,707.00	5.59	294	451,211.00	5.22
From 2001 to 4000	283	9,9,162.42	13.78	286	929,962.00	9.26	231	765,882.00	8.85
From 4001 to 8000	181	1,089,841.78	16.34	272	1,656,067.20	16.49	221	1,355,615.20	15.67
From 8001 to 15,000	118	1,304,812.67	19.56	192	2,142,203.00	21.34	161	1,790,808.00	20.7
From 15,001 to 30,000	70	1,569,408.00	23.52	112	2,309,062.00	23	99	2,061,062.00	23.82
More than 30,000	21	1,211,159.13	18.15	37	2,346,962.00	23.38	33	2,176,368.00	25.16
Total	1135	6,671,307.69	100	1452	10,039,988.86	100	1139	8,651,720.86	100
Average		5877.80			6914.59			7595.89	
Median		3000.00			3500.00			4000.00	
Mode	138	2000.00	12.16	134	2000.00	9.23	119	2000.00	10.45

Source http://notarias.colmex.mx/; Archivo Histórico de Notarías de la Ciudad de México

Table 13.3 Largest credit for each year

Year	Month	Day	Notary	Type of operation	Amount	Rate %	Maturity (days)	Lender
1770	3	1	José Joaquín Arroyo Bernaldo de Quiroz	*Depósito irregular*	1,16,000.00	5		Tesorería General de Bienes
1771	12	16	Andrés Delgado Camargo	*Depósito irregular*	1,00,325.00	5	1825	Juzgado de testamentos, capellanías y obras pías
1772	11	18	Andrés Delgado Camargo	*Censo redimible*	50,000.00	5		Concurso de bienes de Antonio García
1773	8	7	Andrés Delgado Camargo	*Depósito irregular*	1,01,493.00	5	365	Juzgado de testamentos, capellanías y obras pías
1774	2	4	Andrés Delgado Camargo	*Depósito irregular*	40,600.00	5	1825	Juzgado de testamentos, capellanías y obras pías
1775	3	24	Andrés Delgado Camargo	*Depósito irregular*	49,260.00	5	365	Juzgado de testamentos, capellanías y obras pías and Casa Mortuoria and Banco de Plata
1776	1	4	Andrés Delgado Camargo	*Depósito irregular*	61,013.00	5	1825	Juzgado de testamentos, capellanías y obras pías and Casa Mortuoria and Banco de Plata
1777	3	3	Mariano Buenaventura de Arroyo	*Depósito irregular*	1,00,000.00	5	2190	Ana Gómez de Valencia
1778	5	9	Nicolás Francisco Díaz	*Depósito irregular*	1,00,000.00	5		Convento de Nuestra Señora de la Encarnación
1779	5	17	Andrés Delgado Camargo	*Depósito irregular*	40,000.00	5	1095	Juzgado de testamentos, capellanías y obras pías and Casa Mortuoria and Banco de Plata
1819	2	4	Eugenio Pozo	*Depósito irregular*	44,000.00	5	1825	Luis de Escobar

(continued)

Table 13.3 (continued)

Year	Month	Day	Notary	Type of operation	Amount	Rate %	Maturity (days)	Lender
1820	1	12	José María Moya	*Obligación de pago*	1,40,000.00	5	3285	Consortium of Creditors of Don Juan Lucas de Lassaga and Cofradía del Pueblo de Parras
1821	11	14	Ignacio de la Barrera	*Reconocimiento de deuda*	83,000.00	5	1825	Testamentaria de María Josefa González Guerra
1822	5	6	Nicolás de Vega	*Reconocimiento de deuda*	60,000.00	6		Chapter of the Cathedral of Mexico City
1823	8	14	Nicolás de Vega	*Obligación de pago*	60,000.00	5	1825	Cofradía del Santísimo Sacramento en Tepotzotlán
1824	4	21	Ignacio José Montes de Oca	*Depósito irregular*	54,000.00	5	2920	Convento de Jesús María
1825	3	5	Eugenio Pozo	*Obligación de pago*	30,000.00	5		José Gumersindo de la Colina Villanueva y Gutiérrez
1826	9	30	Nicolás de Vega	*Depósito irregular*	40,000.00	5	3285	Cofradía del Santísimo Sacramento en Tepotzotlán
1827	11	14	Francisco Calapiz y Aguilar	*Depósito irregular*	2,40,000.00	5	3285	Consortium of creditors of Sánchez de Tagle
1828	2	15	Francisco Calapiz y Aguilar	*Depósito irregular*	1,24,277.00	5	365	Juan Félix Goyeneche

(continued)

Table 13.3 (continued)

Year	Origin of lender	Notes on lender	Debtor	Origin of debtor	Notes on debtor	Mortgaged property	Value of mort. prop.
1770			Society of Jesus (in exile)			Oficio de Ayador en la Real Casa de Moneda	
1771			Matías de Miramontes	Veracruz	Merchant		
1772			Pedro Alcántara del Valle	Mexico City	Juez de Balanza Propietario de la Real Casa de Moneda		
1773			Santiago Nuñez	Mexico City			
1774			Francisco Ariztimuño y Gorozpe	Mexico City	Alcalde mayor of Cuicatlán and Papalotipac (later, member of the RSBAP and captain of the Acordada)		
1775			Ventura Gutiérrez Via	Mexico City			
1776			Bernardo José Carrillo de Albornoz	Havana	Solicitor of the Reales Consejos and the Real Audiencia, provisto alcalde mayor of Chichicapa and Zimatlán		
1777	Mexico City	Widow	Francisco Ignacio de Iraeta	Guipúzcoa	Merchant		
1778			José Ventura Villanueva	Mexico City	Hacendados of the valley of Toluca, with mayorazgo. Their son was Agustín Villanueva Altamirano y Cervantes	Houses and stores at the Puente del Real Palacio	
1779			Juan Lucas de Lassaga	Spain	Diputado del Real Tribunal de la Minería, contador de menores y regidor perpetuo	Three haciendas in the kingdom of León	350,000.00
1819	Mexico City	Merchant and alcalde	Anselmo and Jorge Rojo	Mexico City	Merchant		

(continued)

Table 13.3 (continued)

Year	Origin of lender	Notes on lender	Debtor	Origin of debtor	Notes on debtor	Mortgaged property	Value of mort. prop.
1820			Juan Goribar and Manuel de Ibarra	Mexico City	Juan Goribar, of Saltillo (Coahuila)	Hacienda of San Lorenzo in Parras (Coahuila)	
1821			Miguel González Calderón	Mexico City	Debtor executor. His father was an important merchant, one of the founders of the *Cofradía del Cristo de Burgos*	Haciendas Rancho Grande and Trujillo in Zacatecas	
1822			Government of Agustín de Iturbide				
1823			Juan Antonio and Gabriel Patricio Yermo	Biscay	Merchant	2/3 of the Hacienda and sugar mill of San Nicolás Tolentino at Izúcar	
1824			Manuela Moreno y Barrios	Mexico City	*Mariscala de Castilla,* Marchioness of Ciria		
1825	Santander	*Caballero de Alcántara*	Consortium of Creditors of Gregorio Sáenz and Ramón Valiente		Son-in-law and father-in-law, merchants, had made several donations to the royalist army in the 1810 s		
1826			Juan Antonio and Patricio Gabriel de Yermo	Biscay		Hacienda and sugar mill of San Nicolás Tolentino	
1827			José María and Cristóbal González Peredo	Mexico City			
1828	Navarre	Merchant	José Rafael Alarid	New Mexico	Colonel (and previously deputy to the constitutional congress of 1824)	Hacienda of San Carlos Borromeo at Yautepec	

Source http://notarias.colmex.mx/; Archivo Histórico de Notarías de la Ciudad de México

By comparison, it never occupied that position in the decade between 1819 and 1828. A nunnery was the largest creditor once in the first series, and the same happened in the second. Individuals and consortiums of creditors provided the largest amount of credit once in the 1770s, but five times in the second series. This shows the extent to which the ecclesiastical institutions, the Court of Chaplaincies in particular, had been affected by the *Consolidación* and the internal upheavals of the 1810s.

When it comes to the largest borrowers, merchants, *hacendados* and public officials dominated the panorama in the 1770s. They were still important in the 1820s, but two new actors had appeared: the military and the state itself. In 1822, the government of Agustín de Iturbide, overwhelmed by the fiscal shortage produced by administrative disorder and tax rebates, took the largest loan of that year from the Chapter of the Cathedral of Mexico. It concluded contracts with the Chapter for a total amount of 115,500 pesos. The funds came from ecclesiastical institutions of all sorts, coordinated by the *Capítulo Metropolitano*—the archiepiscopal seat was vacant at the time. Still as an individual and a member of the military, Iturbide borrowed 2000 pesos in 1820 from the merchant Martín Rafael de Michelena. Names of military officers come up quite often during the 1820s; for example, the influential politician and colonel Alarid took the largest loan in 1828 for nearly 125,000 pesos.

Seven out of ten credits needed no collateral in the 1770s, while six out of ten did during the second period studied: this speaks of greater uncertainty and clearly correlates with the fact that only the more affluent members of society, having real estate as support, had access to credit. Another sign of malaise is the fact that while in the 1770s, nine very large operations were irregular deposits and one was a redeemable *censo*, which meant that the money lent was fresh, only half of the very large operations of the 1820s were irregular deposits. The others were either loans to the government or operations involving the transmission of inheritances or mere renegotiations of outstanding loans.

A quite remarkable finding of our research is the ubiquitous presence of women in all tiers of debtors and creditors. For example, on

12 October 1826, Catalina Vamis lent María Antonia Sandoval García Bravo, a bachelorette, 2000 pesos to be repaid after two years. María Antonia used her house at number 10 Donceles Street in Mexico City as warrant for the credit. The notary José Vicente Maciel clearly stated that they came before him on their own and without representatives. This is certainly a remarkable operation, but further research should clarify the extent to which women were acting independently and what their interests and needs were. When it comes to large loans, rich widows played an important role in economic life (Ramos Medina 2002). In 1777, Ana Gómez de Valencia, by then a widow, lent her husband's former business partner and son-in-law, the thriving merchant Francisco Ignacio de Yraeta, 100,000 pesos. In 1824, Manuela Moreno y Barrios, *Mariscala de Castilla* and Marchioness of Ciria, daughter of the Marquess of Valle Ameno and a widow too after her husband passed away in 1822 (Ladd 2006, p. 279), borrowed 54,000 pesos from the Convent of Conceptionist nuns of Jesús María. She needed the money to put business in order after the death of his husband and committed herself to paying the accustomed 5% and reimburse the principal after 8 years. The reputation of the family and the urban property of their *mayorazgo*, along with the solidarity of guarantors, were enough to secure the credit.

Our Results Vis-à-Vis the General Economic Interpretation of the Period

Our research has enabled us to estimate the real interest rates charged on the transactions made in our sample of the financial sector in the 1770s and in 1819–1828. To the best of our knowledge, there is no previous attempt to build series of real interest rates in the specialised literature. This variable plays an important role in the performance of a given economy: generally speaking, *ceteris paribus*, low real rates of interest stimulate economic growth and vice versa. Therefore, estimating real interest rates allows us to advance some views on the consequences that the developments in the financial sector had on the economic performance towards the end of the viceregal period and at the beginning of Mexico's independence.

For the reasons outlined above, we distinguish between Tables 13.4 and 13.5. Table 13.4 depicts our estimates of the real interest rates for the 1770s and 1819–1828 including all financial transactions recorded in our sources. On Table 13.5, transactions with an interest rate of zero have been excluded. No differences exist between the estimates in both tables for 1770–1779 since there were almost no transactions with zero interest registered in our sample for those years. As expected, the estimates in Table 13.5 for 1819–1828 are somewhat higher than in Table 13.4.

We calculate the real interest rates estimates as the weighted average of nominal interest rates in year T minus the inflation rate in the same year. As a proxy for the inflation rates, we use the yearly per cent change in the silver price of the consumption baskets series built by Challú and Gómez-Galvarriato (2015). Since it is unlikely that the poorest segment of the population participated in the financial sector represented by our sample, we use the "respectable" instead of the "barebones" consumption basket. These authors present two—"low" and "high"—bands of the cost of the "respectable" consumption basket. We are very conscious of the limitations of this exercise. However, we think it makes a step forward into an unexplored field of Mexican economic history.

It is evident that interest rates in our financial market were higher in 1819–1828 than in the 1770s. That could be expected and, *prima facie*, is consistent with both the optimistic perception of the 1770s economy and the pessimistic interpretation of economic performance during the 1810s and 1820s that has so far prevailed among historians and economic historians. There is some debate about when the eighteenth-century-long phase of economic growth started to decline and how deep it fell. Depending on the indicator used (e.g. real wages versus silver production) and the explanation proposed (agrarian crisis, drainage of financial resources by the Crown, decreasing returns in mining, etc.), the picture may differ. Whereas Coatsworth (2008) rejects the idea of any growth from 1700 to 1800, other estimates are positive for most of the eighteenth century (Bolt et al. 2018).

However, a consensus exists that economically, the 1770s were markedly different from 1819 to 1828. It is generally accepted that

Table 13.4 Estimates of real interest rates (%), 1770–1779 and 1819–1828 (all transactions)

	Nominal interest rate	Respectable IPCs		Real interest rates	
		Low	High	Low	High
1770	4.49	7.85	3.07	−3.37	1.41
1771	5	−4.01	−5.11	9.01	10.11
1772	4.56	6.35	5.54	−1.78	−0.98
1773	4.79	−7.28	−5.39	12.07	10.18
1774	4.49	−2.2	2.25	6.69	2.25
1775	4.39	−4.49	−5.87	8.88	10.25
1776	4.28	−2.69	2.8	6.97	1.48
1777	4.5	−5.35	−5	9.86	9.5
1778	4.21	−2.92	−1.59	7.13	5.8
1779	4.38	30.08	16.21	−25.7	−11.83
Average	4.51	1.53	0.69	3	3.8

	Nominal interest rate	Respectable IPCs		Real interest rates	
		Low	High	Low	High
1819	5.05	13.76	−2.59	−8.72	7.63
1820	4.21	−7.34	−8.25	11.55	12.46
1821	4.18	−18.42	−0.83	22.59	5.01
1822	3.79	−3.56	−1.67	7.35	5.46
1823	4.35	9.36	6.89	−5.01	−2.53
1824	4.7	−1.43	−21.41	6.12	26.1
1825	5.2	3.95	3.78	1.25	1.42
1826	4.47	−4.43	13.61	8.9	−9.14
1827	4.26	3.18	−2.78	1.09	7.05
1828	4.68	9.76	6.93	−5.08	−2.25
Average	4.49	0.48	−0.63	4	5.1

Source http://notarias.colmex.mx/; Archivo Histórico de Notarías de la Ciudad de México

Table 13.5 Estimates of real interest rates (%), 1770–1779 and 1819–1828 (all transactions with rates higher than zero)

	Nominal interest rate	Respectable IPCs		Real interest rates	
		Low	High	Low	High
1770	4.49	7.85	3.07	−3.36	1.41
1771	5.00	−4.01	−5.11	9.01	10.11
1772	4.56	6.35	5.54	−1.78	−0.98
1773	4.79	−7.28	−5.39	12.07	10.18
1774	4.49	−2.20	2.25	6.69	2.25
1775	4.39	−4.49	−5.87	8.89	10.26
1776	4.28	−2.69	2.80	6.97	1.48
1777	4.50	−5.35	−5.00	9.86	9.50
1778	4.21	−2.92	−1.59	7.13	5.80
1779	4.42	30.08	16.21	−25.66	−11.79
Average	4.51	1.53	0.69	3.0	3.8

	Nominal interest rate	Respectable IPCs		Real interest rates	
		Low	High	Low	High
1819	5.16	13.76	−2.59	−8.61	7.74
1820	4.88	−7.34	−8.25	12.22	13.12
1821	5.10	−18.42	−0.83	23.51	5.92
1822	5.01	−3.56	−1.67	8.57	6.68
1823	5.11	9.36	6.89	−4.26	−1.78
1824	4.94	−1.43	−21.41	6.36	26.34
1825	5.20	3.95	3.78	1.25	1.42
1826	4.53	−4.43	13.61	8.96	−9.08
1827	4.42	3.18	−2.78	1.24	7.20
1828	4.68	9.76	6.93	−5.08	−2.25
Average	4.90	0.48	−0.63	4.4	5.5

Source http://notarias.colmex.mx/; Archivo Histórico de Notarías de la Ciudad de México

after 1810, a deep and lasting crisis affected New Spain's economy. The length of time it took for the Mexican economy to recover to pre-1810 levels of GDP per capita is debatable. Most estimates indicate an incomplete recovery by 1850, or even later (Coatsworth 1990; Prados 2009; Maddison Project Database 2018). A revision of this pessimistic view is proposed by Marichal (2010).

In any case, it is not necessary to assume that the crisis was equally profound in all regions and sectors after independence in order to accept that general economic conditions in post-1821 Mexico only improved very slowly from the unprecedented disruption and impoverishment of the 1800s and 1810s. This claim has recently received new quantitative support from emergent fields of research. During the second half of the nineteenth century, living standards in Mexico City were still not permanently higher than in the last decades of the eighteenth century (Challú and Gómez-Galvarriato 2015). Human capital formation in independent Mexico, proxied by numeracy, follows a similar U-shaped long-term pattern: it was lower in 1870 than in 1800 (Baten et al. 2012). In other words, gaining independence, and independence itself, were costly for the Mexican economy (Coatsworth 1993). Lucas Alamán himself proved to be only a little less pessimistic than most estimates (Serrano and Vázquez 2010, p. 416).

In particular, our quantitative findings correlate with the dynamics of silver production (see Fig. 13.1). Production of silver has a negative correlation with real interest rates: annual average production was 17.5 and 10.2 million pesos in the 1770s and 1819–1828, respectively. The rise in real interest rates in the second period is also consistent with Cárdenas's (1997, 2003) and Marichal's (1997) views. For those authors, a significant part of the explanation for Mexico's poor economic performance in the decades following independence comes from monetary astringency and financial backwardness, respectively. Both of them are emphatic in their claims.

However, the rise in real interest rates from the 1770s to 1819–1828 does not appear large enough to indicate a "dramatic reduction in the means of payment" (Cárdenas 1997, p. 70). In fact, if we compare the rates of growth of the accumulated stock of silver, "the missing piece in the puzzle" of Mexican economic historiography, in the two periods under analysis, a double perspective arises (see Fig. 13.3).[8] On the one hand, rates of growth were generally lower after 1810 than before. In spite of that, the rate is positive, on average, for both the 1810s and the 1820s. Besides, phases of steep decline in the accumulated stock of silver occurred before 1810. Some of them were not short-lived (e.g. 1770–1785). Significant negative rates of growth are not infrequent either.[9] Looking at the periods under analysis, it seems that the real interest rates charged in our particular financial market are deeply intertwined with the rates of growth of the accumulated stock of silver, if such a relationship, which will be studied in our future research, turns out to be statistically detectable.

On the other hand, the accumulated stock of silver in absolute terms was much larger in 1819–1828 (an annual average of more than 203 million pesos) than in the 1770s (53 million). That difference, even measured in per capita terms or as a ratio of GDP, is very significant. It is also counterintuitive: most specialists would expect a larger stock in a growing economy (1770s) than in a declining or stagnant one (1819–1828). If our estimates are correct, they might have interesting implications for the economic history of New Spain/Mexico that we have not yet explored at this stage of our research. Most Mexicanist scholarship, Cárdenas (1997) being an exception, has paid much more attention to the *production* of silver than to the *accumulated stock* of that precious metal. Actually, our estimates raise an immediate question: Where did all that silver go? A near-continuous accumulation of silver inevitably had economic consequences that deserve examination. The existence of a virtually ever-growing stock contradicts the widespread idea that New Spanish economy suffered from an acute scarcity of money in circulation

---

[8]The calculation of the accumulated stock of silver has been made as follows: we start by estimating for year T the difference between production and exports. Since Mexican sources do not offer information for 1821–1824, we have filled that vacuum of data with those obtained from foreign sources and kindly provided by Sandra Kuntz to whom we are deeply grateful; we proceed by summing up for every year between 1770 and 1849 the accumulated difference between production and exports of silver from 1770 on; as stock of silver for 1769, we use the figure of 25 million pesos given by the *Superintendente* of the *Casa de la Moneda* de la Ciudad for 1772 (Quiroz 2006).

[9]That was the case in 1770, 1773, 1778, 1783, 1785–1787, 1791, 1796, 1802, 1804, 1807 and 1809.

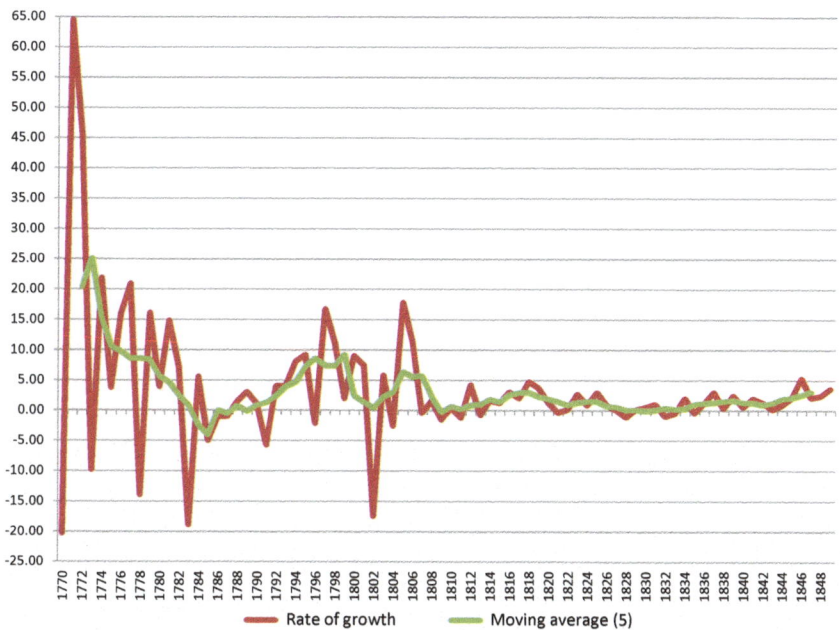

Fig. 13.3 Yearly per cent rates of growth of the accumulated stock of silver, 1770–1849 (*Source* see Footnotes 1 and 8)

(del Valle Pavón 2003; critical: Calderón Fernández and Dobado 2012). Only a gross underestimation of the outflow of silver would imply a significant reduction—that is to say, from abundance to paucity—of the accumulated stock we have calculated. But that does not seem to be the case. Humboldt estimated the accumulated stock of silver in 1803 at 55–60 million pesos. The estimates elaborated by Cárdenas (1997) are, for instance, higher than ours for 1796–1806: 176.1 million pesos versus 71 million. Unfortunately, we cannot further explore this issue here. A discussion in full will appear in a future publication.

Moreover, our estimates of real interest rates are significantly lower than those prevailing in Mexico during the five decades after independence: 12–40% for short-term commercial loans and 30–200% for government loans (Marichal 1997, p. 119). For this author, the economic outcome of such interest rates was that "most potential investors could not be attracted to long-term investment, which offered lower rates" (Marichal 1997, p. 119). Not a minor consequence for economic growth indeed. Even if we take into account that the aforementioned rates *nominal* rather than *real* rates, the difference with our estimates is substantial. Incidentally, the real rates are only slightly higher or lower (roughly a half percentage point) than real rates in 1819–1828 (Tables 13.4 and 13.5).

Therefore, it seems that since the majority of the transactions made under the jurisdiction of the *Juzgado de Capellanías y Obras Pías* and ecclesiastical institutions were long-term loans, see Table 13.1, they did not discourage potential investors from getting involved in long-term business ventures involving the real economy of New Spain/Mexico. It seems that the notion that the "credit activity of the Church contributed to the creation of new capital, increased the supply of available credit funds, prevented its export and retained them within the Viceroyalty, and, ultimately, contributed to providing better possibilities of development to the economy" is basically correct (Martínez and del Valle Pavón 1998, p. 16). Von Wobeser has documented that in the eighteenth century, "72% of credit was destined for productive investment, of which 52% corresponded to trade, 12% to real estate investment and 8% to agriculture and livestock; 16% was channelled toward the payment of debts, and the remaining 12% went to other productive activities and for personal or family matters" (Von Wobeser 1998, p. 195). Therefore, the rapid increase in loans under the *Juzgado* in the 1770s could not but help the growth of New Spain's economy.

Then, the significant reduction between the 1770s and 1819–1828 evident in Fig. 13.4—and it is quite possible that the level of credit attained in the 1790s was significantly higher than in the 1770s[10]—must have negatively contributed to Mexican economic performance. Likely, albeit to an undetermined extent, the reduction in the volume of loans is related to the *Consolidación de Vales Reales* of 1804–1808. According

---

[10]Del Valle Pavón sustains that "From the last third of the eighteenth century, the supply of credit money in the archbishopric of Mexico showed a remarkable expansion as a result of the boom in mining that began shortly after [...] 1770 and the commercial opening of the Pacific [...] in 1774" (del Valle Pavón 2012, p. 30).

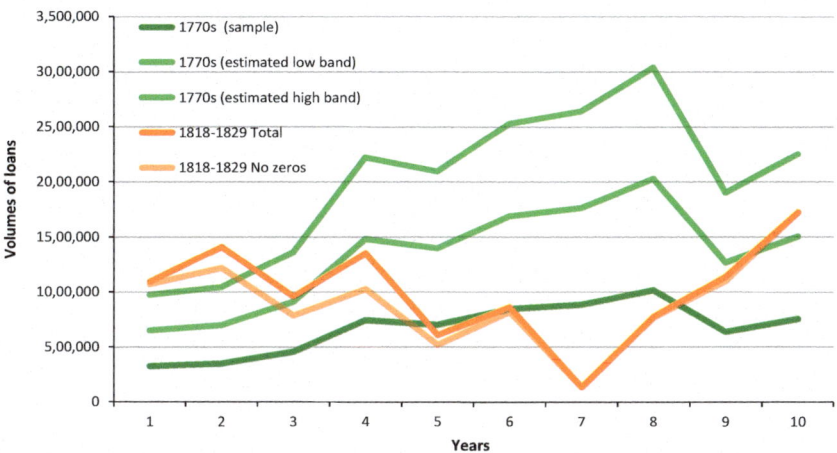

Fig. 13.4 Volumes of loans, 1770s and 1819–1818 (*Source* http://notarias. colmex.mx/; Archivo Histórico de Notarías de la Ciudad de México)

to Humboldt (1822, v. II, p. 463), who used data from the protest manifest sent by the diocese of Valladolid against the *Consolidación* decree, the capital lent by ecclesiastical institutions in 1805 amounted to 44.5 million pesos. The *Consolidación* collected roughly 11 million pesos, that is one quarter of the estimated amount. Thus, Von Wobeser (2003) is probably at least partially right, in her argument that the *Consolidación* resulted in a significant decrease in lending.

However, the effects of civil war and the disruption it caused in New Spain's fiscal and commercial system are much harder to measure. Kizca (1998, p. 60) asserts that "the internal commercial credit systems that Mexico had worked well when the war of independence began in the colony". If this is true, the negative effects of capital flight—estimated by Ward to have been at least 80 million pesos (Romero Sotelo 1997, p. 15)—the destruction of assets and the disruption of the credit system[11] would necessarily be very significant.

----

[11]In 1818, José María Quirós, member of the *Consulado* of Veracruz, "concluded that the Mexican economy was being ruined by the lack of investment" (Ladd 2006, p. 221).

At least in the short to medium run, they could have exceeded the positive effects of the downward revision of the costs of the civil war (1810–1821), the liberalisation of the economy and the subtraction to the logic of transfers of resources within the framework of the imperial order (Sánchez Santiró 2014, p. 216). To sum up, it does seem that the costs of being part of the Spanish Empire had become unbearable by the beginning of the nineteenth century, but also that the costs of independence exceeded the benefits or at least that they needed several decades to come to fruition. The credit market that we have presented in this essay reflects the complicated dynamics of the transition from New Spain to Mexico.

Acknowledgements This research has been funded between August 2016 and August 2018 by a grant of the Bank of Spain research programme, bearing the title *El mundo después del real de a ocho. La plata hispanoamericana tras la independencia, la primera crisis global de la década de 1820 y sus mecanismos de difusión internacional,* 182PR20231. Andrés Calderón thanks the postdoctoral research fellowship programme of Mexico's National University (Programa de Becas Posdoctorales UNAM—DGAPA) that has generously funded his research since August 2018. The authors wish to thank the help of Gerardo Saldaña Domínguez, Violeta Barrientos Nieto, Daniela González Machorro and Claudia Marina Gutiérrez Silveira, our research assistants in Mexico, who have performed an outstanding job assembling the large databases upon which this research is based.

Bibliography

Primary Sources

Humboldt, Alexander Freiherr von, 1822. *Ensayo político sobre el Reino de la Nueva España*, translated by Vicente González Arnao. Paris: Casa de Rosa.
Siliceo, Miguel, 1857. *Memoria de la Secretaría de Estado y del Despacho de Fomento, Colonización, Industria y Comercio de la República Mexicana.* Mexico City: Imprenta de V. García Torres.

Secondary Sources

Arroyo Abad, Leticia; Luiten van Zanden, Jan, 2016. "Growth Under Extractive Institutions? Latin American per Capita GDP in Colonial Times." In: *The Journal of Economic History* 76, 4: 1182–1215.

Ávila, Alfredo; Jáuregui, Luis, 2010. "La disolución de la monarquía hispánica y el proceso de independencia." In: *Nueva historia general de México*, edited by Erick Velásquez García et al. Mexico City: El Colegio de México: 355–396.

Baten, Joerg; Manzel, Kerstin; Stolz, Yvonne, 2012. "Convergence and Divergence of Numeracy: The Development of Age Heaping in Latin America from the Seventeenth to the Twentieth Century." In: *Economic History Review* 65, 3: 932–960.

Bernecker, Walther L., 1992. *De agiotistas y empresarios. En torno a la temprana industrialización mexicana, siglo XIX*. Mexico City: UIA.

Berthe, Jean-Pierre, 1993. "Contribución a la historia del crédito en la Nueva España (siglos XVI, XVII y XVIII)." In: *Prestar y pedir prestado. Relaciones sociales y crédito en México del siglo XVI al XX*, coordinated by Marie-Noëlle Chamoux et al. Mexico City: CIESAS/CEMCA: 25–52.

Bolt, Jutta; Inklaar, Robert; de Jong, Herman; Luiten van Zanden, Jan, 2018. "Rebasing 'Maddison': New Income Comparisons and the Shape of Long-Run Economic Development." In: *GGDC Research Memorandum* 174. https://www.rug.nl/ggdc/html_publications/memorandum/gd174.pdf. Accessed January 13, 2019.

Calderón Fernández, Andrés, 2009. "Una serie de precios de vivienda. Las accesorias del Real Colegio de San Ignacio de Loyola de los Señores Vizcaínos, 1771–1821." In: *Gaceta Vizcaínas* 2, 4: 47–83.

Calderón Fernández, Andrés; Dobado González, Rafael, 2012. "Siete mitos acerca de la historia económica del mundo hispánico." In: *Pintura de los Reinos. Identidades compartidas en el mundo hispánico, siglos XVI–XIX*, coordinated by Rafael Dobado and Andrés Calderón. Mexico City: Fomento Cultural Banamex: 75–103.

Cárdenas, Enrique, 1997. "A Macroeconomic Interpretation of Nineteenth-Century Mexico." In: *How Latin America Fell Behind: Essays on the Economic Histories of Brazil and Mexico, 1800–1914*, edited by Stephen Haber. Stanford: Stanford University Press: 65–92.

Cárdenas, Enrique, 2003. *Cuando se originó el atraso económico de México*. Madrid: Editorial Biblioteca Nueva.

Carrera, Eduardo, 2011. *Las voces de la fe: las cofradías en México, siglos XVII–XIX*. Mexico City: UAM/CIESAS.

Challú, Amílcar; Gómez-Galvarriato, Aurora, 2015. "Mexico's Real Wages in the Age of the Great Divergence, 1730–1930." In: *Revista de Historia Económica/Journal of Iberian and Latin America Economic History* 33, 1: 83–122.

Chamoux, Marie-Nöelle, et al., 1993. *Prestar y pedir prestado. Relaciones sociales y crédito en México del siglo XVI al XX*. Mexico City: CIESAS/CEMCA.

Coatsworth, John, 1990. "La decadencia de la economía mexicana, 1800–1860." In: *Los orígenes del atraso*, edited by John Coatsworth. Mexico City: Alianza Editorial Mexicana: 110–139.

Coatsworth, John, 1993. "La independencia latinoamericana: hipótesis sobre los costes y los beneficios." In: *La independencia americana: consecuencias económicas*, edited by Leandro Prados de la Escosura and Samuel Amaral. Madrid: Alianza Editorial: 17–27.

Coatsworth, John, 2008. "Inequality, Institutions and Economic Growth in Latin America." In: *Journal of Latin American Studies* 40, 3: 545–569.

Del Valle Pavón, Guillermina, 2003. "Historia financiera de la Nueva España en el siglo XVIII y comienzos del XIX, una revisión crítica". In: *Historia Mexicana* 52, 3: 649–675.

Del Valle Pavón, Guillermina, 2012. *Finanzas piadosas y redes de negocios: los mercaderes de la ciudad de México ante la crisis de Nueva España, 1804–1808*. Mexico City: Instituto Mora.

Dobado González, Rafael; Marrero, Gustavo, 2011. "The Role of the Spanish Imperial State in the Mining-Led Growth of Bourbon Mexico's Economy." In: *Economic History Review* 64, 3: 855–884.

Garner, Richard, 1982. "Exportaciones de circulante en el siglo XVIII (1750–1810)." In: *Historia Mexicana* 31, 4: 558–559.

Kicza, John E., 1998. "El crédito mercantil en Nueva España." In: *El crédito en Nueva España*, coordinated by María del Pilar Martínez López-Cano and Guillermina del Valle Pavón. Mexico City: Instituto Mora/El Colegio de Michoacán/El Colegio de México/UNAM-IIH: 33–60.

Kuntz Ficker, Sandra; Tena-Junguito, Antonio, 2017. "Mexico's Foreign Trade in a Turbulent Era (1821–1870): A Reconstruction". In: *Journal of Iberian and Latin American Economic History* 36, 1: 149–182.

Ladd, Doris M., 2006. *La nobleza mexicana en la época de la Independencia, 1780–1826*, translated by Marita Martínez del Río. Mexico City: FCE.

Lerdo de Tejada, Miguel, 1967. *Comercio exterior de México*. México: Banco Nacional de Comercio Exterior.

Lobato López, Ernesto, 1945. *El crédito en México. Esbozo histórico hasta 1925*. Mexico City: FCE.

Ludlow, Leonor; Carlos Marichal, 1998. "Introducción." In: *La banca en México, 1820–1920*, edited by Leonor Ludlow and Carlos Marichal. Mexico: Instituto Mora: 7–30.

Maddison Project Database, 2018. https://www.rug.nl/ggdc/historicaldevelopment/maddison/releases/maddison-project-database-2018. Accessed on August 8, 2019.

Marichal, Carlos, 1997. "Obstacles to the Development of Capital Markets in 19th Century Mexico." In: *How Latin America Fell Behind*, edited by Stephen Haber. Stanford: Stanford University Press: 118–145.

Marichal, Carlos, 1999. *La bancarrota del virreinato. Nueva España y las finanzas del imperio español, 1780–1810*. Mexico City: El Colegio de México/FCE.

Marichal, Carlos, 2010. "El desempeño de la economía mexicana, 1810–1860: de la Colonia al Estado-nación." In: *Historia económica general de México*, coordinated by Sandra Kuntz. Mexico City: El Colegio de México.

Marichal, Carlos; von Grafenstein, Johanna, 2012. *El secreto del imperio español. Los situados coloniales en el siglo XVIII*. Mexico City: El Colegio de México/Instituto Mora.

Marichal, Carlos; Souto Mantecón, Matilde, 1994. "Silver and Situados: New Spain and the financing of the Spanish Empire in the Caribbean in the eighteenth century." In: *The Hispanic American Historical Review* 74, 4: 587–613.

Martínez López-Cano, María del Pilar, 2001. *La génesis del crédito colonial. Ciudad de México, siglo XVI*. Mexico City: UNAM.

Martínez López-Cano, María del Pilar; Speckman Guerra, Elisa; von Wobeser, Gisela, 2004. *La iglesia y sus bienes: de la amortización a la nacionalización*. Mexico City: UNAM.

Martínez López-Cano, María del Pilar; del Valle Pavón, Guillermina, 1998. "Los estudios sobre el crédito colonial. Problemas, avances y perspectivas." In: *El crédito en Nueva España*, coordinated by María del Pilar Martínez López-Cano and Guillermina del Valle Pavón. Mexico: Instituto Mora/El Colegio de Michoacán/El Colegio de México/UNAM-IIH: 13–32.

Martínez López-Cano, María del Pilar; von Wobeser, Gisela; Juan Guillermo Muñoz Correa, 1998. *Cofradías, capellanías y obras pías en la América colonial*. Mexico City: UNAM.

Mejía Torres, Karen Ivett, 2014. *Las cofradías en el valle de Toluca y su relación con el crédito, 1794–1809*. Zinacantepec: El Colegio Mexiquense.

Ortiz de la Tabla Ducasse, Javier, 1978. *Comercio exterior de Veracruz, 1778–1891. Crisis de dependencia*. Sevilla: Escuela de Estudios Hispanoamericanos de Sevilla.

Prados de la Escosura, Leandro, 2009. "Lost Decades? Economic Performance in Post-Independence Latin America." In: *Journal of Latin American Studies* 41, 2: 279–307.

Quiroz, Enriqueta, 2006. "La moneda menuda en la circulación monetaria de la ciudad de México. Siglo XVIII." In: *Estudios Mexicanos* 22, 2: 219–249.

Ramos Medina, Manuel, comp., 2002. *Viudas en la historia*. Mexico City: CEHM Condumex.

Romero Sotelo, María Eugenia, 1997. *Minería y guerra. La economía de Nueva España, 1810–1821*. Mexico City: El Colegio de México/UNAM.

Sánchez Cuen, Manuel, 1958. *El crédito a largo plazo en México. Reseña histórica*. Mexico City: Banco Nacional Hipotecario Urbano y de Obras Públicas.

Sánchez Santiró, Ernest, 2014. "Economía y fiscalidad en la Guerra de Independencia. Nueva España (1810–1821)." In: *Iberoamérica y España antes de las Independencias, 1700–1820. Crecimiento, reformas y crisis*, coordinated by Jorge Gelman et al. Mexico: Instituto Mora/CONACyT/El Colegio de México: 163–224.

Serrano, José Antonio; Josefina Zoraida Vázquez, 2010. "El nuevo orden, 1821–1848." In: *Nueva historia general de México*, edited by Erick Velásquez García. Mexico City: El Colegio de México: 397–442.

Tenenbaum, Barbara, 1985. *México en la época de los agiotistas, 1821–1857*. Mexico City: FCE.

Thompson, Edward Palmer, 2000. *Costumbres en Común*, translated by Jordi Beltrán and Eva Rodríguez. Barcelona: Crítica.

Torales Pacheco, María Cristina, 1985. "II. La conformación de un capital." In: *La compañía de comercio de Francisco Ignacio de Yraeta (1767–1797). Cinco ensayos*. Mexico City: IMCE—UIA.

Von Wobeser, Gisela, 1993. "La postura de la Iglesia católica frente a la usura." In: *Memorias de la Academia Mexicana de la Historia, correspondiente de la Real de Madrid* 36: 121–145.

Von Wobeser, Gisela, 1994. *El crédito eclesiástico en la Nueva España: siglo XVIII*. Mexico City: UNAM.

Von Wobeser, Gisela, 1998. "Los créditos de las instituciones eclesiásticas de la ciudad de México en el siglo XVIII." In: *El crédito en Nueva España*, coordinated by María del Pilar Martínez López-Cano and Guillermina del Valle Pavón, 176–202. Mexico City: Instituto Mora/El Colegio de Michoacán/El Colegio de México/UNAM-IIH.

Von Wobeser, Gisela, 2003. *Dominación colonial: la Consolidación de Vales Reales en Nueva España (1804–1812)*. Mexico City: UNAM.

Von Wobeser, Gisela, 2005. *Vida eterna y preocupaciones terrenales: las capellanías de misas en la Nueva España, 1600–1821*. Mexico City: UNAM.

Von Wobeser, Gisela, 2006. "La consolidación de vales reales como factor determinante de la lucha de independencia en México, 1804–1808." *Historia Mexicana* 56, 2: 373–425.

14

Exchange Rates and Silver Prices at European Fairs, Sixteenth to Eighteenth Centuries

Markus Denzel

International fairs were always more than just places where goods and information were exchanged; these functions could also be fulfilled by great annual markets (*Große Jahrmärkte*), that is, by markets with a relatively large number of foreign participants. International fairs also fulfilled the function of monetary and, more importantly, credit markets (cf. Denzel 2018). Therefore, they needed large amounts of precious metals as a means of payment. When the fairs were located in the vicinity of relevant mining areas, it was appropriate to trade these products at the fairs, thus making some of the necessary precious metals available. Such was the case in Leipzig, where at first the trade with silver from Freiberg and later, in the fifteenth and sixteenth centuries, the *Saigerhandel*[1] from the Erzgebirge was executed (Westermann 2002,

---

[1] Saigerhandel: Trade of silver gained by method of crystalline segregation.

M. Denzel (✉)
Historisches Seminar, Universität Leipzig, Leipzig, Germany
e-mail: denzel@rz.uni-leipzig.de

© The Author(s) 2019   **349**
R. Pieper et al. (eds.), *Mining, Money and Markets*
in the Early Modern Atlantic, Palgrave Studies in Economic History,
https://doi.org/10.1007/978-3-030-23894-0_14

p. 241; cf. Westermann 1983). In return, the trade with these mining products contributed to the gradual ascent of Leipzig's large annual markets to international fairs (Schirmer 1999, p. 89f.).

However, the vast majority of fairs could not rely on such flows of precious metals. For most of the late Middle Ages, until the take-off of precious metal imports from Spanish America, these fairs suffered under the Great Bullion Famine—the epitome of the monetary crisis of the late Middle Ages (Stromer 1981, p. 109; cf. Day 1978; Munro 1972, 1992). In fact, they suffered such that the lack of available precious metals would have effectively made the expansion of business impossible. Without further elaborating on other reasons, the shortage of precious metals was certainly decisive for provoking the development of innovative means of payment, transactions and credit financing at the large international fairs. It also warranted the expansion of business beyond the limitations of available precious metals, both for individual merchant houses and for the business of the fairs as a whole. Thus, by the late twelfth century, merchants at the Champagne fairs started to settle mutual claims and liabilities in a cashless manner. This clearing mechanism had completed its development already by 1180 and was legally guaranteed by the court of the fair (Schneider 1991, I, p. 135f.; Thomas 1977, 1991; Schönfelder 1988; Irsigler and Reichert 2007). Additionally, several forms of cashless payment papers were added, starting with the *lettres de foire* and reaching its peak in the development of the classical bill of exchange, which would dominate the cashless payment system in the following centuries (Denzel 2008, pp. 52–55). In this way, the transfer of remaining claims and liabilities not settled by the clearing until the next fair—wherever it may take place—became possible.

Thus, on the one hand, the large international fairs provided the opportunity to issue money and build credit; on the other, they contributed to international clearing and the dispersion of cashless payment methods from the thirteenth and fourteenth centuries onwards. In other words, they pushed the process of integration towards the emergence of a European exchange market (Denzel 2001, 2011). At the same time, the large international fairs often, though not always, contributed to avoiding financial and, above all, credit crises by means of their regulating impact on monetary, precious metal and credit markets. In case of money shortage, the merchant-bankers active at these fairs provided the needed liquidity, whereas in the case of available surplus money, they would divert it to different monetary markets, mostly other fairs. As a consequence of the process of commercialisation during the late fifteenth and, more importantly, the sixteenth centuries, international fairs would become the actual focal points of financial crises, having developed into centres of cashless payment settlements, particularly between the state and merchant creditors (Denzel 2016). Such was the case at the fairs of Brabant (Van der Wee 1963) and Castile (Ladero Quesada 1982, 1995) as well as at the international fairs of Lyon (Gascon 1971; Boyer-Xambeu et al. 1989) or the Genoese Bisenzone fairs (Gentil da Silva 1969; Felloni 1991). In all cases, these exchange fairs provided for the settlement of payments between the Crown and its creditors. When the Crown was unwilling to pay off its credits with interest, these international fairs resulted in financial crises. Classic examples of such occurrences are the *Grand Parti* in Lyons and the Spanish financial crises of the second half of the sixteenth and the beginning of the seventeenth centuries, which had their origin at the Castilian fairs.[2]

This thought leads to the second point: the mechanisms linking silver prices and exchange rates.[3] Assuming that the currencies accepted at fairs were related to precious metals and that the larger availability of silver made it a means of payment with greater significance than gold, it should be expected that the price of the scarce commodity silver was one of, if not *the* determinant factor in the rise and fall of exchange rates. This, by the way, applies to the fairs as well as to other financial markets of that time. The underlying mechanism is well known and has often been described: increasing scarcity of silver supplies leads to

---

[2]Even though the latter have been re-interpreted recently as short-term liquidity crises rather than as a sign of unsustainable debts (Drelichman and Voth 2010, 2011, 2014).

[3]Here, exchange rates mean rates for bills of exchange drawn in foreign currency. These are *not* money rates for changing one coin in another as examined in Claudia de Lozanne Jefferies' chapter.

higher exchange rates and vice versa (e.g. North 1996).[4] From the second half of the sixteenth century, the Castilian fairs were profoundly influenced by the American silver fleets. Because of the excessive volume of silver flowing to Spain, a financial crisis could emerge.[5] In this case, this was a crisis that stemmed from money being too cheap (*larghezza*), leading to (increased) inflation. But a financial crisis could also emerge when there was too little silver, for example when a fleet did not arrive. If the Crown could not satisfy its creditors, shortage of money (*stretezza*) occurred and the entire European payment structure was shaken; hence, the Castilian fairs were closely related with those of Lyons and "Bisenzone" as well as with Antwerp, Lisbon and the Italian financial markets (Martínez Ruiz 2004; Marsilio 2012, 2018). Incidentally, the silver did not have to be physically present to exert a positive or negative influence: knowledge about its availability in the king's funds was sufficient.

But what do the empirical sources show? First of all, at the international fairs, no immediate prices were quoted for silver. However, the prices of goods were registered *in* silver money, and their increase in the second half of the sixteenth century reflects a phenomenon that is known as the "price revolution". However, it should be stressed that until the eighteenth century, there are no regularly quoted prices for fixed weights of silver at the fairs. In Leipzig, such quotations become regular only after the Seven Years' War. At the fairs of the sixteenth and seventeenth centuries, exchange rates were the only internationally comparable measure between the various currencies.

So, what can be said about the development of the prices of currencies and precious metals on the basis of the quotations of exchange rates? The effects of the silver supply from the New World became emphatically clear in the example of the Castilian fairs in Medina del Campo in the late sixteenth and early seventeenth centuries (Fig. 14.1).

---

[4]We have a quite similar phenomenon in the relationship between silver flows and interest rates in Seville, as Manuel González-Mariscal, Rafael Mauricio Pérez Garcia, Manuel Díaz-Ordóñez and Manuel Fernández Chaves show in their chapter in this volume.

[5]Cf. The contribution by Domenic Hofmann in this volume.

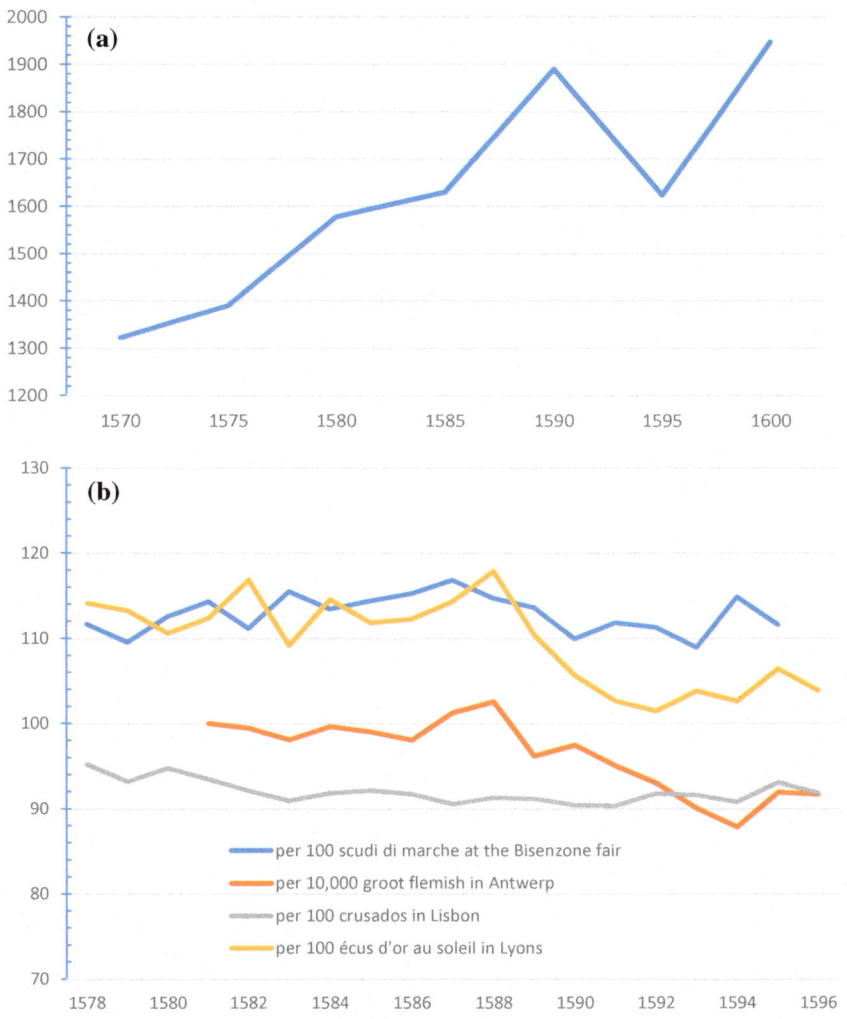

Fig. 14.1 The Price revolution in Spain in the late sixteenth century—which markets are concerned? **a** Value of a basket of commodities, in maravedis, 1570–1600 (*Source* Pieper 1985, p. 139) **b** Exchange rates at Medina Del Campo, in Ducados de Cambio, 1578–1596 (*Source* Denzel 1995, pp. 60–66)

As could be expected, the basket of commodities that Renate Pieper has put together shows a remarkable price increase, whereby the prices were quoted in small silver units—in maravedis. In contrast, several selected exchange rate quotations from the Castilian fairs show a relatively large, but in international comparison not exceptionally high, volatility of the rates. The general trend, however, shows only a small increase and sometimes even no increase at all. This means that the commodity prices registered in small silver coins followed different criteria than the exchange rate quotations. On the basis of which, as it turns out, no price revolution, not even a stronger inflationary tendency, can be detected. A similar result would be obtained if the quotations of other fairs or financial markets at Medina del Campo were consulted.

We could, thus, ask provocatively: Does this mean that there was no price revolution at all? Of course, there was a price revolution in sixteenth-century Europe, but on the commodity markets, not on the exchange markets. How can this finding be explained? The background is that the merchant-bankers at the large international fairs no longer executed their payment transactions on the basis of current silver coins, as used to be the case at the Champagne fairs, but using money of account, which was based on the scarcer precious metal gold. Already at the Geneva fairs of the fifteenth century, a stable unit of accounting was established for all: all prices and rates were quoted in Marc d'Or (Denzel 2008, p. 114; Bergier 1963). This accounting unit would later be adopted by all exchange fairs. From 1439, free determination of the unit price (*conto*) on the basis of the Écu de marc (since 1575 the Écu d'or au soleil) for all foreign, i.e. non-French coins was granted to the *réunion des banquiers* in Lyons. Finally, at the Genoese Bisenzone fairs all exchange rates between different currencies were defined by their relation to a basic coin, the Écu de marc or Scudo di marchi.[6] This was the result of a compensation process which benefitted the Genoese— between the highest valued coins of that year, and more precisely between the Scudo-coins of Spain, Antwerp, Venice, Genoa, Florence and Naples. Unlike the fairs of Lyons, a single national currency was

---

[6]In doing so, the merchant-bankers at the Bisenzone fairs followed the example of their colleagues at the Lyons fairs, where all currencies were compared with the Écu d'or au soleil.

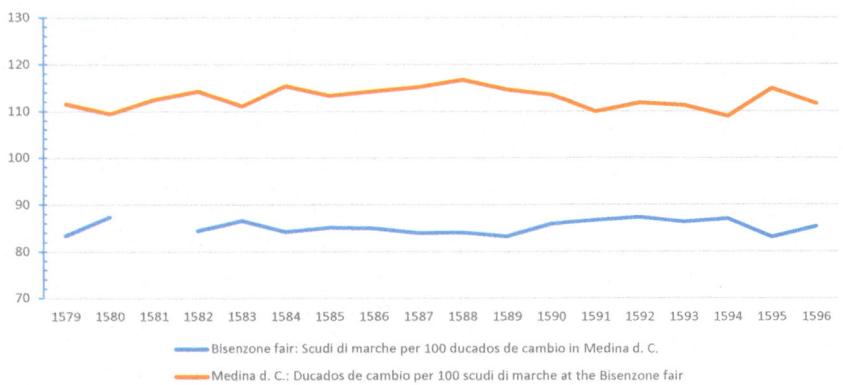

Fig. 14.2 Exchange rates between Medina Del Campo and the Bisenzone fairs, 1579–1596 (*Source* Denzel 1995, pp. 66, 100f.)

no longer the basis for the money of account; the merchant-bankers "calculated" a common currency, so to speak. Between 1552 and 1763, the proportion of this money of account to the average gold weight of the Scudo de oro in oro (Écu d'or en or) remained at 100:101, "which proves the strength of the moneys of account and the flexible ways and methods by which cash payments, expressed in 'scudi di marchi' [sic], could be completed all over the world" (Houtman-de Smedt and Van der Wee 1993, p. 110; cf. Felloni 1984; Racine 1991, p. 155f., 161; Gentil da Silva 1969, p. 81). This stable money of account led to a stable *conto*, the official determination of the exchange rate, which was "always determined by taking into account the rates of the other fairs; a merchant or a banker who drew a bill of exchange at the Genoese fairs or who lent money there, would not have to fear foreign exchange losses" (North 1996, p. 230), which greatly enhanced the attractiveness of these exchange fairs.

At the Castilian fairs, there was also a money of account, the ducado de cambio of 375 maravedis. In the history of the development of different monies of account, this ducat of account should be considered alongside the Geneva or early Lyon fairs; for the merchant-bankers trading at the Castilian fairs, it performed perfectly. It made its users independent from the rise and fall of silver prices and consequently from the inflationary tendencies of the price revolution. However, since it was based solely on the Castilian currency, it was not internationally accepted like the Scudo di marchi of the Bisenzone fairs. This is why quotations in ducados de cambio show somewhat larger fluctuation margins than quotations in scudi di marchi (Fig. 14.2). This, however, was a secondary effect: what mattered was independence from the fluctuations of the silver prices, regardless of their economic or political grounds.

Turning our attention to Central Europe and Leipzig in particular, one may notice that the source base available at Leipzig shows a different finding: in Leipzig, payment in precious metals, namely in current money, continued to be common until the nineteenth century. Exchange business really took off only in the sixteenth century, but it never developed into a business sector independent from the commodity trade. The monetary, credit and exchange markets of Leipzig depended on the commodity business. Unlike Western Europe before the mid-sixteenth century, precious metals were available in sufficient quantities in Leipzig, thanks to the nearby Erzgebirge. When these mines were no longer sufficiently productive, from the seventeenth century onwards, merchants from Poland and later also from Russia and the Ottoman Empire brought coins to the fairs, sometimes in such quantities that in the eighteenth and nineteenth centuries they had to be transported by wagon to Hamburg for clearance. Thus, the payment transactions system of Leipzig was characterised by using bills of exchange for payments to Western European markets like Hamburg, Vienna, Amsterdam, London and Paris, whereas payments to Poland, Russia and the Ottoman Empire were done in current money. In this way, Leipzig formed a hinge between—much simplified—the cashless payment system of the West and the cash payments of Eastern Europe, and this also was a central reason for its existence (Denzel 1999).

As was the case elsewhere, in Leipzig's cashless payment transactions, money of account was used for calculations, but in 1763, after its massive debasement during the Seven Years' War, the Saxon currency was replaced with the Konventionskurant. The so-called *Konventionswechselzahlung*, or exchange money, formed the basis of the regular quotations of exchange and money rates that appeared afterwards. In those quotations, gold and silver prices were also quoted; for silver, the rixdollar (*Reichsthaler*) *Konventionskurant* as the basic unit of

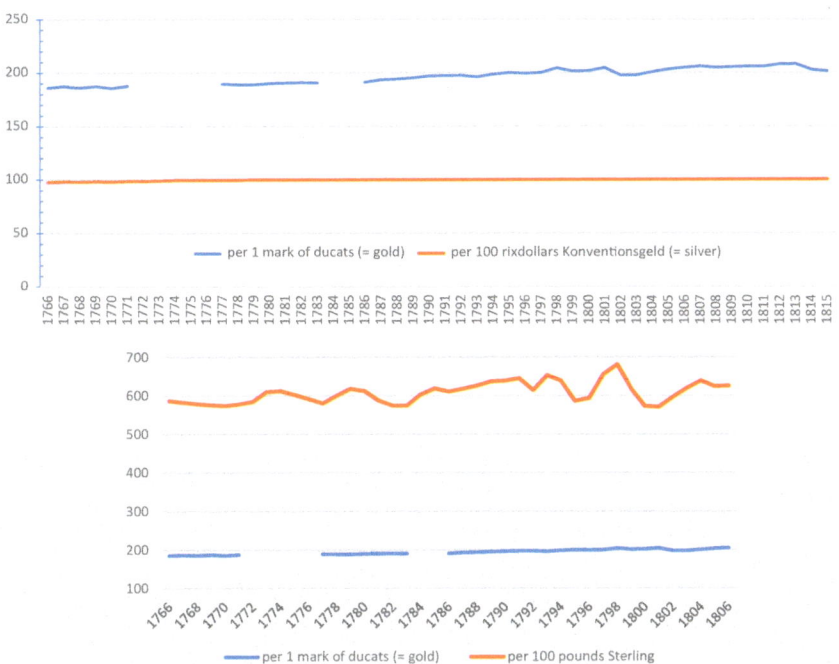

Fig. 14.3 Gold and silver prices at the Leipzig fairs, 1766–1815 (*Source* Denzel 1994, pp. 66, 83f., 114)

currency and for gold the mark of ducats,[7] later in the nineteenth century the mark of fine gold (Fig. 14.3).

If both quotations are compared, an almost continuous parity of the silver currency with the draft payment (bills of exchange payment) becomes apparent, whereas the gold price rose slightly during the war years of the 1790s and 1800s, meaning that silver lost some value in relation to gold. Perhaps this is due to the partial absence of Russian visitors to the fairs, who brought most of the gold ducats to Leipzig. From 1770, the year with the lowest gold price, until 1813, when the highest gold price was quoted, its rise amounted to only about 12%. This was clearly less than the few commodity prices that are available, a detailed analysis of which

---

[7]The basic unit of the ducat weight (*Dukatengewicht*) was the mark of Cologne of 233,855g.

is still to be undertaken. Leipzig's exchange rates were oriented, rather unspectacularly, on the price of silver; in the case of London, because of England's gradual shift towards a gold currency, on the price of gold.

To sum up: what are the results that can be seen? First, the large international fairs in Western Europe needed precious metals as liquid means for cash payments and as security for cashless payment transactions, without themselves being centres of precious metal trade like Amsterdam, London or Hamburg in the seventeenth and eighteenth centuries. Leipzig occupied a peculiar position, on the one hand because of the nearby Erzgebirge, on the other because it functioned as a supplier of precious metals from Eastern Europe to Hamburg. Leipzig owed its special position to its function as a hinge between different types of payment systems in Western and Eastern Europe.

Second, beginning with the Geneva fairs in the fifteenth century, cashless payment transactions at the large fairs were no longer executed on the basis of current money but instead on the basis of money of account, which, in the course of the sixteenth century, came to be based in gold. This innovation, which was developed at the fairs, had a decisive impact on the history of the European payment and financial system. The goal of the merchants was to make their payment transactions independent from the fluctuations of silver prices as well as from the arbitrariness of political leaders in currency policy issues. At the same time, the merchants strove to expanding their money supply in times of money shortage by using cashless methods for payment and credit. Unlike the Western European fairs of the sixteenth century, that development occurred relatively late in Leipzig, only after the currency and financial crisis of the Seven Years' War.

Therefore, a clear distinction must be made between commodity markets, where current money was used, and the markets of high finance for bills of exchange, where crown and state financing also took part. The commodity markets existed in continuous dependence on the availability of silver and therefore on the price of silver at fairs, like anywhere. The financial markets needed silver or gold as security for their transactions and the price of silver interested bankers only in relation to the prevailing money of account. In this respect, the international fairs did not differ from the other financial markets of the early modern period.

Bibliography

Bergier, Jean-François, 1963. *Genève et l'économie européenne de la Renaissance.* Paris: S.E.V.P.E.N.

Boyer-Xambeu, Marie-Thérèse; Deleplace, Ghislain; Gillard, Lucien, 1989. "Goldstandard, Währung und Finanz im 16. Jahrhundert." In: *Geldumlauf, Währungssysteme und Zahlungsverkehr in Nordwesteuropa 1300–1800. Beiträge zur Geldgeschichte der späten Hansezeit*, edited by Michael North. Köln and Wien: Böhlau: 167–181.

Day, John, 1978. "The Great Bullion Famine of the Fifteenth Century." In: *Past and Present 79*: 3–54.

Denzel, Markus A., ed., 1994. *Währungen der Welt X: Geld- und Wechselkurse der deutschen Messeplätze Leipzig und Braunschweig (18. Jahrhundert bis 1823).* Stuttgart: Steiner.

Denzel, Markus A., ed., 1995. *Währungen der Welt IX: Europäische Wechselkurse vor 1620.* Stuttgart: Steiner.

Denzel, Markus A., 1999. "Zahlungsverkehr auf den Leipziger Messen vom 17. bis zum 19. Jahrhundert." In: *Leipzigs Messen 1497–1997. Gestaltwandel. Umbrüche. Neubeginn 1497–1914*, edited by Hartmut Zwahr, Thomas Topfstedt, and Günter Bentele, vol. 1. Köln and Weimar and Wien: Böhlau: 149–165.

Denzel, Markus A., 2001. "Der Beitrag von Messen und Märkten zum Integrationsprozeß des internationalen bargeldlosen Zahlungsverkehrssystems in Europa (13.–18. Jahrhundert)." In: *Fiere e mercati nella integrazione delle economie europee, secc. XIII–XVIII. Atti della 'Trentaduesima Settimana di Studi', 8–12 maggio 2000,* edited by Simonetta Cavaciocchi. Prato: Le Monnier: 819–835.

Denzel, Markus A., 2008. *Das System des bargeldlosen Zahlungsverkehrs europäischer Prägung vom Mittelalter bis 1914.* Stuttgart: Steiner.

Denzel, Markus A., 2011. "Der Beitrag von Messen und Märkten zum Integrationsprozeß des internationalen bargeldlosen Zahlungsverkehrssystems in Europa (13.–18. Jahrhundert)." In: *Globalizing Areas, kulturelle Flexionen und die Herausforderung der Geisteswissenschaften*, edited by Günther Heeg and Markus A. Denzel. Stuttgart: Steiner: 47–60.

Denzel, Markus A., 2016. "The Role of Institutions in Financial Crises: Fairs—Public Banks—Stock Exchanges (Thirteenth to Eighteenth Century)." In: *Le crisi finanziarie: gestione, implicazioni sociali e conseguenze nell'età preindustriale: selezione di ricerche = The Financial Crises: Their Management, Their Social Implications and Their Consequences in*

Pre-Industrial Times: Selection of Essays, edited by Istituto internazionale di storia economica F. Datini. Firenze: Le Monnier: 427–450.

Denzel, Markus A., ed., 2018. *Europäische Messegeschichte, 9.–19. Jahrhundert. Forschungsstand und Forschungsperspektiven.* Köln and Weimar and Wien: Böhlau.

Drelichman, Mauricio; Voth, Hans-Joachim, 2010. "The Sustainable Debts of Philipp II: A Reconstruction of Castille's Fiscal Position, 1566–1596." In: *Journal of Economic History,* 70, 4: 813–842.

Drelichman, Mauricio; Voth, Hans-Joachim, 2011. "Serial Defaults, Serial Profits: Returns to Sovereign Lending in Habsburg Spain, 1566–1600." In: *Explorations in Economic History,* 48: 1–19.

Drelichman, Mauricio; Voth, Hans-Joachim, 2014. *Lending to the Borrower From Hell. Debts, Taxes, and Default in the Age of Philipp II.* Princeton and Oxford: Princeton University Press.

Felloni, Giuseppe, 1984. "Un système monétaire de marc dans les foires de change génoises, XVIᵉ–XVIIIᵉ siècles". In: *Études d'histoire monétaire,* edited by John Day. Lille: Presses universitaires de Lille: 249–260.

Felloni, Giuseppe, 1991. "Kredit und Banken in Italien, 15.–17. Jahrhundert." In: *Kredit im spätmittelalterlichen und frühneuzeitlichen Europa,* edited by Michael North. Köln and Wien: Böhlau: 9–23.

Gascon, Robert, 1971. *Grand commerce et vie urbaine au XVIᵉ siècle. Lyon et ses marchands.* Paris: De Gruyter Mouton.

Gentil da Silva, José, 1969. *Banque et crédit en Italie au XVIIe siècle.* Paris: S.E.V.P.E.N.

Houtman-de Smedt, Helm; Van der Wee, Herman, 1993. "Die Entstehung des modernen Geld- und Finanzwesens Europas in der Neuzeit." In: *Europäische Bankengeschichte,* edited by Hans Pohl. Frankfurt am Main: Knapp: 75–173.

Irsigler, Franz; Reichert, Winfried, 2007. "Les foires de Champagne." In: *Messen, Jahrmärkte und Stadtentwicklung in Europa/Foires, marchés annuels et développement urbain en Europe,* edited by iidem. Trier: Porta Alba: 89–105.

Ladero Quesada, Miguel Ángel, 1982. "Las ferias de Castilla. Siglos XII a XV." In: *Cuadernos de historia de España* 67/8: 315–322.

Ladero Quesada, Miguel Ángel, 1995. *Las ferias de Castilla. Siglos XII a XV.* Madrid: Comité español de ciencias históricas.

Marsilio, Claudio, 2012. "The Genoese and Portuguese Financial Operator's Control of the Spanish Silver Market (1627–1657)." In: *Journal of European Economic History* 41, 3: 69–89.

Marsilio, Claudio, 2018. "The Italian Exchange Fairs and the International Payment System (XVI–XVII Centuries)." In: *Europäische Messegeschichte, 9.–19. Jahrhundert. Forschungsstand und Forschungsperspektiven*, edited by Markus A. Denzel. Köln and Weimar and Wien: Böhlau: 169–180.

Martínez Ruiz, José Ignacio, 2004. "The Credit Market and Profits from Letters of Exchange. *Ricorsa* Exchange Operations Between Seville and the 'Besançon' International Fairs (1589–1621)." In: *Journal of European Economic History* 33: 331–355.

Munro, John H., 1972. *Wool, Cloth and Gold. The Struggle for Bullion in Anglo-Burgundian Trade, 1340–1478*. Bruxelles: Édition de l'université de Bruxelles.

Munro, John H., 1992. *Bullion Flows and Monetary Policies in England and the Low Countries, 1350–1500*. Hampshire: Ashgate.

North, Michael, ed., 1989. *Geldumlauf, Währungssysteme und Zahlungsverkehr in Nordwesteuropa 1300–1800. Beiträge zur Geldgeschichte der späten Hansezeit*. Köln and Wien: Böhlau.

North, Michael, 1996. "Von den Warenmessen zu den Wechselmessen. Grundlagen des europäischen Zahlungsverkehrs in Spätmittelalter und Früher Neuzeit." In: *Europäische Messen und Märktesysteme in Mittelalter und Neuzeit*, edited by Peter Johanek and Heinz Stoob. Köln and Weimar and Wien: Böhlau: 223–238.

Pieper, Renate, 1985. *Die Preisrevolution in Spanien (1500–1640) Neuere Forschungsergebnisse*. Wiesbaden: Steiner.

Racine, Pierre, 1991. "Messen in Italien im 16. Jahrhundert: Die Wechselmessen von Piacenza." In: *Frankfurt im Messenetz Europas—Erträge der Forschung*, edited by Hans Pohl, vol. I. Frankfurt am Main: Union Druckerei und Verlag: 155–170.

Schirmer, Uwe, 1999. "Die Leipziger Messen in der ersten Hälfte des 16. Jahrhunderts. Ihre Funktion als Silberhandels- und Finanzplatz der Kurfürsten von Sachsen." In: *Leipzigs Messen 1497–1997. Gestaltwandel. Umbrüche. Neubeginn, 1497–1914*, edited by Hartmut Zwahr, Thomas Topfstedt; and Günter Bentele, vol. 1. Köln and Weimar and Wien: Böhlau: 87–107.

Schneider, Jürgen, 1991. "Messen, Banken und Börsen (15.–18. Jahrhundert)." In: *Banchi pubblici, banchi privati e monti di pietà nell'Europa preindustriale. Amministrazione, tecniche operative e ruoli economici*, vol. 1. Genova: Società Ligure di Storia Patria: 133–169.

Schönfelder, Alexander, 1988. *Handelsmessen und Kreditwirtschaft im Hochmittelalter—Die Champagnemessen*. Saarbrücken-Scheidt: Dadder.

Stromer, Wolfgang von, 1981. "Hartgeld, Kredit und Giralgeld. Zu einer monetären Konjunkturtheorie des Spätmittelalters und der Wende zur Neuzeit." In: *La moneta nell'economia europea, secoli XIII–XVIII. Atti della „Settima Settimana di studio (11–17 aprile 1975)*, edited by Vera Barbagli Bagnoli. Firenze: Le Monnier: 105–125.

Thomas, Heinz, 1977. "Beiträge zur Geschichte der Champagne-Messen." In: *Vierteljahrschrift für Sozial- und Wirtschaftsgeschichte*, 49: 433–467.

Thomas, Heinz, 1991. "Die Champagnemessen." In: *Frankfurt im Messenetz Europas—Erträge der Forschung*, edited by Hans Pohl, vol. I. Frankfurt am Main: Union Druckerei und Verlag: 13–36.

Van der Wee, Herman, 1963. *The Growth of the Antwerp Market and the European Economy (Fourteenth–Sixteenth Centuries)*. Louvain: Springer.

Westermann, Ekkehard, 1983. "Silbererzeugung, Silberhandel und Wechselgeschäft im Thüringer Saigerhandel von 1460 bis 1620. Tatsachen und Zusammenhänge, Probleme und Aufgaben der Forschung." In: *Vierteljahrsschrift für Sozial- und Wirtschaftsgeschichte*, 70: 192–214.

Westermann, Ekkehard, 2002. "*Die Nürnberger Welser und der mitteldeutsche Saigerhandel des 16. Jahrhunderts in seinen europäischen Verflechtungen.*" In: *Die Welser—Neue Forschungen zur Geschichte und Kultur des oberdeutschen Handelshauses*, edited by Mark Häberlein and Johannes Burkhardt. Berlin: Akademie: 240–264.

Name Index

© The Editor(s) (if applicable) and The Author(s),
under exclusive license to Springer Nature Switzerland AG 2019
R. Pieper et al. (eds.), *Mining, Money and Markets
in the Early Modern Atlantic*, Palgrave Studies in Economic History,
https://doi.org/10.1007/978-3-030-23894-0

Place Index

Subject Index

Printed by Printforce, the Netherlands